11th International Conference on X-ray Microscopy

(XRM2012)

Journal of Physics: Conference Series Volume 463

Shanghai, China
5-10 August 2012

Editors:

Hongjie Xu
Renzhong Tai
Ziyu Wu

ISBN: 978-1-62993-282-8
ISSN: 1742-6588

Printed from e-media with permission by:

Curran Associates, Inc.
57 Morehouse Lane
Red Hook, NY 12571

Printed by Curran Associates, Inc. (2013)

For permission requests, please contact the Institute of Physics
at the address below.

Institute of Physics
Dirac House, Temple Back
Bristol BS1 6BE UK

Phone: 44 1 17 929 7481
Fax: 44 1 17 920 0979

techtracking@iop.org

Additional copies of this publication are available from:

Curran Associates, Inc.
57 Morehouse Lane
Red Hook, NY 12571 USA
Phone: 845-758-0400
Fax: 845-758-2634
Email: curran@proceedings.com
Web: www.proceedings.com

TABLE OF CONTENTS

Preface

The Eleventh International Conference on X-ray Microscopy (XRM2012) was held on 5–10 August 2012 at the Hope hotel in Shanghai. Historically, for the first time the XRM conference took place in China. The conference was jointly hosted by the Shanghai Synchrotron Radiation Facility (SSRF) and the National Synchrotron Radiation Laboratory (NSRL). The series of XRM conferences dates back to 1983 in Göttingen, Germany. Since the Zürich conference, XRM2008, it has been held every two years, showing its increasing popularity among the x-ray microscopy communities around the world. Research in the area of x-ray microscopy is advancing very fast with the development of synchrotron radiation techniques, especially the emergence of third generation light sources with low natural emittance which has significantly pushed forward the development of technologies and applications in this area. This has been fully demonstrated in presentations from this and previous XRM conferences.

XRM2012 was attended by 295 people including 21 invited speakers, 53 contributing speakers, 55 students, and 13 industry exhibitors. Over 232 abstracts were submitted for oral or poster presentation and 56 original, peer-reviewed papers are published in these proceedings. The conference was sponsored by the Chinese Academy of Sciences (CAS) and the National Natural Science Foundation of China (11210301016/A0802), and three gold sponsors active in industrial and technological fields related to x-ray microscopy. An exhibition booth was offered free to Australia synchrotron, the host for XRM2014, to promote the next conference which will be held in Melbourne, Australia in 2014. An unforgettable memory for most conference participants might be the charming night cruise along Pujiang river which was part of the welcome reception on the first evening.

The Werner Meyer-Ilse Award (WMIA) prize this year was awarded to Irene Zanette (TU-München) and Stephan Werner (HZB-Berlin), the former for her pioneering new work in high-sensitivity phase-contrast and dark-field (small-angle x-ray scattering) images, the latter for his outstanding accomplishments in the field of on-chip zone plate stacking. The location for the XRM2016 conference was voted for by all conference attendees, and Diamond Light Source in the United Kingdom was chosen. A tour of SSRF was attended by nearly 100 participants who got a close look at the seven beamlines in operation before the close of the meeting.

We would like to thank all the participants for their active involvement and excellent contributions during the conference and poster session, and the International Program Committee for their guidance, and outstanding support of local organizating committee for a successful XRM2012.

Hongjie Xu, Ziyu Wu and Renzhong Tai
Proceeding Editors

The 11th of International Conference on X-ray Microscopy
August 05-10, Shanghai, China

Journal of Physics: Conference Series **463** (2013) 011001 doi:10.1088/1742-6596/463/1/011001

International Program Committee(IPC)

Sadao Aoki (retired), University of Tskuba, Japan
David Attwood, UC Berkeley, USA
Henry Chapman, University of Hamburg and DESY, Germany
Hans Hertz, Royal Institute of Technology, Sweden
Adam Hitchcock, McMaster University, Canada
Chris Jacobsen (Chair), Northwestern University and Argonne National Laboratory, USA
Janos Kirz (Emeritus), Lawrence Berkeley National Laboratory, USA
Maya Kiskinova, ELETTRA, Italy
Ian McNulty, Argonne National Laboratory, USA
Graeme Morrison, University College London, UK
Keith Nugent, University of Melbourne, Australia
Christoph Quitmann, Paul Scherrer Institut, Switzerland
Gerd Schneider, Bessy, Germany
Günter Schmahl (Emeritus), Universität Göttingen, Germany
Hyun-Joon Shin, Pohang Accelerator Laboratory, South Korea
Jean Susini, ESRF, France
Wenbing Yun, Xradia Inc., USA
Ziyu Wu, USTC and NSRL, China

Journal of Physics: Conference Series **463** (2013) 011001 doi:10.1088/1742-6596/463/1/011001

Local Organizing Committee (LOC)

Hongjie Xu, Chair, SINAP
Ziyu Wu, Co-chair, NSRL/USTC
Renzhong Tai, Executive Chair, SSRF/SINAP
Xiaohan Yu, Scientific Program, SSRF/SINAP
Peiping Zhu, Scientific Program, BSRF
Yangchao Tian, Scientific Program, NSRL
Shensheng Han, Scientific Program, SIOM
Zhanshan Wang, Scientific Program, Tongji University
Aiguo Li, Scientific Program, SSRF/SINAP
Lijuan Zhang, Secretary in General, SSRF/SINAP
Zhiyun Pan, Registration and Abstracts, NSRL
Ling Zhang, Registration and Posters, SSRF/SINAP
Qingru Mi, Web Developer and Registration, SSRF/SINAP
Ying Zou, Sponsoring/Exhibition and Conference Editor, SSRF/SINAP
Ying Fan, SSRF/SINAP
Kai Zhang, BSRF/SINAP
Wanxia Huang, BSRF
Qingxi Yuan, BSRF
Kun Gao, NSRL
Zhili Wang, NSRL
Yi Xia, NSRL

Journal of Physics: Conference Series **463** (2013) 011001 doi:10.1088/1742-6596/463/1/011001

List of Participants

Name	Institute/Company
Abbati Gennaro	University of Applied Sciences Koblenz, Remagen/Germany
Abbey Brian	La Trobe University, Melbourne/Australia
Adams Daniel	JILA University of Colorado, Boulder/USA
Akihiro Suzuki	Osaka university, Osaka/Japan
Altapova Venera	Karlsruhe Institute of Technology (KIT), Karlsruhe/Germany
Amati Matteo	Sincrotrone Trieste S.C.p.A., Trieste/Italy
Anastasi Peter	Silson Ltd, Northampton/UK
Aoki Sadao	University of Tsukuba, Tsukuba/Japan
Attwood David	University of California Berkeley, Berkeley/USA
Baik Jaeyoon	Pohang Accelerator Lab.(PAL), Pohang/South Korea
Barre Sebastien	Institute of Anatomy, University of Bern, Bern/Switzerland
Bauer Sondes	Karlsruhe Institute of Technology (KIT), Karlsruhe/Germany
Baumbach Stefan	University of Applied Sciences Koblenz, Remagen/Germany
Benedetto Bozzini	Salento University, Lecce/Italy
Liu Biqiu	Institute of Ceramics, Chinese Academy of Sciences (CAS), Shanghai/China
Bittencourt Carla	University of Mons, Mons/Belgium
Bohic Sylvain	European Synchrotron Radiation Facility (ESRF), Grenoble/France
Boone Matthieu	Ghent University, Gent/Belgium
Brandstetter Stefan	Dectris Ltd., Baden/Switzerland
Bruyndonckx Peter	Bruker-MicroCT, Kontich/Belgium
Burvall Anna	Royal Institute of Technology (KTH), Stockholm/Sweden
Cai Feng	Haikou People's Hospital, Xiangya Medical School, Centeral South University, Haikou/China
Charalambous Pambos	ZONEPLATES.COM, Longdon/UK
Chen Bo	University of Melbourne, Parkville/Australia
Chen Bo	London Centre for Nanotechnology, UCL, London/UK
Chen Shan	National Synchrotron Radiation Laboratory (NSRL), Hefei/China
Chen Weilong	Shanghai NTI Co., Ltd., Shanghai/China
Chen Yue	Graduate University of CAS, Beijing/China
Cheng Yin	Karlsruhe Institute of Technology (KIT), Karlsruhe/Germany
Choi Jaeho	Dankook University, Cheoana/South Korea
Civita Maria	London Centre for Nanotechnology/UCL, London/UK
David Christian	Paul Scherrer Institut (PSI), Villigen/Switzerland
Deng Junjing	Northwestern University, Evanston/USA
Diete Wolfgang	Bruker ASC GmbH, Bergisch Gladbach/Germany
Ducic Tanja	Deutsches Elektronen-Synchrotron (DESY), Hamburg/Germany
Duke Elizabeth	Diamond Light Source (DLS), Oxfordshire/UK
Eberl Christian	Institut für Materialphysik, University of Göttingen, Göttingen/Germany
Eisebitt Stefan	Technische Universität Berlin (TU-Berlin), Berlin/Germany
Ejima Takeo	Tohoku University, Sendai/Japan
Enders Bjoern	Technische Universität München (TU-München), München/Germany
Ewald Johannes	University of Applied Sciences Koblenz, Remagen/Germany
Eymard Philippe	Synchrotron Soleil, Saint-Aubin/France
Li Faguo	Shanghai Jiao Tong University, Shanghai/China
Falaschetti Christine	Feinberg School of Medicine, Northwestern University, Chicago/USA
Falkenberg Gerald	Deutsches Elektronen-Synchrotron (DESY), Hamburg/Germany
Fan Chunhai	Shanghai Institute of Applied Physics (SINAP), Shanghai/China
Feng Renfei	Canadian Light Source (CLS), Saskatoon/Canada
Ferrara Joseph	Rigaku Americas Corp, Woodlands/USA
Feser Michael	Xradia, Pleasanton/USA
Fink Rainer	Universität Erlangen-Nürnberg, Erlangen/Germany
Fujimori Susumu	NTT Advanced Technology Corporation,Kanagawa Prefecture/Japan
Gao Kun	National Synchrotron Radiation Laboratory (NSRL), Hefei/China
Gianoncelli Alessandra	ELETTRA Sincrotrone Trieste, Trieste/Italy
Giewekemeyer Klaus	European XFEL, Hamburg/Germany
Gil Kyehwan	Pohang Accelerator Laboratory, Pohang/South Korea
Gleber Sophie	Argonne National Laboratory(ANL), Argonne/USA
Gorniak Thomas	Karlsruhe Institute of Technology (KIT), Karlsruhe/Germany
Grevent Corinne	MPI Intellingente Systeme, Stuttgart/Germany
Guan Dongxing	Nanjing University, Nanjing/China
Guo Fangzhun	CHI-VAC Research & Development Co., Ltd., Dalian/China
Gustafson Debbie	Energetiq Technology, Inc., Woburn/USA

Guttmann Peter	Helmholtz-Zentrum Berlin fuer Materialien und Energie GmbH, Berlin/Germany
Hansson Bjorn	Excillum AB, Kista/Sweden
Hansson Emelie	Royal Institute of Technology (KTH), Stockholm/Sweden
Hatano Tadashi	Tohoku University, Sendai/Japan
Hayes Kerry	Australian Synchrotron,Clayton/Australlia
Hertz Hans	Royal Institute of Technology (KTH), Stockhom/Sweden
Higashiguchi Takeshi	Utsunomiya University, Utsunomiya/Japan
Eba Hiromi	Tokyo City University, Tokyo/Japan
Hofer Pascal	Dectris Ltd., Baden/Switzerland
Holt Martin	Center for Nanoscale Materials / ANL, Argonne/USA
Holzner Christian	Xradia Inc., Pleasanton/USA
Hornberger Benjamin	Xradia, Inc., Pleasanton/USA
Huang Qiushi	Tongji University, Shanghai/China
Hwang Chan-Cuk	Pohang Accelerator Laboratory, Pohang/South Korea
Hwang Jeongeun	Pohang University of Science and Technology (POSTECH), Pohang/South Korea
Jacobsen Chris	Argonne National Laboratory, Argonne/USA
Jiang Huaidong	Shandong University, Jinan/China
Jiang Licai	Rigaku Innovative Technologies, Inc., Auburn Hills/USA
Jiao Zhang	Shanghai Jiao Tong University, Shanghai/China
Li Jiaxin	CHI-VAC Research & Development Co.,Ltd., Dalian/China
Johansson Ulf	MAX IV Laboratory, Lund/Sweden
Jung Jiwon	Pohang University of Science and Technology (POSTECH), Pohang/South Korea
Jung Sung Yong	Pohang University of Science and Technology (POSTECH), Pohang/South Korea
Kado Masataka	Japan Atomic Energy Agency, Kyoto/Japan
Kagoshima Yasushi	University of Hyogo, Hyogo/Japan
kai zhang	Institute of High Energy Physics (IHEP),CAS, Beijing/China
Karvinen Petri	Paul Scherrer Institut (PSI), Villigen/Switzerland
Kaulich Burkhard	Diamond Light Source Ltd (DLS), Oxfordshire/UK
Kaznatcheev Konstantine	Photon Sciences, Brookhaven National Laboratory (BNL), Upton/USA
Keskinbora Kahraman	Max Planck Institute for Intelligent Systems, Stuttgart/Germany
Kewish Cameron	Synchrotron Soleil, Saint-Aubin/France
Kim Chan	Gwangju Institute of Science and Technology, Gwangju/South Korea
Kim Hae Koo	Pohang University of Science and Technology (POSTECH), Pohang/South Korea
Kim Yoonhee	Gwangju Institute of Science and Technology, Gwangju/South Korea
Kiskinova Maya	Elettra Laboratory, Trieste/Italy
Kosior Ewelina	European Synchrotron Radiation Facility (ESRF), Grenoble/France
Koyama Takahisa	Japan Synchrotron Radiation Research Institute (JASRI), Hyogo/Japan
Kraemer Markus	AXO DRESDEN GmbH, Dresden/Germany
kuniko takemoto	Kansai Medical University, Hirakata/Japan
Kurta Ruslan	Deutsches Elektronen-Synchrotron (DESY), Hamburg/Germany
Lagomarsino Stefano	Istituto per i Processi Chimico-Fisici Consiglio Nazionale delle Ricerche (IPCF-CNR), Rome/Italy
Lai Barry	Argonne National Laboratory, Argonne/USA
Larsson Daniel	Royal Institute of Technology (KTH), Stockholm/Sweden
Lee Sang Joon	Pohang University of Science and Technology (POSTECH), Pohang/South Korea
Lerotic Mirna	2nd Look Consulting, Hongkong/China
Li Junyun	Oxford Instruments, Shanghai/China
Li Zhilong	Tongji University, Shanghai/China
Liese Tobias	Institut für Materialphysik / University of Göttingen, Göttingen/Germany
Lim Jae-Hong	Pohang Accelerator Laboratory, Pohang/South Korea
Lim Jun	Pohang Light Source, Pohang/South Korea
Liu Ocean	Oxford Instruments, Shanghai/China
Liu Xianping	Tongji University, Shanghai/China
Liu Yijin	Stanford University, Stanford/USA
Lu Yingshen	Canadian Light Source, Saskatoon/Canada
Lundstrom Ulf	Royal Institute of Technology(KTH), Stockholm/Sweden
Luo Hongwei	University of Science and Technology of China(USTC), Hefei/China
Luo Jun	Nanjing University, Nanjing/China
Mak Rachel	Northwestern University, Evanston/USA
Mao Wendy	Stanford University, Stanford/USA
Mayo Sheridan	Commonwealth Scientific and Industrial Research Organisation (CSIRO), Melbourne/Australia
McNulty Ian	Argonne National Laboratory, Argonne/USA
Medjoubi Kadda	Synchrotron Soleil, Saint-Aubin/France
Menzel Andreas	Paul Scherrer Institut (PSI), Villigen/Switzerland
Modregger Peter	Paul Scherrer Institut (PSI), Villigen/Switzerland
Moerner William E.	Stanford University, Stanford/USA
Mokso Rajmund	Paul Scherrer Institut (PSI), Villigen/Switzerland

Morrison Graeme	University College London, London/UK
Nelson Johanna	SLAC National Accelerator Laboratory, Menlo Park/USA
Nho Hyun Woo	Hanyang University, Seoul/South Korea
Nilsson Daniel	Royal Institute of Technology (KTH), Stockholm/Sweden
Ohchi Tadayuki	NTT Advanced Technology Corp., Kanagawa Prefecture/Japan
Ohigashi Takuji	Institute for Molecular Science, Okazaki/Japan
Ohldag Hendrik	SLAC National Accelerator Laboratory, Menlo Park/USA
Otsuka Takamitsu	Utsunomiya University, Utsunomiya/Japan
Park Soyeong	Pohang University of Science and Technology (POSTECH), Pohang/South Korea
Paterson David	Australian Synchrotron, Clayton/Australlia
Peele Andrew	Australian Synchrotron, Clayton/Australlia
Yu Pengfei	Shanghai Institute of Ceramics, CAS, Shanghai/China
Pereiro Eva	ALBA synchrotron, Barcelona/Spain
Poitry-Yamate Carole	Ecole Polytechnique Fédérale de Lausanne (EPFL), Lausanne/Switzerland
Raabe Joerg	Paul Scherrer Institut (PSI), Villigen/Switzerland
Rau Christoph	Diamond Light Source Ltd., Oxfordshire/UK
Ren Xiaoxu	Diamond Light Source Ltd., Oxfordshire/UK
Rose Volker	Argonne National Laboratory, Argonne/USA
Roy Sujoy	Lawrence Berkeley National Laboratory (LBNL), Berkeley/USA
Rutishauser Simon	Paul Scherrer Institut (PSI), Villigen/Switzerland
Ryan Chris	Commonwealth Scientific and Industrial Research Organisation (CSIRO), Melbourne/Australia
Salditt Tim	Institute for X-Ray Physics, Göttingen/Germany
Sasov Alexander	Bruker-MicroCT, Kontich/Belgium
Satoshi Matsuyama	Osaka University, Osaka/Japan
Schlotter Bill	SLAC Linac Coherent Light Source (LCLS), Menlo Park/USA
Schneider Gerd	Helmholtz Zentrum Berlin (HZB), Berlin/Germany
Schroer Christian	Technische Universität Dresden (TU-Dresden), Dresden/Germany
Selin Maarten	Royal Institute of Technology (KTH), Stockholm/Sweden
Shand Matthew	Silson ltd., Northampton/UK
Shen Qun	Brookhaven National Lab, Upton/USA
Shi Xiaowen	University College London, London/UK
Shi Yingxia	Stanford University, Stanford/USA
Shin Hyun-Joon	Pohang Accelerator Laboratory, Pohang/South Korea
Shorne stephen	Energetiq Technology Inc., Woburn/USA
Shu Deming	Argonne National Laboratory, Argonne/USA
Snigirev Anatoly	European Synchrotron Radiation Facility (ESRF), Grenoble/France
Snigireva Irina	European Synchrotron Radiation Facility (ESRF), Grenoble/France
Somogyi Andrea	Synchrotron Soleil, Saint-Aubin/France
Song Xiangxia	University of Science and Technology of China (USTC), Hefei/China
Stahlhut Philipp	Lehrstuhl für Röntgenmikroskopie Universität Würzburg, Würzburg/Germany
Stampanoni Marco	Paul Scherrer Institut / ETH Zürich, Villigen/Switzerland
Stockmar Marco Kurt	Technische Universität München (TU München), München/Germany
Suhonen Heikki	European Synchrotron Radiation Facility (ESRF), Grenoble/France
Sun Tao	Argonne National Laboratory, Argonne/USA
Sun Tianxi	College of Nuclear Science and Technology, Beijing Normal University, Beijing/China
Suzuki Yoshio	SPring-8, Hyogo/Japan
Takahashi Yukio	Osaka University, Osaka/Japan
Takano Hidekazu	University of Hyogo, Kamigori/Japan
Takeuchi Akihisa	JASRI / SPring-8, Hyogo/Japan
Takuya Tsuji	JASRI/SPring-8, Hyogo/Japan
Tang Mau-Tsu	National Synchrotron Radiation Research Center (NSRRC), Hsinchu(Taiwan)
Thieme Juergen	Brookhaven National Laboratory, Upton/USA
Tkachuk Andrei	Xradia, Pleasanton/USA
Uhlén Fredrik	Royal Institute of Technology (KTH), Stockholm/Sweden
Vartiainen Ismo	Paul Scherrer Institut (PSI), Villigen/Switzerland
Verman Boris	Rigaku Innovative Technologies, Inc., Auburn Hills/USA
Vila Comamala Joan	Argonne National Laboratory, Argonne/USA
Vogt Stefan	Argonne National Laboratory, Argonne/USA
Vogt Ulrich	Royal Institute of Technology (KTH), Stockholm/Sweden
Wan Xiaoming	Institute of Geographic Sciences and Natural Resources Research, CAS, Beijing/China
Wang Chenyang	Shanghai Institute of Ceramics, CAS, Shanghai/China
Wang Haipeng	Shanxi University, Taiyuan/China
Wang Hongchang	Diamond Light Source, Oxfordshire/UK
Wang Huiqiong	Xiamen University, Xiamen/China
Wang Jian	Canadian Light Source Inc., Saskatoon/Canada
Wang Jun	Brookhaven National Laboratory, Upton/USA

Wang Zhanshan	Physics Department, Tongji University, Shanghai/China
Watanabe Norio	Faculty of Pure and Applied Sciences, University of Tsukuba, Ibaraki/Japan
Watts Benjamin	Paul Scherrer Institut (PSI), Villigen/Switzerland
Wei Der-Hsin	National Synchrotron Radiation Research Center, Hsinchu(Taiwan)
Weigand Markus	Max Planck Insitute for Intelligent Systems, Stuttgart/Germany
Werner Stephan	Helmholtz-Zentrum Berlin für Materialen und Energie, Berlin/Germany
Wessels Philipp	Institute for Experimental Physics, University of Hamburg, Hamburg/Germany
Wiesemann Urs	Bruker ASC GmbH, Bergisch Gladbach/Germany
Woloschak Gayle	Northwestern University, Evanston/USA
Wu Danlin	Tongji University, Shanghai/China
Wu Nae-Lih	National Taiwan University, Taipei(Taiwan)
Xiong Gang	London Centre for Nanotechnology, London/UK
Xu Feng	Karlsruhe Institute of Technology (KIT), Karlsruhe/Germany
Yabashi Makina	RIKEN SPring-8 Center, Sayo/Japan
Yamauchi Kazuto	Osaka University, Osaka/Japan
Yang Shengfeng	NTT-AT Shanghai representative Office, Shanghai/China
Yang Yang	Karlsruhe Institute of Technology (KIT), Karlsruhe/Germany
Yang YS	Commonwealth Scientific and Industrial Research Organisation (CSIRO), Melbourne/Australia
Liu Gang	National Synchrotron Radiation Laboratory (NSRL), University of Science and Technology of China, Hefei/China
Guan Yong	National Synchrotron Radiation Laboratory (NSRL), Hefei/China
Hong Youli	Institute of High Energy Physics, CAS, Beijing/China
Yuan Qingxi	Institute of High Energy Physics, CAS, Beijing/China
Yun Sylvi	Xradia, Pleasanton/USA
Yun Wenbing	Xradia, Inc., Pleasanton/USA
Zanette Irene	Technische Universität München (TU-München), München/Germany
Zhang Grace	Oxford Instruments, Shanghai/China
Zhang Guanhua	Dalian Institute of Chemical Physics, Dalian/China
Zhao Piqi	College of Materials Science and Engineering, Tongji University, Shanghai/China
Zhao Roger	Oxford Instruments, Shanghai/China
Zheng Jincheng	Xiamen University, Xiamen/China
Zhou Tunhe	Royal Institute of Technology (KTH), Stockholm/Sweden
Zhu Caizhen	Institute of Chemistry, CAS, Beijing/China
Zhu Jing	School of Materials Science and Engineering, Dalian University of Technology, Dalian/China
Zhu Peiping	Institute of High Energy Physics, CAS, Beijing/China
Zhu Yu-Min	Tongji University, Shanghai/China
Wu aiguo	Ningbo Institute of Materials Technology & Engineering (NIMTE), CAS, Ningbo/China

Journal of Physics: Conference Series **463** (2013) 011002 doi:10.1088/1742-6596/463/1/011002

Peer review statement

All papers published in this volume of *Journal of Physics: Conference Series* have been peer reviewed through processes administered by the proceedings Editors. Reviews were conducted by expert referees to the professional and scientific standards expected of a proceedings journal published by IOP Publishing.

Diffractive optics for laboratory sources to free electron lasers

U Vogt, J Reinspach, F Uhlén, D Nilsson, H M Hertz and A Holmberg

KTH/Royal Institute of Technology, Biomedical and X-ray Physics, Albanova University Center, SE – 106 91 Stockholm, Sweden

Email: uvogt@kth.se

Abstract. In this contribution we present our recent results in the field of diffractive optics for both soft and hard x-ray radiation, and for laboratory sources to x-ray free electron lasers (XFEL). We developed a laboratory soft x-ray microscope that uses in-house produced zone plate optics as high-resolution objectives. We continuously try to improve these optics, both in terms of efficiency and resolution. Our latest development is the manufacturing of tungsten soft x-ray zone plates with outermost zone widths of 12 nm and 90 nm high structures. For hard x-rays, we investigated the possibility to use metal zone plates on a diamond substrate for nano-focusing of the European X-ray Free Electron Laser. The simulations show that the heat conduction is efficient enough to keep a zone plate well below melting temperature. However, metal zone plates will experience large and rapid temperature fluctuations of several hundred Kelvin that might prove fatal. To test this, we manufactured tungsten on diamond prototype zone plates and exposed them to radiation from the LCLS XFEL. Results show that metal zone plates can survive the XFEL beam.

1. Introduction

Zone plates are high-quality diffractive optical elements which can be used for diffraction-limited x-ray focusing and imaging down to the 10 nm spatial resolution range [1]. In the soft x-ray range, zone plates are the method of choice used in all existing high-resolution x-ray microscopes. For focusing of hard x-ray radiation other alternatives in the form of refractive and reflective optics are widely in use, mainly because of the difficulty to produce high-efficiency hard x-ray diffractive optics. However, recent progress in nanofabrication improved the performance of hard x-ray zone plates considerably, so their use becomes more and more common. In this paper we present the progress of the Stockholm x-ray optics group in the field of diffractive optics for both hard and soft x-rays, for applications in the laboratory and at free-electron laser facilities.

Zone plates are circular diffraction gratings with decreasing line widths towards the outer diameter. Two major parameters characterize the performance of these optical devices, namely the height or thickness of the grating lines and the width of the outermost line. The latter defines the numerical aperture of the optic and therefore the achievable spatial resolution, which is approximately equal to the smallest line width. The thickness defines the efficiency of the zone plate, i.e., how much of the incoming radiation can be found in the first diffraction order which is normally used for imaging. With typical values for optimal efficiency ranging from 100 nm to several microns and outer line width below 100 nm, the fabrication of these optics can only be done with state-of-the-art nanofabrication techniques.

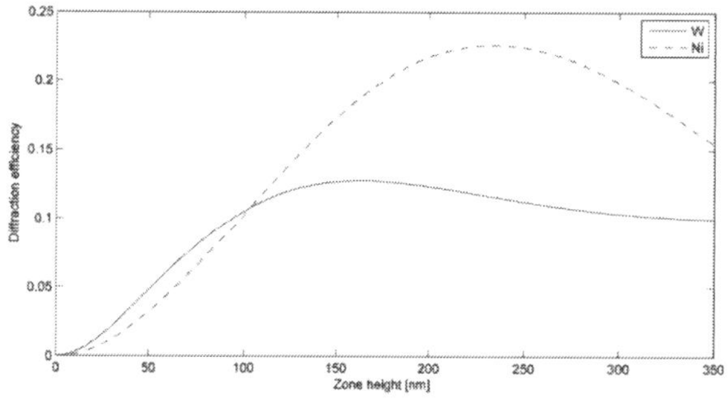

Figure 1. The first order theoretical diffraction efficiency of tungsten (W, solid line) and nickel (Ni, dashed line) as a function of zone height for 2.48 nm wavelength.

In the following we describe our own work towards zone plates with better resolution and higher efficiency for soft x-ray microscopy and hard x-ray zone plates for high-heat-load XFEL applications.

2. Improved soft x-ray zone plates for laboratory x-ray microscopy

The Stockholm group has pioneered the field of laboratory soft x-ray microscopy in the water-window spectral region. Instead of synchrotron radiation, x-rays emitted from laser-produced plasmas are used. Zone plates act as high-resolution objectives in the microscope, and the need for these optics was the reason to start our own nanofabrication of diffractive optics.

Nickel is the optimum material for the 2.48 nm operation wavelength of our microscope, as can be seen in figure 1. Our standard zone plates have outermost zone width of Δr = 25-30 nm and a zone height of 150-200 nm, close to the optimum value for highest efficiency. For details about the manufacturing process, see Ref. [2]. In an attempt to improve the resolution of the optics, i.e., to achieve smaller Δr, we could manufacture nickel zone plates with Δr = 13 nm [3] at a zone height of 35 nm, yielding a theoretical diffraction efficiency of below 2%. This height limit is mainly due to collapsing of higher zones in the electroplating step, and therefore we investigated alternative approaches that could lead to better efficiency.

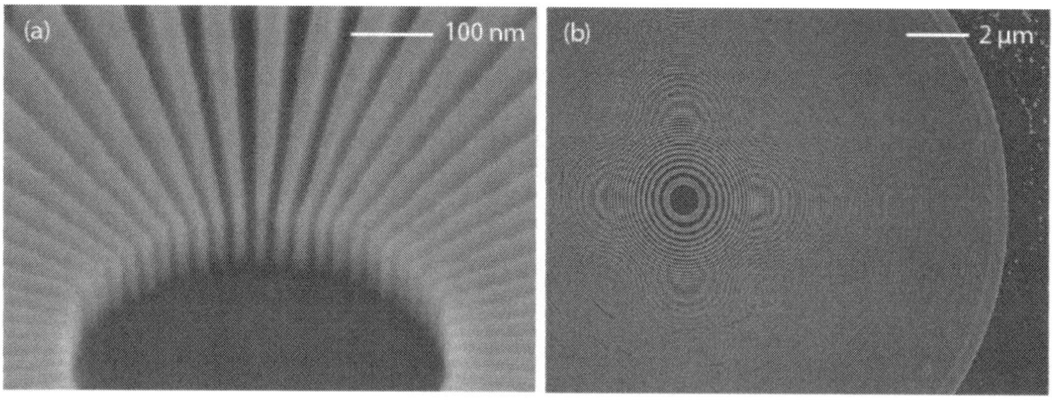

Figure 2. Fabricated high-resolution tungsten structures for soft x-ray microscopy. (a) Inner part of a Siemens star test pattern with innermost half-period of 12.5 nm. (b) Zone plate with Δr = 12 nm. The structure height is 90 nm in both cases.

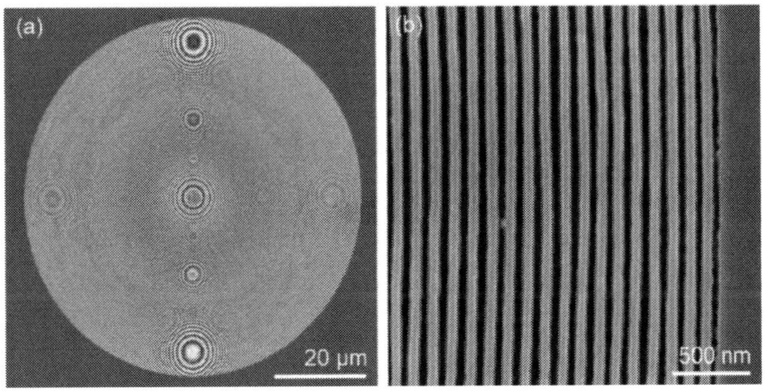

Figure 3. Fabricated prototype tungsten-on-diamond zone plates for FEL applications. (a) Overview of zone plate with 75 µm diameter. (b) Outer part of zone plate with $\Delta r = 50$ nm. Tungsten height is 500 nm.

Tungsten is a material used for hard x-ray zone plates [4], and the efficiency curve for 2.48 nm can also be found in figure 1. As it acts as a pure absorbing material for soft x-rays, it cannot achieve the same maximum efficiency as in the case of Ni, but below a thickness of 100 nm both materials have similar performance. However, in contrast to Ni it can be structured directly by reactive ion beam (RIE) etching, avoiding the electroplating step. Accordingly, higher structures and higher efficiencies would be possible.

We successfully developed a novel fabrication process for high-resolution soft x-ray zone plates in tungsten. Details of the nanofabrication process can be found in Ref. [5]. The key step is the pattern transfer into W by RIE in an $SF_6/0_2$ plasma at a cryogenic temperature of -50 °C. Figure 2 shows the scanning electron microscope (SEM) images of a produced zone plate with $\Delta r = 12$ nm and a tungsten height of 90 nm, resulting in a theoretical diffraction efficiency of 9.6%, and a star test pattern with the same height and 12.5 nm wide lines.

3. Hard x-ray zone plates for XFEL radiation

Many application experiments using hard x-ray free electron laser radiation at energies around 10 keV would benefit from nanofocused x-ray beams with diameters around 100 nm. However, focusing to these dimensions is far from a standard task and only recently attempts were made using a Kirkpatrick-Baez mirror system and a set of compound refractive lenses. We have investigated the use of zone plates instead. They are easy to handle since the single-element optic can be quickly aligned in the beam. Moreover, they offer the prospect of focal spot sizes well below 100 nm, as described in the previous section.

An important question is if zone plates can withstand the high-intensity XFEL beam, since a considerable amount of the incoming radiation will be absorbed in the relatively thin zone plate material layer. Therefore we developed a simulation tool for the heat transfer in zone plate optics irradiated with XFEL beams, based on the finite element method. Calculations showed that the single-pulse temperature increase in zone plates made out of heavy-metal materials placed in the direct unfocused XFEL beam is well below the melting temperature of the material [6, 7].

Due to our experiences with tungsten, we fabricated prototype zone plates in this material [8], see also figure 3. The optics use a diamond substrate for optimal heat conduction. The zone plates were then used in two types of test experiments [9]. First, we irradiated the optics with focused, pulsed high-intensity visible laser pulses that should give similar temperature profiles as compared with XFEL radiation. No damage to the zone plate nanostructures could be observed after several million laser pulses. Second, we placed the same zone plates in the direct, unfocused beam (6 keV photon energy) of the LCLS XFEL in Stanford, USA. Also here no damage to the optics could be observed after 100000 pulses. The same was true in a similar, independent study [10]. These are very promising results that justify possible future use of zone plates for XFEL nanofocusing.

4. Conclusion

We successfully fabricated soft and hard x-ray zone plates made out of tungsten. In the soft x-ray range, zone plates with outermost zone width as low as 12 nm could be manufactured with a considerable height of 90 nm for high diffraction efficiency. In the hard x-ray range, tungsten-on-diamond zone plates were fabricated for use with XFEL radiation. These optics were successfully tested at the LCLS XFEL, where no radiation damage to the nanostructures could be observed.

Acknowledgements

This work has been funded by the Swedish Research Council, the Swedish Foundation for Strategic Research and the Göran Gustafsson Foundation. We thank C. David for the help with the LCLS measurements.

References

[1] Attwood D 1999 *Soft x-rays and extreme ultraviolet radiation* (Cambridge: Cambridge University Press)
[2] Holmberg A, Rehbein S and Hertz H M 2004 *Microel. Engin.* **73-74** 639
[3] Reinspach J, Lindblom M, von Hofsten O, Bertilson B, Hertz H M and Holmberg A 2009 *J. Vac. Sci. Technol. B* **27(6)** 2593
[4] Charalambous P 2000 *AIP Conf. Proc.* **507** 625
[5] Reinspach J, Uhlén F, Hertz H M and Holmberg A 2011 *J. Vac. Sci. Technol. B* **29(6)** 06FG02
[6] Nilsson D, Holmberg A, Sinn H and Vogt U 2010 *Nucl. Instrum. Meth. A* **621** 620
[7] Nilsson D, Holmberg A, Sinn H and Vogt U 2011 *Proc. SPIE* **8077** 80770B
[8] Uhlén F, Lindqvist S, Nilsson D, Reinspach J, Vogt U, Hertz H M and Holmberg A 2011 *J. Vac. Sci. Technol. B* **29(6)** 06FG03
[9] Nilsson D, Uhlén F, Reinspach J, Hertz H M, Sinn H and Vogt U 2012 *New J. Phys.* **14** 043010
[10] David C *et al* 2011 *Scientific Reports* **1** 57

A dedicated micro-CT beamline for the Australian Synchrotron and the Remote-CT project

S C Mayo[1], T E Gureyev[1], Y I Nesterets[1], D A Thompson[1], K K W Siu[2] and K Wallwork[2]

[1]CSIRO Materials Science and Engineering, Private Bag 33, Clayton South, VIC 3169, Australia
[2]Australian Synchrotron, 800 Blackburn Rd, Clayton, VIC 3168

E-mail: Sherry.Mayo@csiro.au

Abstract. A dedicated micro-CT beamline is planned for the Australian Synchrotron which will extend the synchrotron's imaging and tomography capability down to the smaller scale, incorporating phase-contrast and absorption-contrast, and an additional focussing-based mode for high-resolution. The beamline will use multi-layer mirror monochromators for enhanced flux, and will focus particularly on dynamic and high throughput studies in both monochromatic and pink-beam mode. Together with the existing Imaging and Medical beamline, this beamline will produce numerous large datasets of 10 GB or more, providing a significant data-processing challenge. The Remote-CT project addresses this by combining the "MASSIVE" supercomputing GPU cluster with XLI / X-TRACT software, developed at CSIRO. This software has extensive functionality for both processing and simulation of absorption and phase-contrast tomography data and has now been modified for parallel operation on a GPU cluster to take maximum advantage of the speed-up this enables.

1. Introduction

Australia has an active micro-CT community active in lab-based micro-CT [1] and already using overseas synchrotrons such as the APS, SLS and ESRF for conducting micro-CT experiments which cannot be performed in the lab. In many cases this involves in-situ or dynamic experiments. At the Australian Synchrotron the Imaging and Medical Beamline (IMBL), which opens to users in November 2012, will provide a facility targeted primarily at larger-scale imaging and tomography together with radiotherapy. The planned micro-CT beamline is envisaged as a complementary beamline to IMBL, focused on higher-resolution, smaller-scale applications and with a particular emphasis on high-speed CT for dynamic, in-situ and high-throughput applications.

2. A dedicated micro-CT beamline on a bending magnet

The micro-CT beamline will use a bending magnet source with multi-layer broad-band monochromation to maximize monochromatic flux, together with pink/white-beam operation when maximum flux is required for speed. It will have two end stations within one extended hutch and will support in-line and grating-based phase-contrast modes with fields of view ranging from 20mm down to ~1mm with a submicron pixel size and resolution. It will also have a secondary focusing-based mode for higher resolution hard x-ray micro-CT with a resolution of <100nm. The main operational modes and operating parameters of the beamline are summarized in Table 1.

Table 1. Main parameters of the planned micro-CT beamline

Energy Range	8-40keV	
Source type	Bending magnet 6BM1	
Monochromation type	White beam, Pink beam, Multilayer broadband monochromator (bandwidth 2%), Si monochromators (bandwidth 0.02%), possible option of extra broadband multi-layer (bandwidth 8-10%)	
End stations	Station 1 (upstream)	a) Parallel-beam imaging and micro-CT with option of in-line phase contrast. Including high-speed with automated sample exchange. b) Focussing-based high-resolution imaging/MCT mode. (with Stn.2)
	Station 2 (downstream)	a) Grating-based phase-contrast imaging/tomography b) Focussing-based high-resolution imaging/MCT mode. (with Stn.1)
Field of view	Station1	4mm V x up to 2cm H (standard) 50-100microns (focussing mode) (can cover larger areas by stitching)
	Station 2	6 mm V x up to 3cm H (can cover larger areas by stitching)
Pixel size range & max resolution	In-line phase-contrast/absorption contrast	Pixels: 0.2 - 10μm Max resolution: ~0.7μm
	Grating-based phase-contrast	Pixels: 1- 10μm Max resolution: ~4μm
	Focussing-based high-resolution mode	Pixels:20nm Max resolution: (~100nm imaging, ~200nm tomography)
Maximum tomographic data collection speed	Inline phase-contrast/ absorption- contrast	Subsecond in high-speed mode, 5-10min at maximum resolution
	Grating-based phase-contrast	15-45 min
	Focussing-based high-resolution mode	15-30min

As noted above a major focus of this beamline will be on high-speed CT for in-situ/dynamic studies, and for high throughput. This will require high-speed stages and detectors, and robotic sample exchange. In addition it is planned to offer a number of sample environments (heating, cooling, pressure, fluid flow) as standard to facilitate in-situ studies. In developing these aspects of the beamline we are fortunate to be able to draw on the considerable experience at high-speed micro-CT beamlines at SLS, ANKA and ESRF [2].

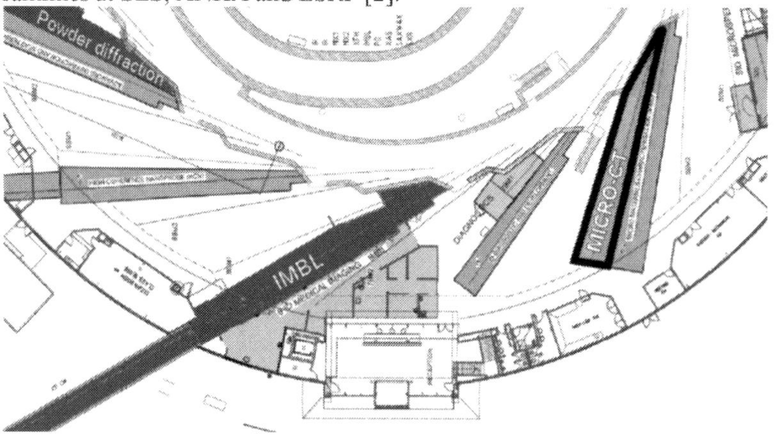

Figure 1. Australian Synchrotron layout showing existing beamlines in black, the micro-CT beamline in grey with a black border and other planned beamlines in grey.

For high-speed micro-CT the aim is to enable data collection within a second or faster if required using pink-beam. This will require high-speed CMOS detectors and reconstruction algorithms able to provide a satisfactory reconstruction with limited numbers of views. This will reduce the volume of data enabling a higher number of continuous scans to be acquired, or else enabling scans to be acquired more rapidly for a given camera speed. Standard air-bearing stages are already available capable of 20 revolutions per second so camera speed and signal-to-noise are the primary limitations. Current state of the art CMOS cameras such as the PCO Dimax enable over 4000 frames per second to be acquired at 1000x1000 resolution, however detectors are rapidly advancing and even faster detectors may become available prior to beamline construction.

High-throughput reconstruction will produce quality data with larger image sizes, and more views than fast CT, at a slower speed perhaps 10-20 minutes per dataset. Automated sample exchange will enable 100 datasets a day to be acquired thus, as with high-speed CT, huge quantities of data will be generated. The following section outlines how we will meet the challenge of processing these quantities of data.

3. The MASSIVE high-performance cluster and the Remote-CT project.

Computational requirements at the Australian Synchrotron are being met by the MASSIVE supercomputer, one of two such machines installed at the Australian Synchrotron and Monash University as part of a Victorian Government funded collaborative project. MASSIVE is a GPU cluster with 42 nodes, each with 2x 6-core X5650 CPUs, 2 nVidia M2070 GPUs and 48GB of RAM. MASSIVE also has a 58TB, GPFS file system capable of 2GB/s sustained write and is connected via a high-performance network (Mellanox IS5200 Infiniband QDR 4X switch).

The existing Imaging and Medical beamline (IMBL) has requirements for high-speed reconstruction of tomographic data up to $(8K)^3$ voxels in size, and both the proposed micro-CT beamline (as outlined above) and IMBL will also be capable of generating many tomographic scans in a relatively short amount of time. The resulting need for high-speed tomographic reconstruction and phase-retrieval is being addressed using X-TRACT software, developed by CSIRO and adapted for parallelized operation on the MASSIVE GPU cluster [3]. Tomographic reconstruction is a readily-parallelizable computational problem and can also benefit greatly from speed-up associated with GPU-based algorithms. Running on the MASSIVE cluster enables X-TRACT to operate in parallel over many CPUs and GPUs, with minimal penalties associated with file input-output (IO) thanks to the high-speed file system.

3.1. X-TRACT software and the Remote-CT project

X-TRACT is a Microsoft Windows application which has been under development in CSIRO for a number of years. It incorporates tools for tomographic reconstruction, phase-retrieval, simulation of phase-contrast/tomographic data, and a large number of general image processing tools which can be accessed and combined into scripts using the X-TRACT Image Calculator.

The Remote-CT project has extensively modified X-TRACT to run in parallel on MASSIVE for high-speed operation with large datasets, and to be accessed remotely over the network. A fully-featured PC version of X-TRACT on a user's computer can be used as a stand-alone application, but it can also operate as a remote client for the server version of X-TRACT running on the cluster. The user only needs to tick a check box for 'remote mode' in the dialog they are using and X-TRACT will 'see' and operate on the cluster file system, dispatching the tasks to the X-TRACT server running on MASSIVE. Currently MASSIVE has dual-booting Windows/Linux nodes to accommodate X-TRACT but a Linux port of the X-TRACT server is currently underway.

The IMBL is connected to MASSIVE via a dedicated optic-fibre link to ensure rapid transfer of data to the cluster. A similar linkage is envisaged for the micro-CT beamline. X-TRACT has been in operation on IMBL for a number of months during commissioning and will be available to users for the start of the user program in late 2012. Users will be able to continue working on their data using the cluster in remote mode after they finish their experiments.

3.1.1. Processing speed-up on MASSIVE

The majority of tasks involved in image processing, phase-retrieval and tomographic reconstruction are 'embarassingly parallel' in that many tasks are carried out file by file and as such can be distributed over many worker threads in order to speed up processing of the full dataset. Initially the time taken goes in inverse proportion to the number of worker threads used, however, for some cluster architectures the file IO starts to limit the increase in speed available by recruiting increasing numbers of worker-threads. In the case of the MASSIVE cluster, however, the high-speed file-system mitigates this problem enabling larger numbers of threads to be used to speed up a task without significant penalty from file IO. This can be seen in Figure 2 which shows a comparison of reconstruction speeds versus number of workers for X-TRACT implemented on the CSIRO GPU cluster and on the MASSIVE cluster. Due to the latter's high-performance IO system, reconstruction time continues to reduce as more workers are used, whereas for the CSIRO cluster reconstruction time levels off for 50 or more workers.

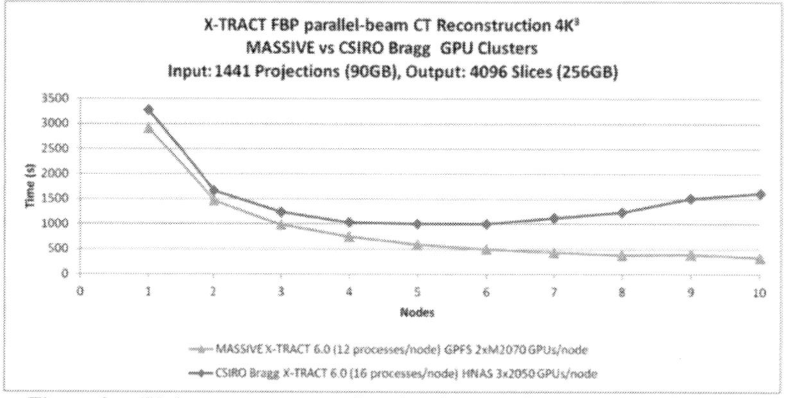

Figure 2 – 4K dataset reconstruction time in seconds vs number of worker threads for the CSIRO GPU cluster and the high-speed IO MASSIVE cluster

4. Conclusions

The micro-CT beamline combined with the IMBL and the supporting high-performance computing facilities will provide high-quality and rapid CT for investigation of dynamic processes and a wide range of samples, with reconstruction tools commensurate with the speed of data collection. Future computational challenges lie in addressing the need for rapid segmentation and analysis of the datasets in order to extract the maximum value from them on a manageable timescale.

References

[1] Trinajstic K, Long J, Johanson Z, Young G, Senden T 2012 *J. Vertebr Paleontol.* **32**, 4 757-780
Wang, Y., Wertheim, D., Jones, A., Chang, H., Coombes, A. 2010. *J. Pharm. Sci* **99**(6), 2854-2862.

[2] Rack A, Weitkamp T, Bauer Trabelsi S, Modregger P, Cecilia A, dos Santos Rolo T, Rack T, Haas D, Simon R, Heldele R, Schulz M, Mayzel B, Danilewsky A N, Waterstradt T, Diete W, Riesemeier H, Müller B R, Baumbach T 2009 *Nucl. Inst. and Meth. B*: **267**, 1978-1988
Mader, K, Marone, F, Hintermuller, C, Mikuljan, G, Isenegger, A, Stampanoni, M 2011 *J. Synchrotron Rad.* **18**, 117
http://www.esrf.eu/UsersAndScience/Experiments/StructMaterials/ID15/Technical_overview/ultra-fast-micro-tomography

[3] Gureyev T E, Nesterets Y I, Thompson D, Wilkins S, Stevenson A, Sakellariou A, Taylor J, & Ternovski D 2011 *Proceedings of SPIE*, vol. **8141**, pp.81410B-1 - 81410B-14,
Thompson D, Nesterets Ya I, Gureyev T E, Sakellariou A, Khassapov A and Taylor J A 2011 MODSIM 2011, Perth, Australia, http://www.mssanz.org.au/modsim2011/A8/thompson.pdf
XLI / X-TRACT, http://www.ts-imaging.net/Services/AppInfo/X-TRACT.aspx.

Direct mapping of ^{19}F in ^{19}FDG-6P in brain tissue at subcellular resolution using soft X-ray fluorescence

C Poitry-Yamate[1,4], A Gianoncelli[2], G Kourousias[2], B Kaulich[3], M Lepore[1,4], R Gruetter[1,4,5] and M Kiskinova[2]

[1]Biomedical Imaging Center, Ecole Polytechnique Fédérale de Lausanne, Switzerland
[2]ELETTRA – Sincrotrone Trieste S.C.p.A., Trieste, Italy
[3]Diamond Light Source Ltd, Harwell Science and Innovation Campus, Oxfordshire, UK
[4]Radiology Department, University of Lausanne, Switzerland
[5]Radiology Department, University of Geneva, Switzerland

E-mail:carole.poitry-yamate@epfl.ch

Abstract. Low energy x-ray fluorescence (LEXRF) detection was optimized for imaging cerebral glucose metabolism by mapping the fluorine LEXRF signal of ^{19}F in ^{19}FDG, trapped as intracellular ^{19}F-deoxyglucose-6-phosphate (^{19}FDG-6P) at 1μm spatial resolution from 3μm thick brain slices. ^{19}FDG metabolism was evaluated in brain structures closely resembling the general cerebral cytoarchitecture following formalin fixation of brain slices and their inclusion in an epon matrix. 2-dimensional distribution maps of ^{19}FDG-6P were placed in a cytoarchitectural and morphological context by simultaneous LEXRF mapping of N and O, and scanning transmission x-ray (STXM) imaging. A disproportionately high uptake and metabolism of glucose was found in neuropil relative to intracellular domains of the cell body of hypothalamic neurons, showing directly that neurons, like glial cells, also metabolize glucose. As ^{19}F-deoxyglucose-6P is structurally identical to ^{18}F-deoxyglucose-6P, LEXRF of subcellular ^{19}F provides a link to *in vivo* ^{18}FDG PET, forming a novel basis for understanding the physiological mechanisms underlying the ^{18}FDG PET image, and the contribution of neurons and glia to the PET signal.

1. Introduction

Among the biological and medical applications combining low energy (soft) x-ray fluorescence (LEXRF) and scanning transmission x-ray microscopy (STXM) [**1**], brain energy metabolism has not been addressed. It is well established that glucose, the brain's principle energy fuel substrate, is essential to maintaining mature brain function; however, its cell-specificity has not yet been completely established. This is regrettable since this issue is key to understanding the mechanism of the tripartite synapse, i.e. presynaptic and postsynaptic neurons and glial processes, quantitating the physiological mechanisms underlying positron emission tomography (PET) images [**2**], and assessing the contribution of neurons and glia to the PET signal.

^{18}FDG (Fluoro-Deoxy-D-glucose) PET is the gold standard in the non-invasive imaging and quantitative evaluation of brain glucose metabolism, both in clinical and research settings. Use of the glucose analogue Deoxy-D-glucose (DG) isolates DG transport into the cell through specific glucose transporter proteins and its intracellular phosphorylation to 2-deoxy-D-glucose-6-phosphate (DG-6P), subsequently trapped in the metabolically active cell in proportion to the rate of glycolysis and at concentrations that can reach μmol/g, detectable by NMR [**3**].

In the present study we have used the structurally identical PET glucose analogue, albeit nonradiolabeled, i.e., ^{19}F-fluoro-deoxy-D-glucose (^{19}FDG) to map ^{19}F in ^{19}FDG-6P at subcellular resolution.

Synchrotron-based X-ray fluorescence imaging is based on secondary (fluorescent) photon emission, unique for each element of the periodic table irradiated with a high flux of photons with sufficient energy that results in the removal of an element's core level electrons. The XRF spectrum therefore contains information on the abundance of a specific elemental constituent in the specimen. Using an X-ray microprobe and scanning the specimen allows 2-dimensional mapping of the sample's elemental distribution at submicron spatial resolution [1]. Since the cross-section of fluorescence emission -- a process based on the photoelectric effect -- depends on the primary photon energy for light elements, such as C, O, N, F etc, sufficient signal can be obtained using soft X-rays. Using low energy XRF (LEXRF), the fluorescent emission of ^{19}F, monitored at 676 eV, is directly related to the cellular compartment where DG-6P is trapped in the brain cell.

We recently demonstrated the feasibility of direct mapping of cerebral glucose metabolism *in situ* at subcellular resolution using LEXRF detection of ^{19}FDG-6P by reproducing the expected localization and gradients of glucose metabolism in glial cells from mammalian retina and periventricular hypothalamus [**4**]. Here, the LEXRF measurements were extended to brain structures more closely resembling the general cerebral cytoarchitecture, following the upgrade of the Twinmic instrument with 4 more fluorescence detectors; introducing endogenous low Z elemental mapping, combined with absorption and phase contrast images, further extended the identification of morphological and structural landmarks and features of the preparation.

2. Methods

2.1 Metabolic ^{19}FDG experiments

The hypothalamus was rapidly excised from brains of rats, decapitated under pentobarbital (65mg/kg) anesthesia. Brain slices ≤ 300μm in thickness were prepared and transferred to glucose (10mM) containing-bicarbonate-buffered Ringer's solution (BBRS, 37^{0}C, pH, 7.4). After 6o min slices were washed of D-glucose and then superfused for 50 min with a mixture of 4 mM ^{19}FDG and 1mM D-glucose. Metabolism was stopped and non-phosphorylated FDG was removed by washing the slices in 3x600μl chilled glucose-free BBRS. Brain slices were stored in 4% formalin in Dulbecco's phosphate buffered saline at 5^{0}C for at least 3 days.

2.2 Specimen preparation and production of planar surfaces

Over the course of 5 days, and subsequent washing to remove formalin, tissue slices were dehydrated in a series of increasing acetone (in %: 30x4, 50x4 70x5 90x5 and 100x5, 20 min each), followed by epon infiltration (acetone:epon by volume: 3:1, 1:1, 1:3 and 0:3). Epon was prepared for maximum hardness to minimize radiation damage [**4**]. Brain slices were reoriented in 100% epon and epon polymerization was performed at 70^{0}C for 2 days. Tissue blocks were shaped and then sectioned to a thickness of 3 μm on a conventional ultra-microtome, removing undesirable topographical information at the sample surface [4] and to enable scanning transmission x-ray imaging of the specimen in parallel with LEXRF detection.

2.3 LEXRF set-up and analysis

LEXRF measurements were performed at 909 eV with a 1 μm diameter X-ray microprobe, delivered by a Fresnel zone plate, providing an increase in the incident photon flux (10^{11} ph/s) by 10-fold compared to an earlier study [**4**]. A significant undertaking of the beamline upgrade was the doubling of silicon drift detectors (SDDs) from 4 to 8 SDDs in an annular configuration. Because X-ray fluorescence is isotropic, the number of detected fluorescence photons is proportional to the number of SDD's, thus enabling a reduction in dwell times from 50s to 10s per 1μm x 1μm pixel. The Kα emissions of O, N and F were monitored and expressed in LEXRF maps as the number of fluorescent photons collected per pixel without beam correction since the synchrotron current was maintained constant in top-up mode, a significant improvement from working in decay mode used earlier [**4**]. The LEXRF set-up was coupled to STXM imaging in both absorption and phase-contrast modes. Deconvolution of the fluorescence spectrum per pixel was performed using the PyMCA Hypermet algorithm [5] and constant baseline subtraction.

3. Results

Subsequent to the upgrade of the Twinmic beamline and further optimization of ^{19}F detection, we aimed to explore the potential of LEXRF to image the stable isotope of fluorine in phosphorylated FDG (^{19}FDG-6P) at 1 μm^2 spatial resolution from brain structures closely resembling the general cerebral cytoarchitecture.

3.1 Metabolic and structural mapping of hypothalamic neurons

The ventral medial hypothalamus (VMH), a putative glucose sensing region of the brain, was chosen for illustrative purposes because of the large size of neurons and their dense clustering *in situ*, 4 of which are visible in Figure 1a. To differentiate structures, differences in the refractive index of biological matter (Fig. 1a) and LEXRF mapping of nitrogen (likely reflecting protein content) (Fig. 1b) were exploited. Nitrogen displayed high specificity for the neuropil (N), and 2 intracellular compartments in neurons: cytoplasm (*) and putative endoplasmic reticulum (ER, open arrows). As shown in Fig. 1c, ^{19}F in ^{19}FDG-6P was disproportionately high in the neuropil, punctuated by "hotspots" (closed arrows), reminiscent of dendritic swellings. A remarkable finding was the unequivocal localization of trapped ^{19}FDG-6P in the two aforementioned intracellular compartments of the cell body, best rendered when N and F maps were merged (Fig. 1d). Merging both Z elemental maps, moreover, unambiguously revealed two additional neuron cell bodies (outlined in yellow) -- not discernible by x-ray transmission imaging -- which have been sectioned through different planes as judged by cell body diameter

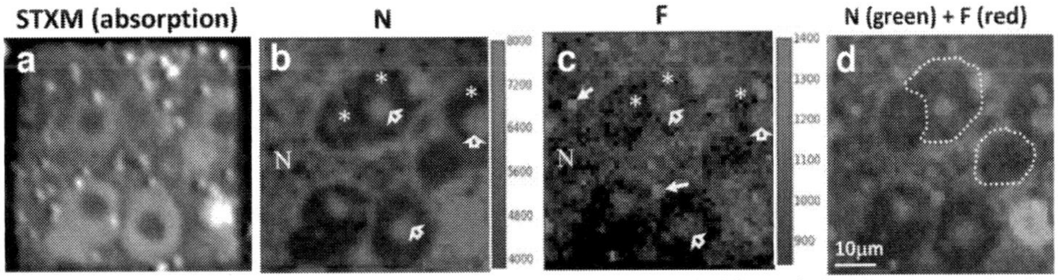

Fig. 1 Direct evidence of glucose metabolism by ventral medial hypothalamic neurons *in situ* and the use of endogenous nitrogen to map brain cytoarchitecture. Scale: photon counts.

and the presence or absence of putative ER. The distribution of ^{19}FDG-6P within the yellow outlines clearly illustrates the glycolytic capacity of these neurons (Fig. 1c). Together, such results, reproducible in 4 preparations, show that glucose metabolism occurs in VMH neurons and even moreso in the neuropil -- the region containing dendritic and axonal processes, synaptic connections, and glial endfeet processes.

3.2 Further structural mapping of hypothalamic neurons by LEXRF mapping of oxygen

In the example illustrated in Figure 2, LEXRF measurements were pursued within the ventral medial hypothalamus. Three neurons, identical to those shown in Fig. 1a, were readily identifiable in the absorption image (Fig. 2a, left hand side), while structures elsewhere were indiscernible. In striking contrast, the full-field structural content of the VMH of Fig. 2a was revealed in fine detail when the maps of oxygen (likely reflecting complex lipids) and nitrogen (likely reflecting protein content) were co-registered (Fig. 2b). Note the exquisite rendering of dendritic/axonal processes emanating from the cell bodies of 2 fusiform-shaped neurons located

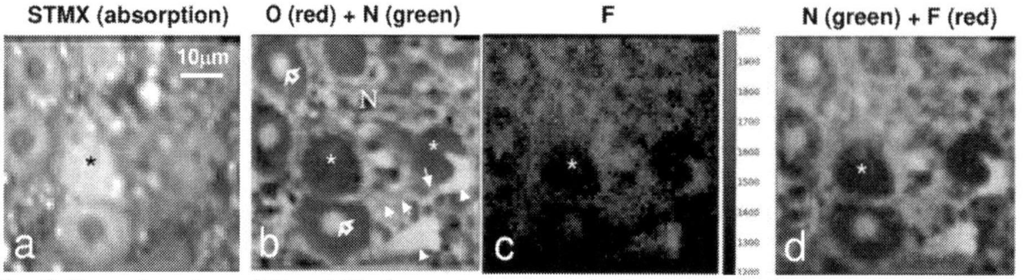

Fig. 2 Co-registration of O and N places glucose metabolism by 2 distinct neuronal cell types within a rich morphological and cytoarchitectural framework.

in the map at the lower right. The dendritic/axonal processes are indicated at the white arrows, while the white arrow heads point to the neuronal cell bodies. Co-registration of oxygen and nitrogen maps additionally delineated the neuropil (N) from the cell cytoplasm (*) and putative ER (open yellow arrows) of the large diameter neurons to a greater extent than seen in Fig. 1b, enabling the identification of 7 large diameter neurons sectioned through different planes of the cell body. This result exemplifies the inherent power of imaging multiple Z elements linked to endogenous organic molecules of the specimen to differentiate cellular structures used as landmarks. Like the neurons in Fig. 1c, the ^{19}FDG-6P distribution in large neurons situated in the upper half of Fig. 2c correlates well with nitrogen content (Fig. 2d). This relationship was modified in the microenvironment of the glycolytically active fusiform-shaped neurons.

4. Discussion

Imaging cerebral ^{18}FDG-6P in practice (Fig. 3) has been extended here to imaging ^{19}FDG-6P in one selected brain region with subcellular resolution. We exploited x-ray fluorescence imaging of F, O and N, where well-defined fluorescence cross-sections and the high brightness of the synchrotron source allowed obtaining element-specific information in the micron domain, showing that neurons from a functionally key region of the central nervous system resembling the general cerebral cytoarchitecture take up and phosphorylate glucose.

We first discuss the importance of LEXRF from the perspective of an ^{18}FDG PET neuroimaging scientist. We then discuss why LEXRF-based ^{19}F detection is extremely challenging, and how it might be extended even further. Finally, we discuss why LEXRF imaging of multiple Z elements potentially offers outstanding alternatives to conventional cell stains and dyes, and how the ^{19}FDG-6P results impact on current neuroscientific thought.

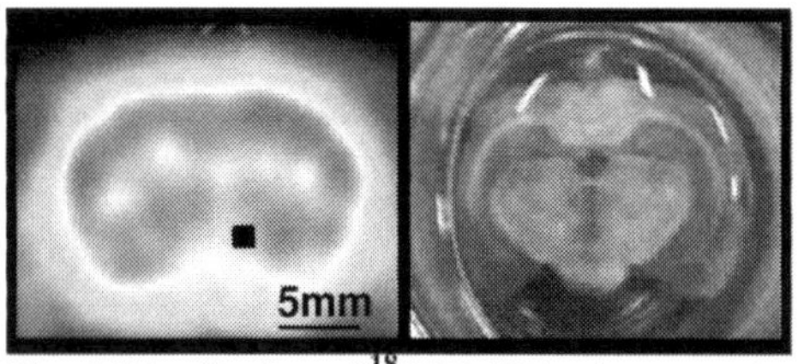

Fig. 3 Distribution of ^{18}FDG-6P in the coronal plane at Bregma -1.8mm of rat brain, placed next to the equivalent brain slice post-mortem.

Positron diffusion by virtue of ^{18}F decay limits the ultimate resolution attainable by PET and is the principle reason why the distribution of ^{18}FDG-6P (Fig. 3, at left) at steady-state appears as a blurry image of the rat brain (at right) at comparable stereotaxic brain coordinates. Given that the region of interest (black square) in the PET image covers the hypothalamus of the right hemisphere, non-invasive ^{18}FDG PET imaging allows, at best, quantitation of glucose metabolism from one brain region to another. Clearly, tackling the questions addressed in the Introduction requires micron resolution, unattainable with ^{18}FDG PET methodology. The commercial availability of non-radiolabeled FDG, i.e. ^{19}FDG, and the fact that the low Z element fluorine is not endogenous to brain tissue per se -- subsequently confirmed experimentally, see Supplementary Data -- were together an attractive combination that led us to assess the application of synchrotron-based x-ray fluorescence to address cellular brain energy metabolism that underlies brain function.

Both methodologies are complimentary and synergistic, and each has its advantages and disadvantages, summarized in Figure 4.

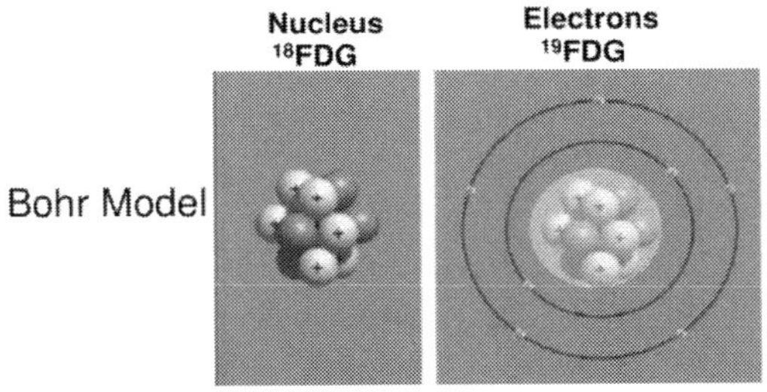

Physical process	positron emission (β^+)	photoelectric effect
Signal emitted	2 high energy γ-photons (511keV)	fluorescent x-ray photons (250-750eV)
Methodology	**PET**	**synchrotron-based LEXRF**
[FDG]	[nM]	> [1mM]
Spatial resolution	1-2 mm	≤ 1μm
Preparation	in vivo	dry, fixed tissue
Object thickness	2-3 cm	≤ 3μm
Morphology	CT	Z elements, STXM; no stains/dyes;
Partial vol. effects	spill-in/spill-out	endogenous F indistinguishable from detector noise
Multiple detection	no, all at 511keV	yes, multiple Z elements

Fig. 4 Salient features of ^{18}FDG positron emission tomography (PET) and synchrotron-based LEXRF of ^{19}F in ^{19}FDG-6P. In both methodologies F replaces the hydroxyl group at the second carbon of the glucose molecule. While the physical process is confined to the nucleus of the F atom in PET, that in LEXRF is confined to the electron orbitals of the F atom. To date, no other Z element conventionally linked to intracellularly trapped DG-6P, e.g. ^{3}H or ^{14}C [**4**] has the ability to be resolved as a stable element (at the second carbon) of the DG/DG-6P molecule from that linked to endogenous organic molecules comprising biological matter.

Three inter-related x-ray reasons and 3 metabolic-related reasons make LEXRF measurements of ^{19}F(DG-6P) extremely challenging. In the first case: (1) the low-energy incident photons needed to remove fluorine's core level electrons have low penetrating power; (2) F has inherently low fluorescence yields [**6**] due to low interaction cross-section, hence low probability of secondary photon emission; and (3) potential self-absorption in brain samples due to low energy, i.e, 676 eV fluorescence emission [**7**]. In the second case: (1) the transport of ^{19}FDG (at non-tracer amounts) into the cell was in direct competition with D-glucose, decreasing the amount of phosphorylated ^{19}FDG; (2) there is possible loss of DG-6P (^{19}FDG-6P), known to be soluble and diffusible during tissue fixation [**8**]; and (3) the distribution of trapped ^{19}FDG-6P cannot be a priori known as it depends on the distribution and affinity of glucose transporter proteins, brain region and stimulus conditions; and the level through which cells of the specimen were microtome-sectioned. The X-ray challenges have been largely overcome by the upgrade of the beamline and optimization of diffractive optics; challenges related to brain energy metabolism are multiple, and increasingly complex. Nonetheless, we are actively pursuing the metabolic experiments under physiological conditions in the live animal, and using a redesigned and high performance tissue fixation procedure.

Cellular brain nitrogen distribution displayed high specificity for the neuropil; and what we believe is the endoplasmic reticulum: currently the beamline set-up requires the removal of tissue water, mainly cytoplasmic, which was displaced by epon that is itself poor in N (see Supplementary data). Likewise, nucleoplasm, of similar low protein content as cytoplasm, was also likely displaced by epon. Nevertheless, N and O mapping is structural, architectural mapping of nitrogen and oxygen linked to organic compounds found in living matter; their detection is independent of the dimensions of cellular and extracellular compartments—a serious impediment for the penetration of cell stains and dyes.

The high uptake of ^{19}FDG in the neuropil implies highly active glucose metabolism in astrocytes (glia cells) which individually occupy non-overlapping domains ~50 μm in diameter, to form a spatially patterned mosaic [**9**]. Given the low level of dendritic arborisation [**10**] and the importance of glutamate uptake into neighboring astrocytes [**11**], it is likely that glucose-metabolism is very high in active astrocytes occupying a large part of the cortical neuropil in an interwoven fashion. However, the expected distribution ^{19}FDG-6P in multiple cellular compartments, shown here for neurons, indicates that glial cells are not the exclusive site of glucose metabolism (and ^{18}FDG PET signal).

5. Conclusion

The present study describes the adaptation, optimization and use of synchrotron-based LEXRF imaging of fluorine in ^{19}FDG trapped as ^{19}FDG6P in biological tissue subsequent to the upgrade of the TwinMic LEXRF set-up to 8 SDDs. We conclude that LEXRF of F is a novel imaging approach that allows studying glucose metabolism at subcellular resolution *in situ*. We furthermore conclude that glucose metabolism occurs both in neurons and glial processes in brain structures resembling the general cerebral cytoarchitecture. Mapping endogenous N and O, in addition to transmission x-ray imaging, enabled placing F maps in a stunning morphological and cytoarchitectural context. Coupled to the subcellular resolution capabilities and the quantitative nature of elemental detection, synchrotron-based LEXRF and STXM together hold remarkable potential for assessing cerebral FDG metabolism. Extending this technology to high Z elements, i.e. $Z > 50$ and high-energy XRF on a suitable experimental x-ray beamline is expected to be more sensitive, as photon yield and penetration depths increase with Z [**12**].

Supported by the Integrated Infrastructure Initiative "European Light Sources Activities – Synchrotrons and Free Electron Lasers," contract RII3-CT-2004-506008 (C. P-Y), the CIBM and Swiss National Science Foundation (131087, RG). We thank Professor Giorgio Margaritondo for the financial contribution to the purchase of 4 SDDs, and Dr Christian David at the PSI, Villigen, Switzerland for the Fresnel zone plate.

References

[1] Gianoncelli A, Kaulich B, Alberti R, Klatka T, Longoni A, De Marco A, Marcello A, and Kiskinova M 2009 Nucl Instr. Meth. Phys. Res. S **608** 195
[2] Frackowiak R S J, Magistretti P J, Shulman R G, Altman J S, and Adams M 2002 Strasbourg, Human Frontier Science Program Organization
[3] Cohen D M, Wei J, O'Brian Smith E, Gao X, Quast M J, and Sokoloff L 2002 Magn Reson Med **48** 1067
[4] Poitry-Yamate C, Gianoncelli A, Kaulich B, Kourousias G, Magill A W, Lepore M, Gajdosik V, and Gruetter R 2012 JNR (in press).
[5] Solé V A, Papillon E, Cotte M, Walter Ph, and Susini J 2007 Spectrochim Acta B **62** 63
[6] Hubbell J H, Trehan P N, Singh N, Chand B, Mehta D, Gard M L, Garg R R, Singh S and Puri S 1994 J Phys. Chem. Ref. Data **23** 339
[7] Hoefler H, Streli C, Wobrauschek P, Ovari M and Zaray Gy 2006 Spectrochimica Acta B **61** 1135
[8] Nehlig A, Wittendorp-Rechenmann E, and Lam C D 2004 JCBFM **24** 1004
[9] Halassa M, Fellin T, Takano H, Dong J-H, and Haydon P G 2007 J Neurosci. **27** 6473
[10] Mishchenko Y, Hu T, Spacek J, Mendenhall J, Harris K M, and Chklovskii D B 2010 Neuron **67** 1009
[11] Poitry-Yamate C, Vutskits L, and Rauen T 2002 J Neurochem **82** 987
[12] Paunesku T, Vogt S, Maser J, Lai B, and Woloschak G 2006 J Cell Biochem **99** 1489

Supplementary Data

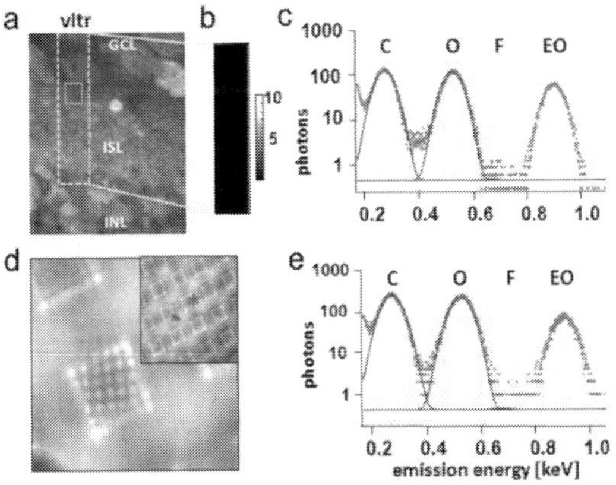

Supplementary Fig. 1 Fluorine is not endogenous to the specimen matrix nor to brain tissue. Modified from [4]. Inset: EM grid with epon matrix out of focus. * shows location of the LEXRF scan.

^{19}F mapped from the 3 innermost and glial-rich stratified layers of retinae (Suppl Fig. 1a), subsequent to exposure to 5mM D-glucose in place of ^{19}FDG, was absent (Suppl Fig. 1b), and normalized photon counts at 676 eV were indistinguishable from detector noise (Suppl Fig. 1c). Similar results were obtained from the specimen matrix, epon (Suppl Fig. 1d and e). Together, these results led us to conclude that the ^{19}F detected is derived from ^{19}F in ^{19}FDG. Since any non-phosphorylated ^{19}FDG was removed by washing [4] the fluorine detected was ^{19}F in ^{19}FG-6P.

Life science applications and research potential of the TwinMic spectromicroscopy station at ELETTRA

A. Gianoncelli[1], G Kourousias[1], L Pascolo[2], C Rizzardi[3], G Ceccone[4], B Kaulich[5], M Kiskinova[1]

[1]ELETTRA – Sincrotrone Trieste S.C.p.A., Trieste, Italy
[2]Institute for Maternal and Child Health, IRCCS Burlo Garofolo, Trieste, Italy
[3]Department of Anatomical Pathology, Department of Pathology and Forensic Medicine, University of Trieste, Trieste, Italy
[4]European Commission, Joint Research Centre, IHCP, Ispra, Italy
[5]Diamond Light Source Ltd, Harwell Science and Innovation Campus, Oxfordshire, UK

E-mail: alessandra.gianoncelli@elettra.trieste.it

Abstract. Since its opening to the users in late 2007, the TwinMic beamline at the Elettra synchrotron has been attracting the interest of a broad life science community. The recent development and implementation of a low energy x-ray fluorescence detection mode have increased the capability and the versatility of the scanning mode of the TwinMic microscope. Now using simultaneously the x-ray transmission and emission detection systems one can perform elemental, absorption and phase contrast imaging that provides complementary chemical and morphological information for the sample under investigation. The recent achievements in the performance of the instrument using combined acquisition modes for characterization of nano and micro-structured biomaterials will be illustrated with selected exemplary results.

1. Introduction

Understanding the biochemistry and the processes occurring in complex biological systems at their natural subcellular length scales requires an interdisciplinary research approach. X-ray microscopy using a synchrotron light that combines high resolution x-ray imaging with element sensitive spectroscopies, namely x-ray fluorescence (XRF) and x-ray absorption spectroscopy (XAS) is one of the powerful techniques that has been gaining the life scientists interest in the recent decade [1, 2]. Both type of X-ray microscopes, full-field imaging and scanning, are in operation at most of the synchrotron facilities using soft or hard X-rays. The major advantage of the microscopes using hard x-rays is their high penetration depth and micro-XRF ability to detect trace elements but for soft matter the absorption of hard x-rays is very weak and for low Z elements the fluorescence signal using hard x-rays is extremely low. This paper emphasizes the potential of soft x-ray microscopy, combing x-ray imaging and low energy x-ray fluorescence (LEXRF) imaging and microspectroscopy for analysis of materials at submicron scale. In particular, we present short description of the instrument and methodology and selected results from the ongoing research related to the cellular distribution of $CoFe_2O_4$ nanoparticles in fibroblast cells and composition and structure of the asbestos body developed in human lung tissues after inhalation of asbestos, which illustrate its potential in bio-medical research.

2. The TwinMic spectromicroscopy station

TwinMic instrument [3] is a soft x-ray microscope operated at the Elettra synchrotron (Trieste, Italy) in the x-ray range 400 and 2200 eV. It consists of two modules that host a full-field transmission x-ray microscope (TXM) and in scanning x-ray microscope (SXM), respectively, with easy switch between the two [3]. Here we report some recent results obtained using the SXM, where the sample is raster scanned under a microprobe formed by zone plate diffractive optics. The transmitted x-rays providing absorption and differential phase contrast images and also micro-XAS spectra are collected by using a fast readout CCD camera or a photodiode [4]. The X-ray fluorescence photon emission, which provides elemental images and LEXRF micro-spot spectra can be simultaneously detected by 8 silicon drift detectors (SDDs) in an annular configuration [5]. The LEXRF mode of TwinMic detects the K lines of elements from B to P and the L lines of chemical elements with emission in the 400-2100 eV energy range (such as Fe, Ni, Cu, and Zn) with submicron spatial resolution [6]. The LEXRF set up has been extensively requested for life science applications since it permits elemental mapping of light elements, namely C, N, O, Mg, Na, that are the main constituents of life matter and the elemental maps can be correlated to the morphology of the specimen revealed by the absorption and phase contrast images.

3. Results

The TwinMic microscope has been successfully used in different research fields including fuel cell technology [7], biotechnology, biomaterials, food science and nanotoxicology, neuroscience and clinical medicine. Recent research efforts have provided new insights into (i) the morphology distribution and correlation of the elements as result of plant's growth under various environmental conditions and in presence of toxic substances [8, 9], (ii) concentration dependence of penetration of magnetic nanoparticles in different cell organelles and changes in the nanoparticle chemistry inside the cells [10,11], (iii) cellular internalisation and degradation of magnetic nanoparticles in digestive gland epithelium [12], (iv) reaction of lung tissue in the presence of inhaled asbestos fibres [13], (v) glucose metabolism in neurons [14], and many others. In the paper we report two examples of ongoing bio-related studies that illustrate the potential of the instrument in relevant research domains.

3.1 Cellular distribution of $CoFe_2O_4$ nanoparticles in fibroblast cells

One of the ongoing nanotoxicology-related research efforts focuses on the analysis of the effects of magnetic nanoparticles on cells. In particular, we have investigated Balb3T3 mouse fibroblast cells grown on Si_3N_4 windows after exposure for 24 hours to different concentrations (40 and 1000 μM) of $CoFe_2O_4$ nanoparticles (NPs) with average NP diameter of 30 nm. We analysed at least 3 cells for each condition. These results represent the first attempt of evaluating the behavior of $CoFe_2O_4$ NPs at sub-cellular level by x-ray spectromicroscopy. The information contained in LEXRF maps and micro-spot XRF spectra [10], acquired after treatments using different NPs concentration exposure have revealed that, for concentrations below 500 μM, Co and Fe are localized in the cytosol, i.e. in the perinuclear region, whereas at higher NPs concentrations, although the most of the NPs are still localized in the cytosol, Fe and Co XRF peaks were also observed in the nuclear region of the cell. Moreover, the Fe and Co XRF intensities retained the stoichiometric ratio of the intact NPs, whereas the Fe/Co ratio in the nucleus indicated a sensible accumulation of Co. As an example Figure 1 shows the absorption, phase contrast, and elemental XRF maps of a fibroblast cell grown in a medium containing $CoFe_2O_4$ NPs with concentration of 1000 μM. While the C signal delineates the cell structure and morphology, the Fe and Co maps indicate the distribution of the NPs. Since oxygen is also present in the cell constituents the O signal inside the cell becomes stronger where the particles are located. The absorption and phase contrast images clearly show the presence of a white corona delimitating the cell nucleus that becomes particularly visible at high NPs concentrations, which is associated with the formation of liquid droplets [10].

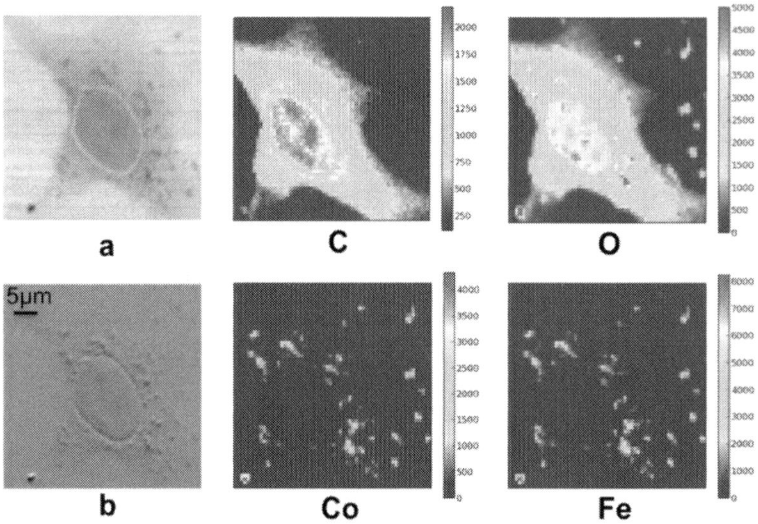

Fig. 1 Absorption (a) and phase contrast (b) images (40μm x 40μm) of a fibroblast cell exposed to 1000 μM concentration of $CoFe_2O_4$ NPs and corresponding elemental maps of C, O, Fe and Co, acquired at a photon energy of 1.1 keV and with a spot size of 670 nm.

3.2 Asbestos body morphology and composition in human lung tissues

Asbestos toxicity and its carcinogenic effect are known since a few decades but the mechanism involved in the inflammation and toxicity process are still not well understood.

The results obtained with TwinMic [12] represent the first attempt for exploring the morphology and composition of an asbestos body in the lung of exposed patients using untreated paraffinated histological lung tissues by a combination of soft x-ray microscopy approaches. The observed specimen morphology and differences in the local distribution of specific chemical elements (namely Si, Fe, O and Mg) have provided distinct fingerprints for the core asbestos fibre, where Si is the main constituent, and for the ferruginous bodies contained in the tissues, characterized by a strongly increased content of Fe, Mg and O compared to the surrounding tissue. The results have shown that, along with Fe, Mg also plays an important role in the formation mechanisms of asbestos bodies that can be related to processes arising as a response of the lung tissue to the asbestos toxicity [13].

Summary of the asbestos-relevant results is shown in Figure 2. The absorption and phase contrast images of an asbestos body in lung tissue coupled with the corresponding elemental maps of Si, Mg and Fe, reveal that the darkest, highly absorbing regions are the one containing Fe, which is the principle constituent of the body developed around the fibre that is well visible in the Si map. The presence of Mg coincides with the location of Fe but some accumulation of Mg can be clearly found also in the tissue in the vicinity of the body.

This first study was carried out at low energy (2 keV) by using the TwinMic XRF setup. Further combined investigations (TwinMic beamline in Elettra and ID21 beamline in ESRF) at higher energy will be presented in a future manuscript in preparation.

4. Conclusion

The present paper briefly describes the research potential of the TwinMic spectromicroscope in the life science field by concentrating on two case studies of on-going research. The combination of x-ray imaging (absorption and phase contrast) with elemental analysis (XRF) have provided new insights in the biological events occurring at submicron scale. The planned future developments aim at complementing the XRF with XANES spectroscopic analyses, also available at TwinMic microscopy station (see for instance ref. 7), where speciation will add new information to morphological and elemental analyses.

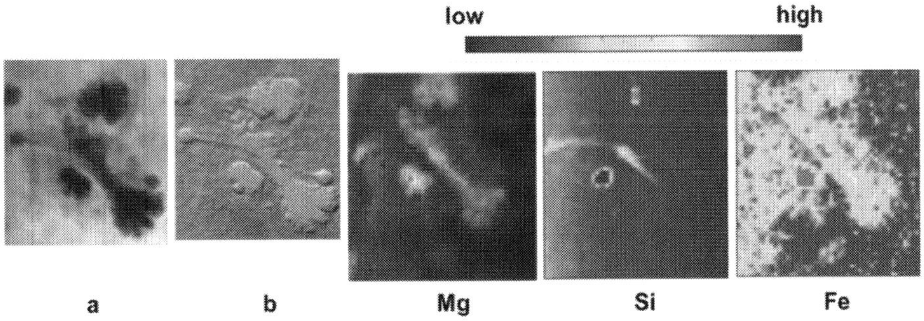

Fig. 2 Absorption (a) and phase contrast (b) images (25μm x 32μm) of an asbestos body in lung tissue. The corresponding elemental maps show the distribution of Si, Mg and Fe in an area surrounding the asbestos body (29μm x 35μm, 1.95 keV, 500nm beam diameter).

References

[1] Howells H, Jacobsen C and Warwick A Principles and Applications of Zone Plate X-ray Microscopes, Springer, Berlin 2006

[2] Ide-Ektessabi A. Application of Synchrotron Radiation: Microbeams in Cell Microbiology and Medicine. Spinger-Verlag, Berlin 2007

[3] Kaulich B, et. Al., in Proc. 8th Int. Conf. X-ray Microscopy, edited by S. Aoki, Y. Kagoshima & Y. Suzuki (IPAP,Tokyo,2006) **2006** 7: 22

[4] Gianoncelli A, Kaulich B, Alberti R, Klatka T, Longoni A, De Marco A, Marcello A and Kiskinova M 2009 *Nucl Instrum Meth A,* **608** 195

[5] Gianoncelli A, Morrison GR, Kaulich B, Bacescu D and Kovac J *Appl Phys Lett* 2006 **89**.

[6] Kaulich B, Gianoncelli A, Beran A, Eichert D, Kreft I, Pongrac P, Regvar M, Vogel-Mikus K and Kiskinova M *J R Soc Interface* 2009 **6** S641

[7] Bozzini B, Gianoncelli A, Kaulich B, Kiskinova M, Prasciolu M and Sgura I *ChemSusChem* 2010 **3** 846

[8] Regvar M, Eichert D, Kaulich B, Gianoncelli A, Pongrac P, Vogel-Mikus K and Kreft I *Journal of Experimental Botany* 2011 **62** 3929

[9] Tolrà R, Vogel-Mikuš K, Hajiboland R, Kump P, Pongrac P, Kaulich B, Gianoncelli A, Babin V, Barceló J, Regvar M and Poschenrieder C *Journal of Plant Research* 2011 **124** 165

[10] Marmorato P, Ceccone G, Gianoncelli A, Pascolo L, Ponti J, Rossi F, Salomé M, Kaulich B and Kiskinova M 2011 *Toxicology Letters* **207** 128

[11] Gianoncelli A, Marmorato P, Ponti J, Pascolo L, Kaulich B, Uboldi C, Rossi F, Makovec D, Kiskinova M and Ceccone G *X-ray spectrometry*, in press.

[12] Novak S, Drobne D, Golobič M, Zupanc J, Gianoncelli A, Kiskinova M, Kaulich B, Pelicon P, Vaupetič P, Jeromel L, Ogrinc N and Makovec D submitted to *Nanotoxicology* (under review)

[13] Pascolo L, Gianoncelli A, Kaulich B, Rizzardi C, Schneider M, Bottin C, Polentarutti M, Kiskinova M, Longoni A and Melato M *Particle and Fibre Toxicology* 2011 7

[14] Poitry-Yamate C, Gianoncelli A, Kaulich B, Kourousias G, Magill A W, Lepore M, Gajdosik V and Rolf Gruetter R 2012 *JNR*, in press

New Developments in Hard X-ray Fluorescence Microscopy for In-situ Investigations of Trace Element Distributions in Aqueous Systems of Soil Colloids

Sophie-Charlotte Gleber[1], Britta Weinhausen[2], Sarah Köster[2], Jesse Ward[1], David Vine[1], Lydia Finney[1], and Stefan Vogt[1]

[1] X-ray Science Division, Argonne National Laboratory, Argonne, Il 60304, USA

[2] Institut für Röntgenphysik, Georg-August-Universität Göttingen, 37077 Göttingen, Germany

E-mail: gleber@aps.anl.gov

Abstract. The distribution, binding and release of trace elements on soil colloids determine matter transport through the soil matrix, and necessitates an aqueous environment and short length and time scales for their study. However, not many microscopy techniques allow for that. We previously showed hard x-ray fluorescence microscopy capabilities to image aqueous colloidal soil samples [1]. As this technique provides attogram sensitivity for transition elements like Cu, Zn, and other geochemically relevant trace elements at sub micrometer spatial resolution (currently down to 150 nm at 2-ID-E [2]; below 50 nm at Bionanoprobe, cf. G. Woloschak et al, this volume) combined with the capability to penetrate tens of micrometer of water, it is ideally suited for imaging the elemental content of soil colloids. To address the question of binding and release processes of trace elements on the surface of soil colloids, we developed a microfluidics based XRF flow cytometer, and expanded the applied methods of hard x-ray fluorescence microscopy towards three dimensional imaging. Here, we show (a) the 2-D imaged distributions of Si, K and Fe on soil colloids of Pseudogley samples; (b) how the trace element distribution is a dynamic, pH-dependent process; and (c) x-ray tomographic applications to render the trace elemental distributions in 3-D. We conclude that the approach presented here shows the remarkable potential to image and quantitate elemental distributions from samles within their natural aqueous microenvironment, particularly important in the environmental, medical, and biological sciences.

1. Introduction

The trace element distribution on soil colloids determines matter transport through the soil matrix. However, these binding and release processes on the nanometer scale are not fully understood yet, despite the huge impact in various areas of basic human interest: For example, in agriculture, optimization of nutrient uptake would increase plant production and lower fertilization costs, and in healthcare the spread of toxicants towards drinking water reservoirs needs reliable prediction and prevention. Hard X-ray fluorescence microscopy techniques enable unique possibilities to study these colloidal interactions. As shown recently, colloid fractions of soil samples can be investigated in an aqueous environment at the hard x-ray microprobe at beamline 2-ID-E of the Advanced Photon Source, revealing the heterogeneous distribution of trace elements on soil colloids forming clusters [1]. To make full use of this capability for trace element detection on aqueous samples, a manipulative sample mount was introduced allowing for in-situ manipulation to follow the soil dynamics due to changes in the pH of the soil surrounding water. With the implementation of a 4 element fluorescence detector (Vortex-ME4), the fast fly scan mode became more powerful. Dwell times per pixel in the millisecond range and no overhead time resulting when step scanning the sample stage in x direction enable on-the-fly imaging of most samples in significantly shorter time even at smaller step sizes and thus reasonable oversampling. This is crucial for general operation of in-situ manipulation as well as tomography using X-ray fluorescence (XRF).

2. In-situ manipulation of aqueous samples

For sample manipulation at the X-ray fluorescence microprobe, the sample holder described in [1] was used. Fig. 1 demonstrates its application on a cluster of soil colloids in an aqueous Pseudogley sample: Two scans of the same cluster (imaged 5 min apart) are shown, prior to (left) and after (right) the pH of the surrounding media was increased by the addition of urea in-situ. Most of the cluster remains unchanged, indicating the structural stability despite the slight flow during in-situ injection. However, two initially prominent particles of high carbon content and medium iron content detach from the outer area of the cluster, and the arrangement in the upper area changes slightly as well. The change of location of the large Si particle demonstrates clearly the mobility of particles between both Si_3N_4 membranes of the sample holder.

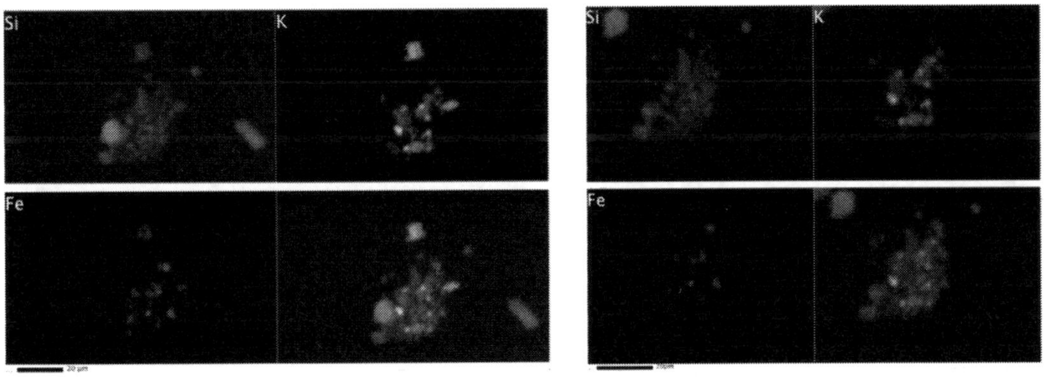

Figure 1. In-situ manipulation of an aqueous Pseudogley sample. 3 element colocalization of the Si fluorescence signal (red), K fluorescence signal (green), Fe fluorescence signal (blue) and the overlay of the 3 elements for projections before (left) and after (right) in-situ addition of urea. Note the changes in cluster arrangement. Scale bar 20 µm. Scans taken at 10keV with 0.5 µm step size and 20 ms dwell time per pixel in fly scan mode. Data analysis: MAPS [3].

3. X-ray fluorescence flow cytometer

A photograph of the XRF flow cytometer at the bending magnet beamline 8-BM-B is shown in Fig. 2. An ideally horizontally focused x-ray beam illuminates a vertical channel of an XRF microfluidic device with a typical width of tens of micrometers and 20 µm height. A syringe pump feeds one of the channels of a microfluidic device with a solution of particles. The sample flow gets adjusted such that particles pass the beam individually and an XRF spectrum gets collected per particle. A camera mounted to a zoom lens records the particle flow, so the fluorescence signal and single particles can be matched and statistically interpreted as a function of the elemental distribution within a population of particles.

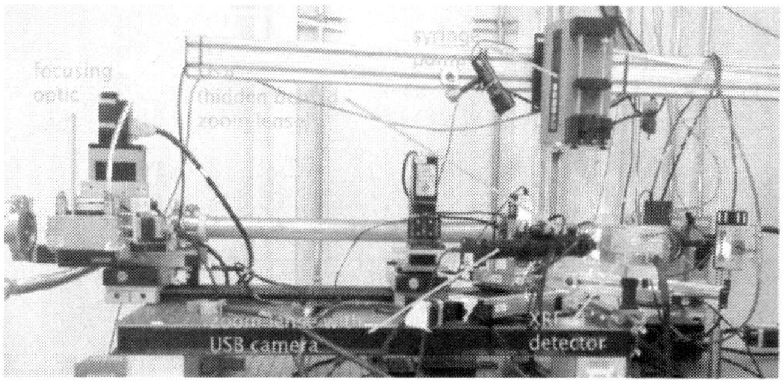

Figure 2. XRF flow cytometer. Main components: focusing optic (Fresnel zone plate or compound refractive lens), OSA, XRF detector at -90° to incident beam, zoom lens with USB camera, and syringe pump feeding microfluidic device at flow rates adjusted to the flow properties and fluorescence signals of the sample particles.

4. X-ray fluorescence tomography

As shown in Fig. 3, a new tomography stage has been implemented to the 2-ID-E microprobe, allowing for robust conditions for both two and three dimensional scanning operation. This setup enabled acquiring the first XRF tomography data set from a cluster of soil colloids of Pseudogley in its aqueous environment enclosed by a chamber constructed of silicon glue-sealed stack of Si_3N_4 membranes. Two projections are shown in Fig. 4, spanning a rotation angle of 24° and revealing a coarse impression of depth within the sample. The entire data set ranges from -60° to 135° in 3° or 6° rotation step sizes. Due to the total frame thickness of 400 µm of the stack of Si_3N_4 windows, the wedge where the fluorescence signal or the incident beam get blocked is larger than for dry samples.

Figure 3. Sample environment. **Left:** He environment sample box at 2-ID-E. Tomography stage included, holding 2-ID-E standard sample mounts with Al stick. In the back: Order Sorting Aperture (OSA) and at -90° to incident beam, fluorescence detector snout with collimator. For standard 2-d scans, sample at -15° to fluorescence detector (as shown). **Right:** Sample mount. Al stick holding sample support is inserted to kinematic mount, shown with stack of two Si_3N_4 windows for aqueous samples.

5. Conclusion

New, exciting capabilities for dynamic, three-dimensional and quantitative XRF studies on sub micrometer length scales have been demonstrated. They are also very valuable for other research areas besides the investigation of soil colloidal interactions: While in-situ manipulation of aqueous samples is not applicable for samples very sensitive to radiation damage, the new XRF tomography setup has been used successfully on samples from material science, biology and biomedical research, e.g. chemically fixed or embedded cell samples were often stable over for a tomographic data set of 60 projections. This achievement is also due to the lower dosage on the sample with a scanning setup compared to full field imaging. The flow cytometer is also well suited for biological samples, as it provides trace element content per particle without high

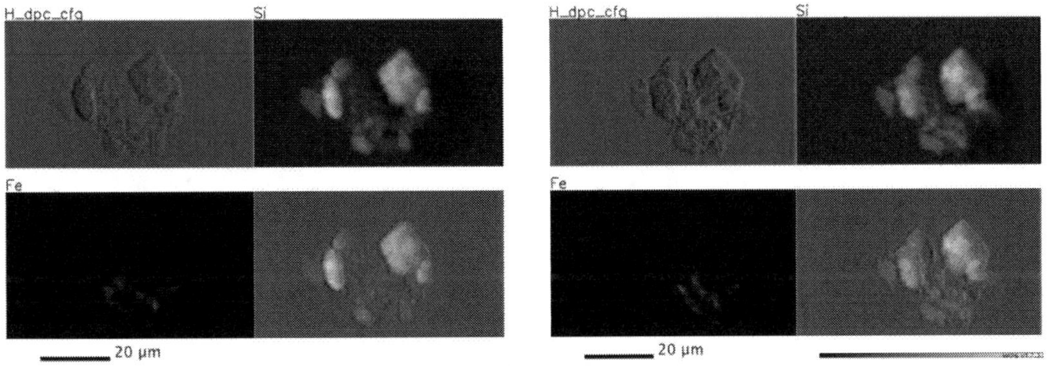

Figure 4. Two projections from tomographic data set of aqueous Pseudogley. 3 element colocalization of horizontal differential phase contrast (red), Si fluorescence signal (green), Fe fluorescence signal (blue) and overlay of the 3 elements for projections at sample rotation angles -21° (left) and -45° (right). Scale bar indicates 20 µm. Scans taken at 10 keV with 0.3 µm step size and 10 ms dwell time per pixel in fly scan mode. Data analysis: MAPS [3].

spatial resolution. Even if microscopic structural damage of a particle occurs, it is while the particle is in the focused beam and the XRF data is already getting recorded. For samples which consist of single particles and remain stable in a liquid medium, this method provides statistically valuable data complementing time consuming high resolution XRF data on single particles. To further improve these techniques, we are focusing on the development of quantitatively correct tomographic reconstruction algorithms as well as the optimization of microfluidic devices for x-ray fluorescence to study dynamic processes under controlled conditions of mixing [4].

Acknowledgments

We would like to thank C. Roehrig, E. Maxey, D. Legnini, and R. Bradford (ANL) for their competent technical support, M. Rivers and T. Lanzirotti (UChicago) for fruitful discussion on fluorescence detector and tomography software, and F. DeCarlo, P. Shevchenko and B. Lai (ANL) for generous supply with optics. Use of the Advanced Photon Source at Argonne National Laboratory was supported by the U.S. Department of Energy, Office of Science, Office of Basic Energy Sciences, under Contract No DE-AC02-06CH11357. The work was partly funded by the Deutsche Forschungsgemeinschaft under contract no GL681/1-1 and the frame work of SFB755 'Nanoscale Photonic Imaging'. B. Weinhausen was supported by the Deutsche Telekom Stiftung.

References

[1] Gleber S C, Vogt S, Niemeyer J, Finney L, McNulty I and Thieme J 2011 *The 10th International Conference on X-Ray Microscopy* (*AIP conference proceedings* vol 1365) ed McNulty I, Eyberger C and Lai B (American Institute of Physics) pp 365–368

[2] Twining B S, Baines S B, Fisher N S, Maser J, Vogt S, Jacobsen C, Tovar-Sanchez A and Sañudo-Wilhelmy S A 2003 *Analytical Chemistry* **75** 3806–3816

[3] Vogt S, Maser J and Jacobsen C 2003 *X-Ray Microscopy 2002, 7th Intl. Conf. on X-Ray Microscopy* (*Journal de Physique IV Proceedings* vol 104) ed Susini J, Joyeux D and Polack F (EDP Sciences, Les Ulis) pp 617–622

[4] Brennich M E, Nolting J F, Dammann C, Nöding B, Bauch S, Herrmann H, Pfohl T and Köster S 2011 *Lab on a Chip* **11** 708–716

Construction of the Scanning Transmission X-ray Microscope Beamline at UVSOR

T Ohigashi[1], H Arai[1], T Araki[2], N Kondo[1], E Shigemasa[1], A Ito[1,3], N Kosugi[1] and M Katoh[1]

[1] Institute for Molecular Science, 38 Nishigo-naka, Myodaiji, Okazaki, Aichi 444-8585, Japan

[2] TOYOTA Central R&D Labs., Inc., 41-1, Yokomichi, Nagakute, Aichi 480-1192, Japan

[3] Tokai University, 4-1-1 Kitakaname, Hiratsuka, Kanagawa 259-1292, Japan

E-mail: ohigashi@ims.ac.jp

Abstract. Construction of a scanning transmission x-ray microscope (STXM) beamline is in progress at UVSOR (Okazaki, Japan). To obtain high brilliance with high resolving power for the incident x-rays, the combination of an in-vacuum undulator and a Monk-Gillieson mounting monochromator with a varied line spacing plane grating was adopted. Resolving power, spectrum and size of the incident x-rays of the beamline were discussed through simulations Then the photon flux at a sample of ~10^8 photons/s at the resolving power of 5,000 in the photon energy range from 100 to 700 eV is expected. As the first results, spectrum of calcium in chalk is shown.

1. Introduction

The scanning transmission x-ray microscopy (STXM) is a powerful tool to perform 2-dimensional mapping of the x-ray spectroscopy with spatial resolution of several tens of nm. This method is appropriate for using the soft x-ray region covering the K absorption edges of light elements and the L absorption edges of transition metals. These absorption edges make the STXM useful for study of functional materials, such as polymers and carbon nanotubes. In addition, using the energy range between the K absorption edges of carbon and oxygen (282~539 eV), called the "water window" region, enables us to observe samples in water, such as cells, with high spatial resolution. For these features, the STXM used in the soft x-ray region becomes a unique technique in broad research fields. To achieve a satisfactory performance of the STXM, a tunable x-ray source with high brilliance, such as a synchrotron radiation, is necessary. In these years, several STXM systems have been installed at synchrotron radiation facilities, Advanced Light Source [1, 2], Swiss Light Source [3], BESSY [4], Canadian Light Source [5], National Synchrotron Light Source [6], Shanghai Light Source Facility [7], Elettra [8] and more to come. The UVSOR facility (Okazaki, Japan), is a synchrotron light source (750 MeV, 300 mA) designed for using the photon energy from IR to soft x-rays. In 2012, the storage ring of the UVSOR-II was upgraded to UVSOR-III to increase the brilliance by reducing the emittance from 27 to 15 nmrad. In the upgrade project, an in-vacuum undulator was also installed at the short straight section for the new STXM beamline, BL4U. These improvements and energy specification of the facility are suitable to operate the STXM with a satisfactory performance so that the construction project of the new STXM beamline, BL4U, has been launched.

In this report, the concept of the STXM beamline project, its current status and the perspectives are shown.

2. Construction of the beamline

To achieve a satisfactory performance of the STXM, it is important to use the incident x-rays with high brilliance and stability. In our beamline, the in-vacuum undulator is used as an x-ray source. The undulator has 26 magnetic periods with the period length of 38 mm pitch and the gap width can be changed from 13 to 40 mm. Brightness of the x-ray source with the undulator, calculated by using SPECTRA [9], was plotted in Fig. 1 against the energy with changing the gap width. These plots show that the photon energy range from 100 to 700 eV can be used effectively.

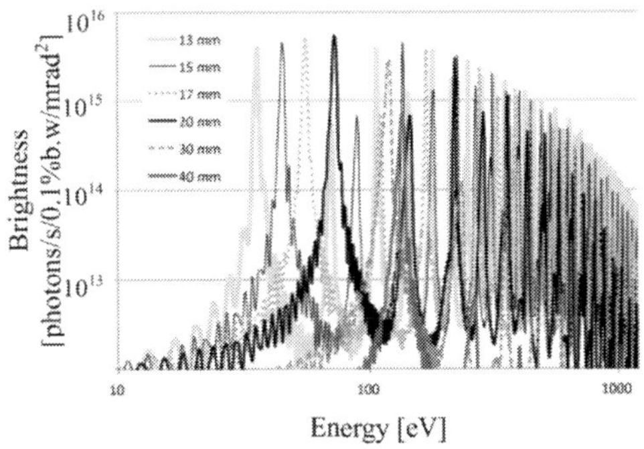

Fig. 1: Calculated brightness of the x-ray source with the in-vacuum undulator with changing the gap

2.1. *Monochromator*

Our beamline employed a variable-included-angle Monk-Gillieson mounting monochromator with a varied line spacing plane grating (VLSPG) [10]. The monochromator consists of 3 mirrors, grating and two slits. A schematic image of the optical system is shown in Fig. 2. The incident x-rays from the undulator enters a toroidal mirror M_0 (R=3725.09 m and ρ=0.2193 m), which is set 6.5 m downstream from the light source with the incident angle of 88° through a slit, S_1. The M_0 mirror focuses the incident x-rays onto a fixed exit slit, S_2, downstream 6.29 m vertically and collimates the x-ray horizontally. The collimated beam is reflected vertically by a plane mirror, M_1, and enters the VLSPG. The line spacing of the VLSPG, N, is given by $N=N_0\cdot(1+a_1\cdot x+a_2\cdot x^2+a_3\cdot x^3)$, where x is a position on the VLSPG and its parameters as follows; N_0=500 lines/mm, $a_1=-3.9534\times10^{-4}$ mm^{-1}, $a_2=1.1668\times10^{-7}$ mm^{-2} and $a_3=-3.1044\times10^{-11}$ mm^{-3}. Sum of the incident angles of the M_1 and the VLSPG varies from 167 to 176°. The beam is focused onto the S_2 horizontally by a cylindrical mirror, M_2 (R=85.963 m) with the incident angle of 88°. Ray-trace simulations of shapes of the beams at the S_2 with changing the resolving power, E/ΔE, of 3,000, 5,000, 10,000 and 20,000 with their vertical FWHMs are shown in Fig. 3. The main photon energy was fixed at 285 eV. These ray-tracing simulations were performed by X.O.P [11]. From Fig. 3, the FHWMs of the vertical beam size (*i.e.* required width of the S_2) become smaller with increasing the E/ΔE and converge at around that of 22 μm. Therefore, the E/ΔE of 10,000~ is considered as the maximum resolving power of our beamline from Fig. 3(c) and 3(d). Practically, it is effective for performance of

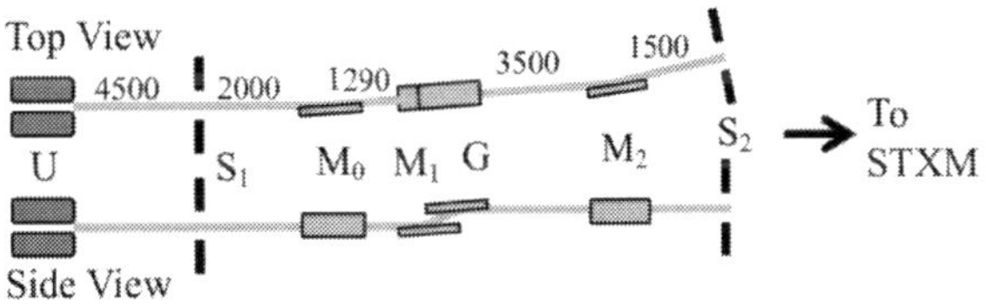

Fig. 2: Schematic images of the optical system of BL4U, where U: the in-vacuum undulator, S_1: the slit, M_0: the toroidal mirror, M_1: the plane mirror, G: the varied line spacing plane grating, M_2: the cylindrical mirror and S_2: the fixed exit slit. The unit of the length is mm.

the STXM to use more photon flux rather than the higher E/ΔE than 5,000 (*i.e.* width of the S_2 is set as 30 μm).

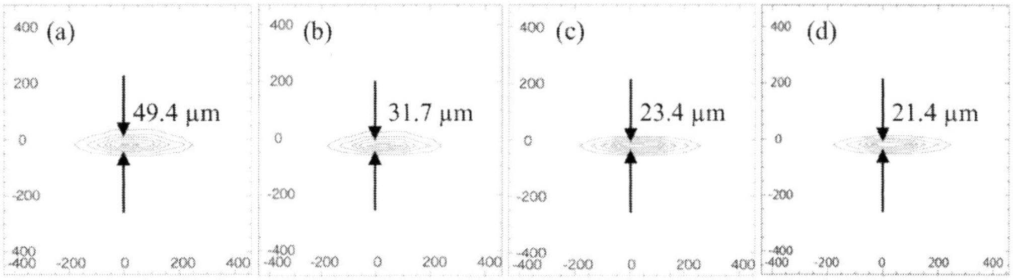

Fig. 3: Ray-trace simulations of the shapes of the monochromatized beams at the fixed exit slit at the E/ΔE of (a) 3,000, (b) 5,000, (c) 10,000 and (d) 20,000 with their vertical FWHMs. The main photon energies are 285 eV. The unit of length is μm.

2.2. Scanning transmission x-ray microscope

A Fresnel Zone Plate (FZP) is used as a focusing optical device. By using the S_2 as a virtual source, the FZP placed at 1,334 mm downstream from the S_2 focuses the x-rays onto the sample through an order sorting aperture. With raster scanning of the sample by piezo stages, the transmitted x-rays are detected by a photon multiplier tube with scintillator.

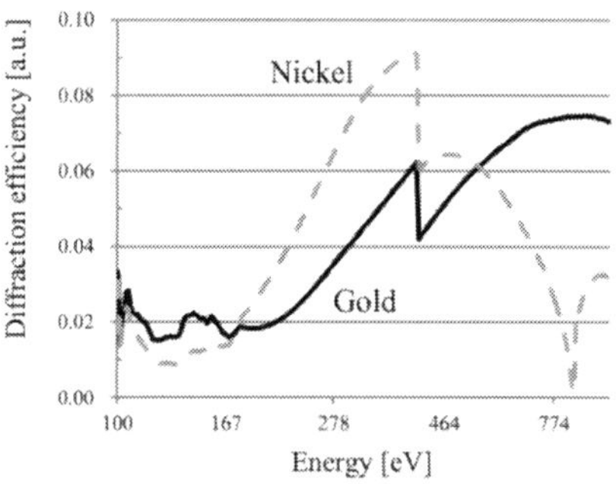

Fig. 4: Calculated efficiencies of the FZPs made of gold and nickel including transmittance of Si_3N_4 substrate membrane

We have been preparing two FZPs, whose parameters are the outermost width of 25 nm, the diameter of 240 μm, the number of the zones of 2,400 and the pattern material of gold or nickel of 100 nm thick. Calculated efficiencies of the FZPs made of gold and nickel, including transmittance of Si_3N_4 support membrane of 100 nm thick, are plotted in Fig. 4. In these plots, the efficiency of the nickel FZP is better than the energy region from 170 to 504 eV so that using the nickel one is more effective to analyze the elements in this region, such as carbon and nitrogen. Totally, by taking into account of efficiency of the FZP, the photon flux of ~10^8 photons/sec at the sample at E/ΔE=10,000 is expected for our beamline.

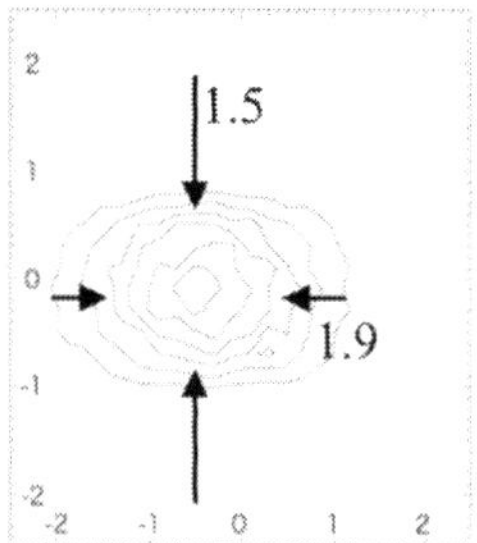

Fig. 5: Simulated shape of the incident beam at the FZP for the diffraction-limited spatial resolution. The unit of length is mm.

To obtain the diffraction-limited spatial resolution for the STXM, the expected focusing spot size of ~30 nm, the FZP should be illuminated by the incident x-rays with longer coherent length than its diameter and with the E/ΔE bigger than number of the zones of the FZP, 2,400. The radius of the coherent area R with coherency of 0.88 is given by $R=0.32z\lambda/D$, where z is the distance between the source and the FZP, λ is the wavelength of the incident x-ray and D is size of aperture of the fixed exit slit, S_2. For the coherent illumination at C K-edge (285 eV), D should be set as 15.5 μm while the E/ΔE becomes more than 10,000. Then, a simulated shape of the incident beam at the FZP is shown in Fig. 5. This beam size of 1.5 (H)× 1.9 (V) mm^2 is large enough to illuminate whole FZP, 240 μm. Therefore, the diffraction-limited spatial resolution is available in this condition.

However, the photon flux is estimated to be about one order less than the condition of S_2 for $E/\Delta E = 10{,}000$.

3. Commissioning of the beamline

In late September 2012, the construction of the beamline was finished and its alignment is started in October. The monochromator is calibrated by measuring K absorption edge spectrum of nitrogen gas (shown in Fig. 6) with setting the S_2 as 30 μm width. From ratio of heights of π^* peak and valley, the $E/\Delta E$ was estimated as around 6,000 and this result was reasonable compared with Fig. 3 (b). From December, the STXM is commissioned and grinded particles of a chalk on microgrid were observed as the first sample by noticing L absorption edge of calcium. 91 Transmitted x-ray images were acquired every 0.1 eV from 346 to 356 eV and the image at 346 eV is shown in Fig. 7 (a) as an example. From area in Fig 7 (a) indicated by a dotted circle, spectrum of optical density of calcium of the chalk is obtained and is shown in Fig. 7 (b). Four peaks, which are assumed to be calcium oxide, are clearly seen on the plot.

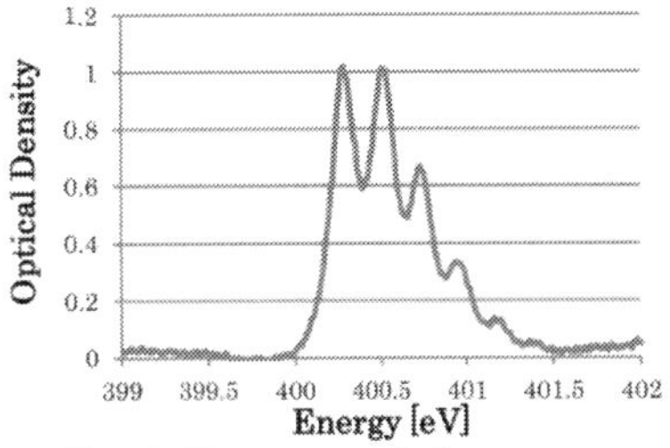

Fig. 6: Spectrum of nitrogen gas

We have been improving our beamline for using wider energy region, stability and calibrations. Our beamiline is due to open for general users in June 2013.

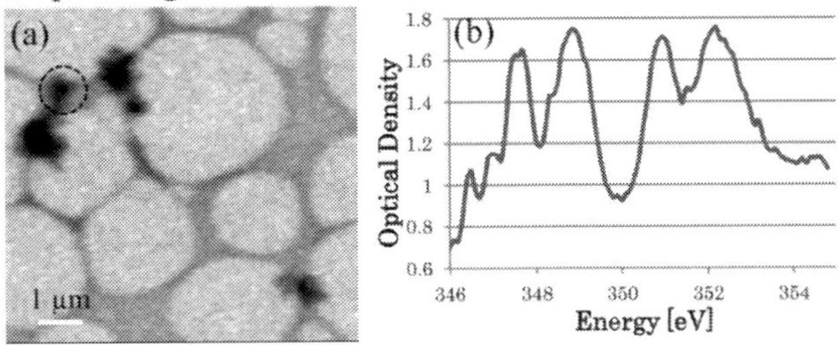

Fig. 7: STXM observation of chalk particles on microgrid (a) transmission image at 346 eV and (b) spectrum of the chalk indicated by a dotted circle in (a)

References

[1] Kilcoyne A L D *et al* 2003 *J. Synchrotron Rad.* **10** 125-136
[2] Bluhm H *et al* 2006 *J. Electron Spectrosc. Relat. Phenom.* **150** 86-104
[3] Raabe J *et al* 2008 *Rev. Sci. Instrum.* **79** 113704
[4] Wiesemann U *et al* 2003 *J. Phys. IV France* **104** 95-98
[5] Kaznatcheev K V *et al* 2007 *Nucl. Instr. Meth. Phys. Res. A* **582** 96-99
[6] Kirz J *et al* 1992 *Rev. Sci. Instrum.* **63** 557-563
[7] Xue C *et al* 2010 *Rev. Sci. Instrum.* **81** 113502
[8] Kaulich B *et al* 2006 *IPAP Conf. Series* **7** pp 22–25
[9] http://radiant.harima.riken.go.jp/spectra/index.html
[10] Koike M and Namioka T 1995 *Rev. Sci. Instrum.* **66** 049501
[11] http://www.esrf.eu/computing/scientific/xop2.1/

Laboratory x-ray microscopy using a reflection target system and geometric magnification

P Stahlhut[1], T Ebensperger[1,2], S Zabler[1,2], R Hanke[1,2]

[1]Chair of X-ray Microscopy, University Wuerzburg, Josef-Martin-Weg 63, 97074 Wuerzburg, Germany
[2]Fraunhofer Development Center X-ray Technology EZRT, Dr.-Mack-Str. 81, 90762 Fuerth, Germany

E-Mail: Philipp.Stahlhut@physik.uni-wuerzburg.de

Abstract: This paper describes a laboratory X-ray microscopy setup, based on geometric magnification. The setup uses a sharp metal tip as a reflection target and a highly focused electron beam of a scanning electron microscope. Here we will describe the structuring process for these metal targets. To demonstrate the capabilities of our system, we show radiographs of test structures corresponding to resolutions below 100nm. There are abilities for 3D imaging in later updates of the system. Further we discuss the first imaging examples for high- and low-absorbing samples.

1. Introduction

The principal limit to the spatial resolution of conventional X-ray sources is set by the focal spot size. In dedicated laboratory setups it can be reduced to less than 1μm [1], e.g. by performing coherent diffraction imaging [2] or by using Fresnel zone-plates [3]. In contrast to these techniques and other which are restricted to synchrotron beamlines, the standard laboratory setup achieves high resolution by geometric magnification in analogy to early X-ray shadow microscopy. This method does not require any optical elements in the beam path, which are both expensive and sensitive to vibrations. Further there are little restrictions concerning X-ray wavelength bandwidth, and one can afford to increase the acceleration voltage of the electron beam up to 30keV without filtering. The physical limit for the resolution of such a system is mainly given by the size of the X-ray source spot. We present a laboratory based X-ray microscope with an ultra small source size and a resolution below 100nm.

2. General considerations

The concept of geometric magnification

$$M = \frac{z_1 + z_2}{z_1}$$

with z_1 for the distance between source and object and z_2 for the distance between objet and a pixelated X-ray detector (shown in fig. 1), is resulting in two main resolution constrains, assuming there is no detector blur. On the one hand the resolution is limited by the physical pixel size P_p of the detector. This leads to a sampling size or an effective pixel size P_s with

$$P_p = M \cdot P_s.$$

On the other hand the sharpness of the picture is limited by the projection of the source s on the imaging plane

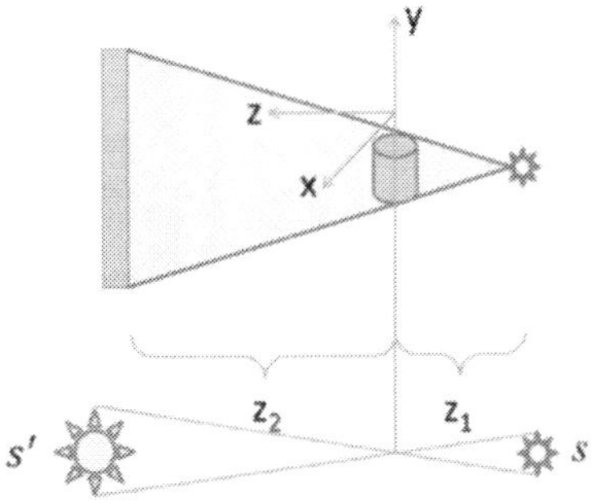

Figure 1. The principle of geometric magnification, by using a pixelated detector. The projection of the source s on the imaging plane s', as well as the effective pixel size are both have an effect on the resolution.

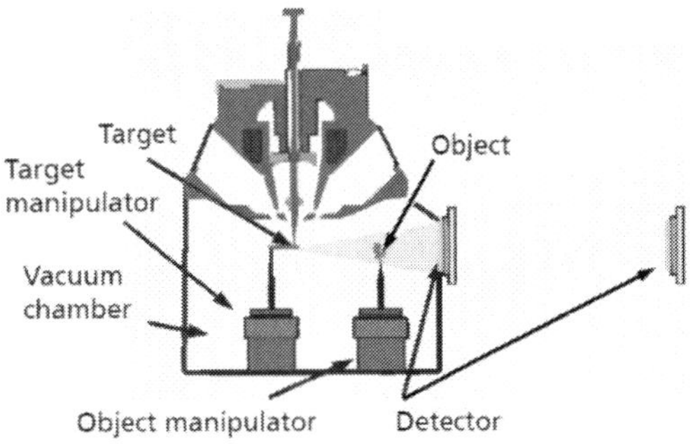

Figure 2. Schematic of the reflection setup. Both target and object are inside the vacuum chamber. The detector is outside and movable.

$$s' = s \cdot \frac{z_2}{z_1} = s \cdot \left(\frac{P_p}{P_s} - 1\right).$$

Putting these two together, plus assuming a spot size $s < 100\text{nm}$ as well as a physical detector pixel size of $P_p = 55\mu\text{m}$, this results into

$$P_{opt} = \frac{s \cdot P_p}{s + P_p} \approx s \qquad ; s \ll P_{p.}$$

Hence the resolution is mainly limited to feature sizes that are in range of the focal spot size.

3. Setup

We modified a JEOL JSM7100-F SEM by replacing the original sample stage with a manipulator for both object and target, which results in a reflection target system (shown in fig. 2). The target is a pointy needle of tungsten or molybdenum with a tip curvature below 100nm, which results in a very small X-ray source spot. The acceleration voltage of the focused electron beam is set to 30keV. To optimize the detected photon flux, we used a 2x3 (hexa assembly) Medipix2 MXR photon counting detector [4]. The sensor layer is composed of 1mm thick cadmium telluride, with 768x512 pixels by a pitch of 55µm. The detector is placed outside of the vacuum chamber. Therefor we used a X-ray window composed of 250µm thick beryllium. This setup allows to work with a small focal spot and achieves large magnifications combined with a source-object distance below 1mm in combination with a reasonable photon flux.

4. X-ray target preparation

Electrochemical etching turned is a fast and reliable method to produce sharp metal tips[5]. As electrolyte we used a 2N sodium hydroxide solution. The two electrodes are identical 500µm thick wires made of either Mo or W. When applying a DC voltage, the etching process is mainly controlled by the electrochemical potential gradient around the tip and the haphazard ionic motion of the electrolyte. With an AC voltage on the other hand the etching rate mainly depends on bubble growth rate and sliding along the tip. The bubble buoyancy during one half of the circle drives the recently etched metal oxides upwards, were they get partially absorbed during the second half of the circle. The result is a higher removal rate of the metal on the lower part of the wire (shown in fig. 3a), in this way the tip shape is formed.

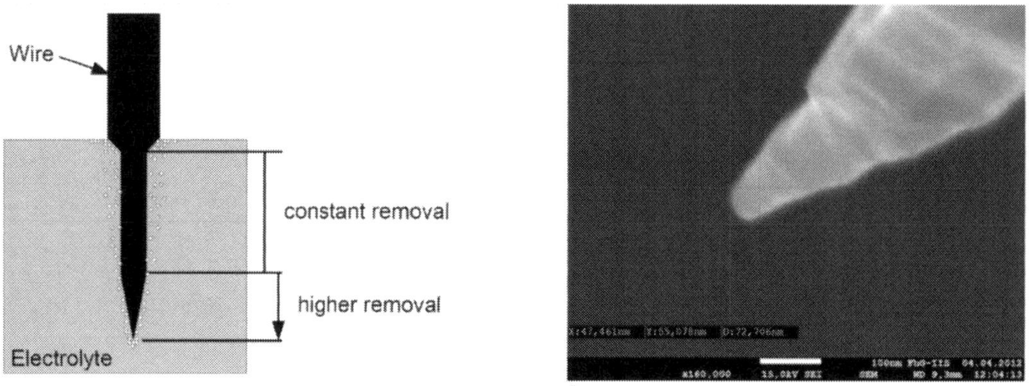

a b

Figure 3. a) Schematic of the electrochemical etching process. The wires made of Mo or W are etched in a 2N NaOH solution. By using an AC voltage the tip is mainly formed by the bubble float, this results in a higher removal rate on the lower part of the wire. **b)** SEM image of a tungsten tip manufactured with the proposed process, with a radius below 70nm.

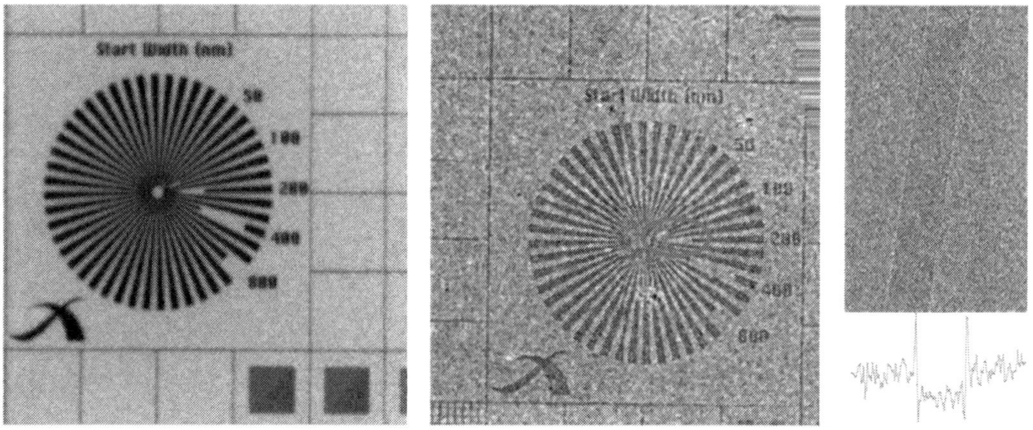

a b c

Figure 4. Image of a test pattern (**a**) imaged with the Xradia UltraXRM L200, 5min exposure, monochromatic Cu K_α radiation courtesy of Peter Krüger, Fraunhofer IZZP, Dresden; **b**) imaged with our setup, 2min exposure, polychromatic 30keV tungsten spectrum); **c**) shows an image of a low absorbing 20μm Al_2O_3 fiber (3min exposure, polychromatic 30keV tungsten spectrum) with an intensity profile beneath. Edge enhancement due to inline phase contrast is clearly visible.

The etching rate can be controlled via AC frequency, peak-to-peak voltage and viscosity of the electrolyte. Therefore we added a small amount of DECON®90 to the solution. By applying a 10Hz sinusoidal ac voltage ($20V_{PP}$), the metal wire becomes a metal tip with a very small radius (down to 70nm) after 20 to 40 minutes of etching (shown in fig. 3b).

5. Results

To determine spatial resolution, a modified Siemens Star pattern with structures down to 50nm was imaged with the presented setup and with a commercially available X-ray microscope, the Xradia UltraXRM L200, which is based of Fresnel zone plates (shown in fig. 4a and 4b). It is clearly visible that with our Setup a resolution of <100nm is possible and both setups are quiet comparable, due to the fact that our setup is still in its early commissioning stage, there is plenty of room for improvements. The contrast in our image is lower due to the fact, that our spectrum has higher energies (30keV tungsten, contrary to the monochromatic Cu K_α of the UltraXRM). A low contrast image with a corresponding intensity profile of a 20μm Al_2O_3 fiber is shown in fig. 4c. It shows that with small spot sizes the spatial coherence of the source is given, and so inline phase contrast imaging is possible.

6. Summary and outlook

The presented laboratory X-ray microscope, using a reflection target source and geometric magnification, is able to achieve a resolution better than 100nm in 2D imaging, as well as inline phase contrast imaging, both at a photon energy up to 30keV. We produced sharp metal tips made of tungsten or molybdenum with a tip radius below 100nm by electrochemical etching, to achieve an ultra small X-ray spot size.

In further updates details study of 3D imaging is planned, as well as damping target and object vibrations, movement precision inside the chamber and focal point stability of the electron beam for longer exposure times.

Acknowledgements

The authors greatfully acknowledge the funding from the Bavarian State Ministry of Economic Affairs, Infrastructure, Transport and Technology which supports the project group "Nano-X-ray Systems for Material Characterization". Further support for the Chair of X-ray Microscopy comes from the Fraunhofer-Gesellschaft and the University of Würzburg.

References

[1] Salomon M, Hanke R, Küger P, Uhlmann N and Voland V 2008 Realization of a computed tomography setup to achieve resolution below 1μm *Nuc. Instr. and Meth.* **591** 50-53

[2] Krüger S P, Neubauer H, Bartels M, Kalbfleisch S, Giewekemeyer K, Wilbrandt P J, Sprung M and Salditt T 2012 Sub-10nm beam confinement by X-ray waveguides: design, fabrication and characterization of optical properties *J. Synchrotron Rad.* **19** 227-236

[3] Chao W, Harteneck B D, Liddle A, Anderson E H and Attwood D T 2005 Soft X-ray microscopy at a spatial resolution better than 15nm *Nature* **435** 1210-1213

[4] Llopart X, Campbell M, Dinapoli R, San Segundo D and Pernigotti E 2002 Medipix2: A 64-k pixel readout chip with 55-μm square elements working in single photon counting mode *Nuc. Sci., IEEE Trans.* **49** 2279-2283

[5] Fotino M 1992 Tip sharpening by normal and reverse electrochemical etching *Rev. Sci. Instrum.* **64(1)** 159

Instrumentation and Analysis Software Development for X-Ray Microfluorescence Experiments at the 4B Beamline of Pohang Light Source-II

Kyehwan Gil[1], Sung Duck Jang[1], Jae-Hong Lim[1], Jung-Yun Huang[1], Hae Koo Kim[2]

Pohang Accelerator Laboratory[1] / Biofluid and Biomimic Research Center[2]
Pohang University of Science and Technology, San 31, Hyoja-dong, Nam-gu, Pohang, Gyeongbuk 790-784, Korea

khgil@postech.ac.kr

Abstract. The 4B beamline of the Pohang Light Source-II performs X-ray microfluorescence experiments together with X-ray microdiffraction experiments, which was upgraded from the old operating conditions of 2.5 GeV and 200 mA to 3.0 GeV and 400 mA. The X-ray microbeam for the microfluorescence experiments is focused to a FWHM size of about 2 μm by adjusting a Kirkpatrick-Baez (KB) mirror system through a LabVIEW application for automatic beam focusing, in the same manner as in the X-ray microdiffraction experiments. A fluorescence detector collects X-ray fluorescence spectra with high-spatial resolution by two-dimensionally scanning a sample attached on an XYZ motorized stage about the focused microbeam. The scanning process is also performed by a LabVIEW application and the collected fluorescence spectra are saved in text format. A microfluorescence analysis software has been newly developed to obtain two-dimensional maps of elements composing a sample from fluorescence spectra. This paper introduces the instrumentation for X-ray microfluorescence experiments at the 4B beamline, describes the details of the developed microfluorescence analysis software, and reports its first analysis results.

1. Introduction

The source of the 4B beamline is a bending magnet in the storage ring of the Pohang Light Source-II with the present operating conditions of 3.0 GeV and 100 mA. The 4B beamline was originally a beamline for X-ray microdiffraction experiments and its main experimental equipments consisted of a hutch slit, two ionization chambers, a shutter, a monochromator, a KB mirror system, a sample stage assembly, and a CCD area detector. Figure 1 shows the setup of the X-ray microdiffraction experimental equipments installed within its hutch. The KB mirror system focuses a microbeam of about 2 μm (FWHM) to a sample 31.5 m far from the source.

Though existing fluorescence analysis softwares such as PyMCA, aXis2000, and GeoPIXE are available [1-3], the development of a microfluorescence analysis software relevant to the 4B beamline was achieved along with additional configurations for the experimental equipments to conveniently perform X-ray microfluorescence experiments as well as X-ray microdiffraction experiments at the 4B beamline.

A newly installed fluorescence detector for X-ray microfluorescence experiments may be used for detecting small metallic pads sputtered on a sample to establish an exact scanning position for X-ray microdiffraction experiments; however, the micrometric elemental mappings of a sample are the fundamental principal object of the fluorescence detector.

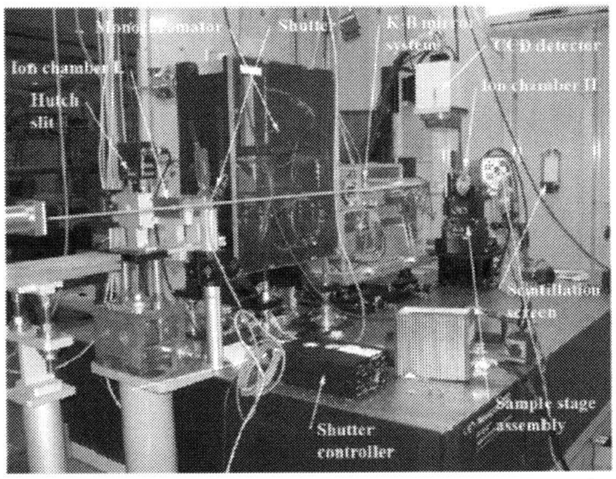

Figure 1. X-ray microdiffraction setup within the 4B beamline hutch.

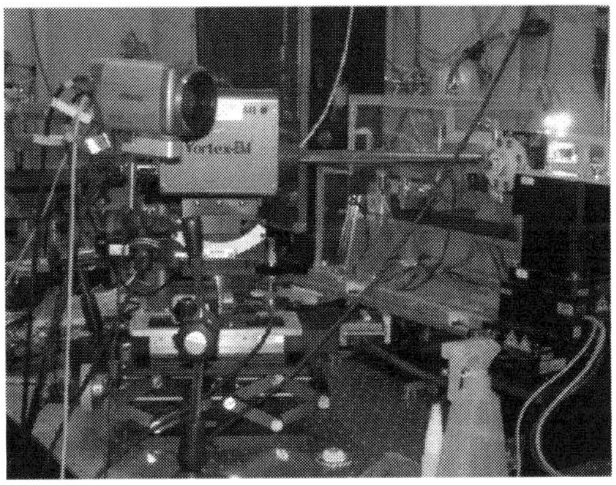

Figure 2. Fluorescence detector and its stage.

2. Instrumentation

As a fluorescence detector, a Vortex-EM from SII NanoTechnology USA Inc. was installed. This fluorescence detector has a cool down time of less than 3 min by thermoelectric cooling and an energy resolution of less than 136 eV. This fluorescence detector is convenient to dispose by virtue of its compact structure that reduces interference with other experimental equipments. Furthermore, its thermoelectric cooling is advantageous for alignment and maintenance.

The fluorescence detector and its stage in operation at the 4B beamline are shown in figure 2. The fluorescence detector stage consists of a linear stage positioned at the lowest part, a yaw stage on the linear stage, and a tilt stage on the yaw stage. The fluorescence detector is positioned on the tilt stage.

Fluorescence samples are attached to a paddle-shaped sample holder shown in figure 3 and are scanned about focused X-rays by a sample stage assembly that is configured into an XYZ stage by assembling three Newport motorized linear stages. Scanning can be performed in three scanning modes. The scanning planes of the three scanning modes are shown in figure 4. LabVIEW scanning applications to collect fluorescence spectra according to the three scanning modes were prepared.

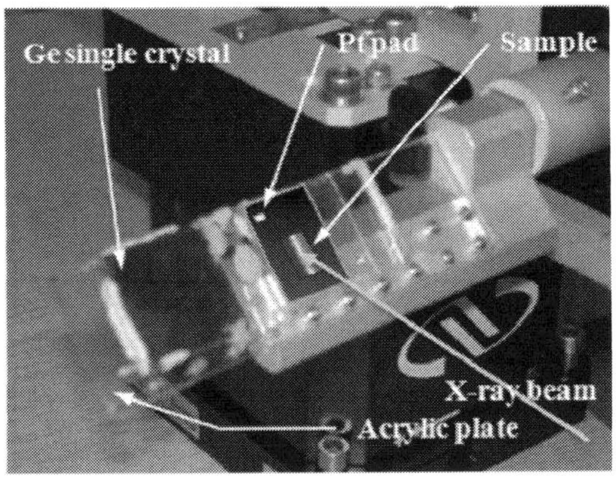

Figure 3. Sample holder on the sample stage assembly.

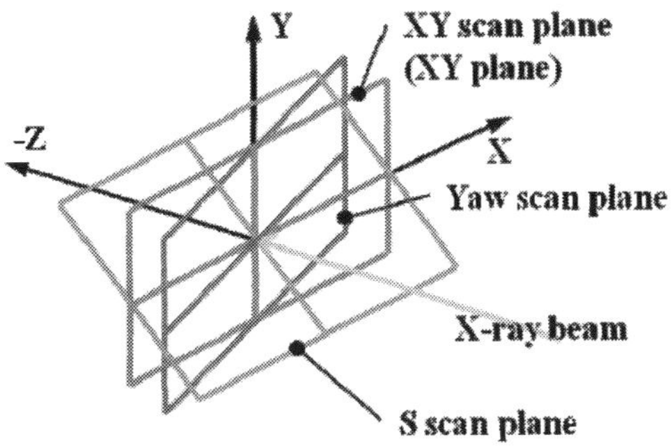

Figure 4. Scanning planes of the three scanning modes.

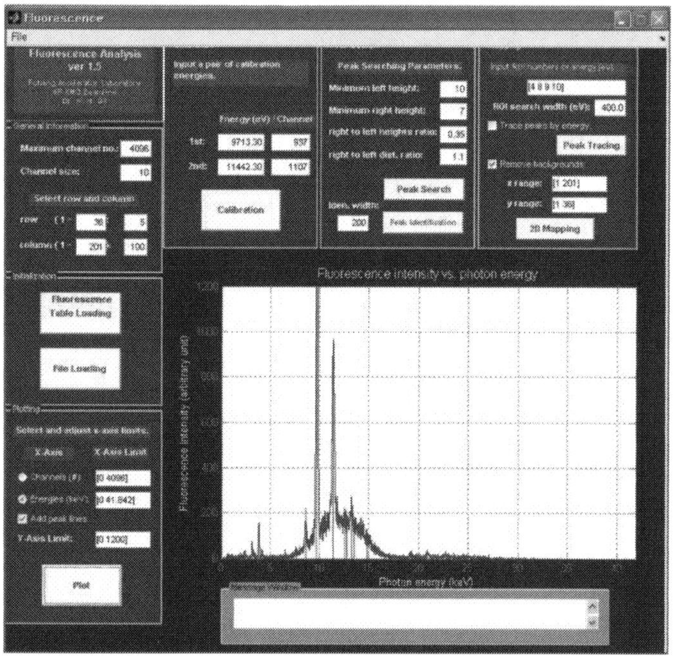

Figure 5. Front panel of the microfluorescence analysis software. The plotting window shows a fluorescence intensity plot with respect to energies along with peak lines.

3. Microfluorescence analysis software

The front panel of the microfluorescence analysis software coded in Matlab is shown in figure 5. Measured fluorescence spectra are saved in a text file. Both the numbers of measured columns and rows are recorded in the first row of the file as two numbers. From the second row, fluorescence spectra measured on each spot of an interesting sample region are recorded one by one in the order of measurement in the following rows. A spectrum consists of an array of photon counts pertaining to the measurement channels of the fluorescence detector. At the back end of each spectrum in one row, the ionization chamber output in A and the live time in msec at the measurement time are recorded.

The main functions of the microfluorescence analysis software are as follows:

• Loading of the energies of X-ray emission lines: This function loads the energies of X-ray emission lines required for the following peak identification function.

• Loading of a fluorescence spectrum file: This function loads a fluorescence spectrum file to analyze through a popup window.

• Plotting of a fluorescence spectrum: This function plots a spectrum pertaining to a measurement spot with specific row and column numbers from a loaded fluorescence spectrum file. Fluorescence spectra are plotted with the x-axis of measurement channels and the y-axis of fluorescence intensities.

• Energy calibration: This function matches the measurement channels with energies. In this research, energy calibration is performed by the following procedure: two clear peaks are determined from a plotted spectrum; then, all measurement channels are converted into energies by matching the central measurement channels of both peaks with the known energies of their emission lines.

• Peak search: This function searches all the nontrivial peaks of a spectrum [4]. Consequently, red verticals are drawn at the peak positions and the serial numbers of the peaks are denoted beside the red verticals on the picture of the fluorescence spectrum. Besides, the serial numbers, energies, and intensities of the searched peaks are tabulated in the command window of Matlab.

• Peak identification: This function identifies the searched fluorescence peaks and presents the source elements of these peaks by comparing the central energy of each searched peak with the energies of emission lines.

• Peak intensity trace: This function computes the intensity distribution of selected specific peaks over the whole measured region. In case the energy of an emission line to trace is known, one can trace its intensity distribution by directly inputting only its energy. In addition, one can compute the intensity distribution of an emission line without any background by using the background removal option.

• Mapping: This function generates the fluorescence maps of the intensity distributions over the whole measured regions. The maps are displayed in two types: 2D color map and 3D bar plot.

4. Discussions and Conclusions

The sample used was an Arabidopsis stem that the uptake of Au nanoparticles had been conducted for. The sample was attached on the sample holder shown in figure 3 and scanned with a live time of 1 sec about the white beam of the 4B beamline with the energy range of 4 – 20 keV. The FWHM size of the X-ray beam was 2.5 (H) × 2.0 (V) μm^2, the horizontal and vertical scanning steps were both 5 μm, and the overall measurement area was 1000 (H) × 250 (V) μm^2. The XY scanning mode was used.

The resulting fluorescence spectrum file was analyzed by using the developed microfluorescence analysis software. Figure 6 shows the fluorescence intensity maps of the Au Lα_1 emission line, displayed in a 2D color map (left) and a 3D bar plot (right). It is shown that the clear distribution of the Au nanoparticles can be effectively observed at a micrometer level.

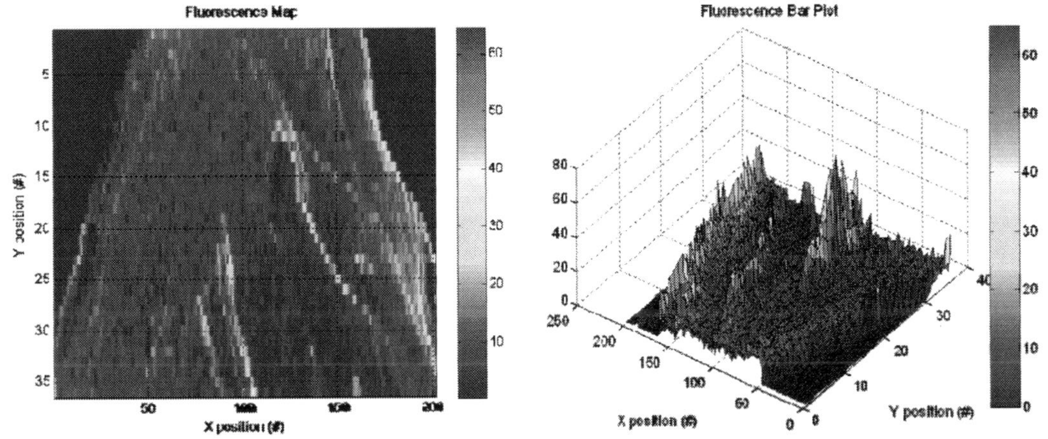

Figure 6. Fluorescence intensity maps displayed in a 2D color map (left) and a 3D bar plot (right).

By installing a fluorescence detector and its stage within the 4B beamline hutch of the Pohang Light Source-II, a setup for X-ray microfluorescence experiments in a micrometer level was established. LabVIEW scanning applications to control these equipments and collect fluorescence spectra according to three scanning modes were completed. In addition, a microfluorescence analysis software was developed in Matlab and successfully applied to analyze an X-ray microfluorescence experiment for an Arabidopsis stem.

As the future objectives of the 4B beamline, we aim to reduce the present measurement time per spot and enable the simultaneous performance of X-ray microfluorescence and X-ray microdiffraction experiments by developing a shutter insertable between the sample and the CCD area detector above it.

Acknowledgments

This work was supported by the Korean Ministry of Education, Science, and Technology.

References

[1] Solé V A, et al. 2007 *Acta Part B* **62** 63
[2] aXis 2000 Software, http://unicorn.mcmaster.ca/aXis2000.html
[3] Ryan C G 2000 *Intl. J. of Imaging Systems and Tech.* **11** (219)
[4] Mariscotti M A 1967 *Nuclear Instruments and Methods* **50** 309

Improvement of cryogenic 3-dimensional observation system of soft x-ray microscope at the SR center of Ritsumeikan University

K Takemoto[1], K Usui[2], T Ohigashi[3, 4], H Fujii[2], M Yoshimura[3], H Namba[2] and H Kihara[3]

1 Kansai Medical University, 18-89 Uyamahigashi, Hirakata, 573-1136, Japan

2 Ritsumeikan University, 1-1-1 Noji-Higashi, Kusatsu, 525-8577, Japan

3 The SR Center, Ritsumeikan University, 1-1-1 Noji-Higashi, Kusatsu, 525-8577, Japan

4 Institute for Molecular Science, 38 Nishigonaka, Myodaiji, Okazaki 444-8585, Japan

E-mail: takemoto@makino.kmu.ac.jp

Abstract. The improvements of a soft x-ray microscope beamline (BL-12) at the SR center of Ritsumeikan University are reported. A wedge-shaped slit and Si plane mirror were newly introduced. The better energy resolution was expected and the +2nd order diffraction from the CZP (1.2 nm at 2.4 nm observation) was suppressed. A new sample holding fixture allows the sample to be replaced quickly and accurately. A new sample cooling system allowed a stable cryogenic x-ray imaging.

1. Introduction

The full-field imaging soft x-ray microscopy (XM) system was installed at BL-12 of the SR center in Ritsumeikan University (575 MeV, 300 mA) [1, 2]. In 2009, the x-ray microscope was improved to perform the nano computed tomography (nano-CT), and we reported the 3-dimensional structures of the brain cortex of the mouse in XRM2010 [3]. After the previous conference, improvements of an illumination optics and a sample holding fixture were carried out, and better nano-CT images were observed.

Cryogenic x-ray microscopy is now accepted as one of the most useful tools for the direct observation of bio-specimens because cryogenic technique promises reduction of their radiation damages. We have developed and installed a cryogenic sample chamber system to the soft XM at BL-12. The temperature of the specimens could be regulated continuously from 273 K down to 173 K. The reduction of the radiation damage was experimentally achieved and the cryogenic images of biospecimens have been taken. However, because of the thermal drift, the spatial resolution of the image was lower than that of the room temperature observation [4, 5]. In addition, the sample stages of the XM at BL-12 have been improved to perform the CT. Therefore, we also developed a new cryogenic sample chamber that improved the cryogenic system. Using this new system, we pursue the goal of the achievement of reducing radiation damage experimentally and the acquirement of x-ray image of biospecimens.

In this paper, we report the overview of the illumination optics, the sample holding fixture, and the new cryogenic system.

2. Improvement of illumination optics

The schematic of the full-field imaging soft XM is shown in Figure 1. In the full-field imaging soft XM with two Fresnel zone plate optics, the combination of a plane mirror, a condenser zone plate (CZP) and a pinhole work not only as an illumination to the sample but also as a linear monochromator [1,2]. The linear monochromator uses difference of focal lengths by chromatic aberration of the CZP. Then, the plane mirror works to suppress the higher diffraction from the CZP, which has the same focal length with the main wavelength. The performance of this optics essentially determines the contrast and the spatial resolution of the image. However, in our XM, the CZP produces +2nd order diffraction of half wavelength of the target wavelength because of its figure error [6]. Illumination optics has been improved to suppress the +2nd order diffraction in order to obtain proper energy resolution for better imaging.

The XM was designed to use the wavelength of the water window region (2.3-4.4 nm) mainly. For the improvement of the performance of the illumination optics in this region, especially the wavelength of 2.4 nm, just below the K-edge of oxygen, a wedge-shaped slit and a silicon (Si) plane mirror were installed. The energy resolution of the linear monochromator is determined by the size of the CZP and the pinhole approximately. Then, the demagnetized light source image by the CZP should be smaller than the size of the pinhole for the expected monochromatic performance [7]. The slit, whose horizontal size varies from 0.37 to 1.9 mm, set downstream the light source was used as a virtual source and reduces the horizontal size of the light source. The energy resolution, $E/\Delta E$, of 300 was expected by using the pinhole of 15 μm diameter.

On the other hand, the Si plane mirror was installed to suppress the +2nd order diffraction from the CZP, which was the wavelength of 1.2 nm in use of 2.4 nm. The cut-off wavelength changes from 1.0 to 1.2 nm by changing the current silicon carbide (SiC) mirror to the Si one. The ratios of +2nd order reflectance to +1st order one is expected to decrease from 0.16 to 0.05 (32%) at the wavelength of 2.4 nm [8].

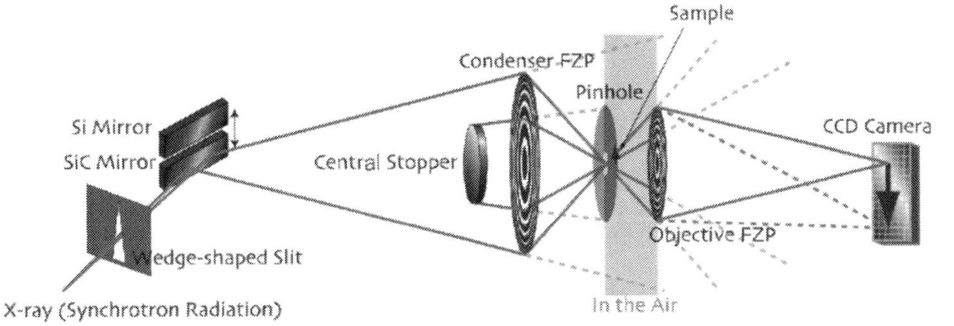

Figure 1. Schematic of the full-field imaging soft XM. The wedge-shaped slit set downstream the light source as a virtual source. Si and SiC plane mirrors are installed. The CZP is a Göttingen KZP 7 type (diameter: 9 mm, outermost zone width: 53.7 nm, number of the zones: 41,890) [9]. The pinhole diameter is 20 μm. The objective ZP (OZP) was fabricated in NTT-AT (diameter: 84 μm, outermost zone width: 38 nm, number of the zones: 550). The images are detected by a back-illuminated CCD camera (C4880-21-24WD, Hamamatsu K.K.).

In order to simplify the mounting of the sample, a sample holding fixture were designed and introduced. Figure 2 shows an XM sample stage and the sample holding fixture head for CT A disciform Neodymium magnet and three steel balls are attached on a backside of the base plate of the head. Three receptacle dimples are formed on the sample stage stainless steel plate (SUS403). Positioning is accomplished by the steel balls and dimples, and rigid fixing is accomplished by the magnet and stainless steel plate. The quick-release fixture allows the sample to be replaced quickly and accurately. The head is also used for wet sample which is contained in glass capillary tube. When 2-dimensional imaging is performed, the head is changed for a dedicated head for it.

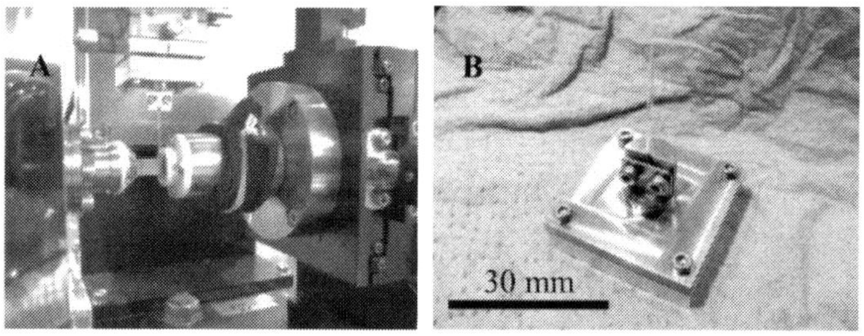

Figure 2. (A) XM sample stage. (B) Sample holding fixture head for CT and wet sample.

3. New cryogenic system

We developed a new cryogenic sample chamber that improved the sample cooling system (SCS). The schematic image and the photo of the SCS are shown in Figure 3. The SCS consists of a sample chamber made of aluminum block and Dewar vessel. Liquid nitrogen stored in the Dewar vessel flows into a metal tube in the aluminum block. In order to avoid frost forming, the sample is mounted in cryogenic oxygen gas environment. The oxygen gas flow rate is about 0.15L/min. The sample chamber is encompassed high heat insulating material (STYRO FORM^{TM}, EK-2, The Dow Chemical Co., the thermal conductivity coefficient: 0.028 W/m·k). The temperature stability was improved by introducing non-contact cooling system with the use of liquid nitrogen and cryogenic oxygen gas. The lowest sample chamber temperature reached -170 °C. X-ray micrographs of polystyrene (PS) latex spheres with x-ray irradiation at room temperature and at cryogenic temperature are shown in Figure 4. Alleviation of radiation damage in PS latex spheres at cryogenic temperatures was verified.

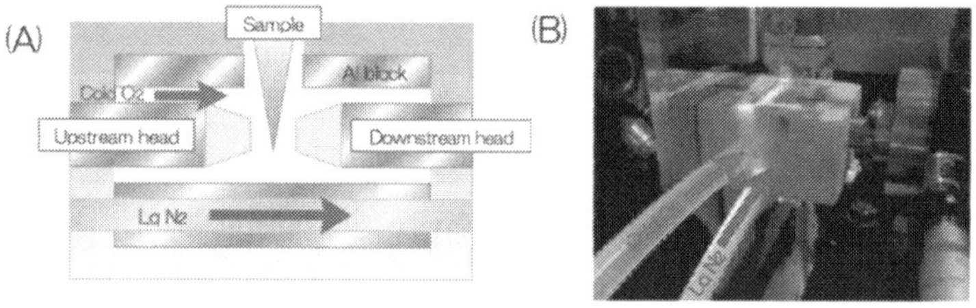

Figure 3. Schematic diagram (A) and photograph (B) of cryogenic sample chamber.

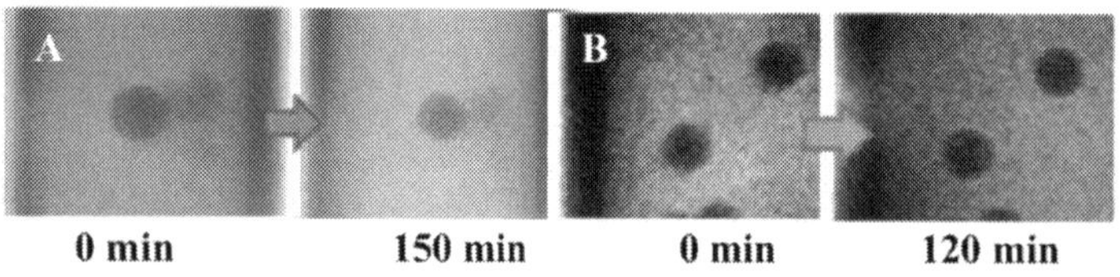

Figure 4. X-ray micrographs of PS latex spheres suspension. (A) was taken at room temperature and (B) was at -160 °C.

4. X-ray images

X-ray images are shown in Figure 5. (A) is air-dried *Scenedesmus* and (B) is air-dried *Skeletonema potamos*. Both images were taken at 2.33 nm wavelength with exposure time of 120 s. The imaging fields are quite good and image contrasts are also good.

Figure 6 presents an x-ray integrated image of cryogenic *S. potamos*. A sealed glass capillary tube 10 μm in diameter containing the laboratory-cultured *S. potamos* was mounted on sample holding fixture head. During exposure time, each cell maintains its position. CT imaging that uses this system is scheduled to be carried out soon.

5. Conclusion

We have improved an illumination optics, sample holding fixture, and new cryogenic system of the soft x-ray microscope beamline (BL-12) at the SR center of Ritsumeikan University. The wedge-shaped slit and the Si plane mirror were newly introduced. The energy resolution, $E/\Delta E$, of 300 was expected and the +2nd order diffraction from the CZP which was the wavelength of 1.2 nm in use of 2.4 nm was suppressed. The new sample holding fixture allows the sample to be replaced quickly and accurately. The new sample cooling system allowed a stable cryogenic x-ray imaging, and alleviation of radiation damage at cryogenic temperatures was also verified.

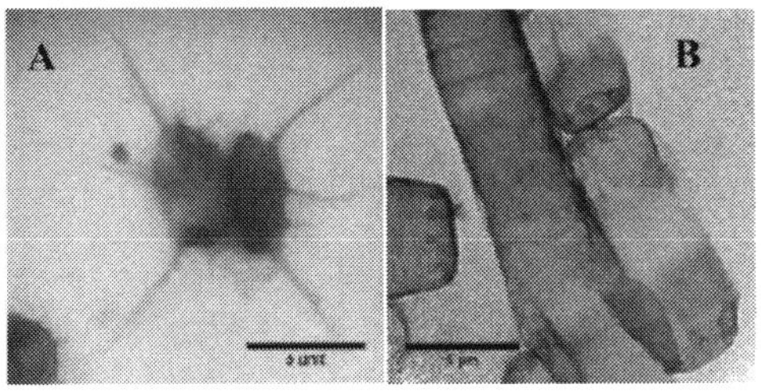

Figure 5. X-ray micrographs of phytoplanktons. (A) *Scenedesmus* and (B) *S. potamos*. Wavelength: 2.0 nm, exp. time: 120 s, air-dried, room temperature.

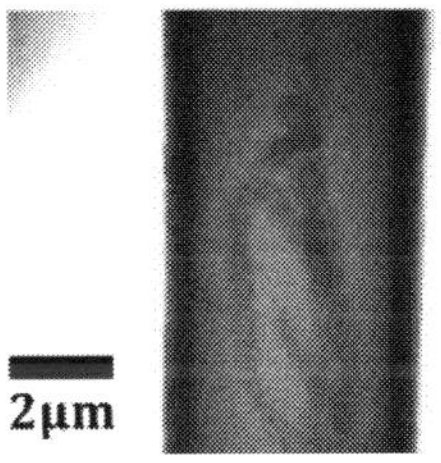

Figure 6. X-ray micrograph of *S. potamos*. Wavelength: 2.33 nm, exp. time: 12 s, wet, cryo-condition. This is an integrated image of three microphotographs.

Acknowledgement

This work was supported in part by Open Research Center Project for Private Universities matching fund subsidy (2007-2011) and Grant-in-Aid for Scientific Research (C) 21540422 from the Ministry of Education, Culture, Sports, Science and Technology, Japan. We thank Mr. Ichise at Lake Biwa Environmental Research Institute for providing us phytoplankton.

References

[1] Hirai A, Takemoto K, Nishino K, Watanabe N, Anderson E, Attwood D, Kern D, Hettwer M, Rudolph D, Aoki S, Nakayama Y and Kihara H 1998 *J. Synchrotron Radiat.* **5** 1102
[2] Hirai A, Takemoto K, Nishino K, Niemann B, Hettwer M, Rudolph D, Anderson E, Attwood D, Kern D P, Nakayama Y and Kihara H 1999 *Jpn. J. Appl. Phys.* **38** 274-278
[3] Ohigashi T, Fujii H, Usui K, Namba H, Mizutani H, Takemoto K and Kihara H 2011 *AIP Conf. Proc.* **1365** 124
[4] Takemoto K, Kimura M, Usui K, Ohigashi T, Fujii H, Nakanishi K, Namba H and Kihara H 2011 2011 *AIP Conf. Proc.* **1365** 419
[5] Kimura M, Takemoto K, Ohashi Y, Ohigashi T, Nakanishi K, Namba H, Kihara H 2011 *Jpn. J. Appl. Phys.* **50** Article No.026601
[6] Niemann B, Rudolph D and Schmahl G 1974 *Opt. Commun.* **12** 160
[7] Ohigashi T, Namba H, Takemoto K, and Kihara H 2011 *Memoirs of SR Center Ritsumeikan Univ.* **13** 201
[8] Schmahl G, Rudolph D, Niemann B, Guttmann P, Thieme J, Schneider G, David D, Diehl M, Wilhein T, 1995 *Optik* **93** 95

Performance of the post-focusing mirror system at the reflectometry beamline BL-11D of the Photon Factory

Tadashi Hatano, Shogaku Aihara, Kentaro Uchida and Toshihide Tsuru

IMRAM, Tohoku University, 2-1-1 Katahira, Aoba-ku, Sendai 980-8577, JAPAN

E-mail: hatano@tagen.tohoku.ac.jp

Abstract. Beamline BL-11D of the Photon Factory was recently opened for the characterization of extreme-ultraviolet and soft X-ray optical components. For reflectometry of multilayers for soft X-ray microscope optics, a small focus size on the sample surface matching the small acceptances of the curved multilayer samples is required. The post-focusing mirror system of BL-11D is composed of horizontally and vertically focusing elliptical mirrors. The performance was evaluated by microscopic beam profile observation, by a knife-edge scan test, and by the Ronchi test. The FWHM beam size was 120 μm (H) × 30 μm (V) with an insignificant spherical aberration, which is smaller than the requirement.

1. Introduction

Multilayer characterization tools are crucial for the development of extreme ultraviolet (EUV) and soft X-ray imaging optics based on multilayer reflectors deposited on the mirror substrates. Beamline 6.3.2 of the Advanced Light Source is a standard facility for these operations [1]. In Japan, although some reflectometry beamlines are in operation, none of them specialize in the characterization of curved multilayer samples. An adequately focused incident beam as well as a multi-axis sample stage are required. In our previous study, periodic thicknesses of concave EUV multilayers were measured at a high accuracy by reflectometry using a 1-mm-diameter parallel beam at BL-12A of the Photon Factory (PF) [2], where the beam size was less critical because of the larger acceptance of the EUV multilayer. In Schwarzschild-type soft X-ray microscope optics, however, a mirror substrate with a 10-mm radius of curvature is used with a multilayer coating having a period number of several hundred, e.g., 600 period Cr/Sc multilayer at 400 eV [3]. The angular acceptance $|\Delta(\cos\phi)|/\cos\phi = \tan\phi\ \Delta\phi$, which is equivalent to the bandwidth $\Delta\lambda/\lambda$, is approximately the inverse of the period number, and $\Delta\phi$ is 1° when ϕ is 5°. For a substrate with a 10-mm radius of curvature, a 200-μm beam size in the tangential plane, or 800-μm beam size in the sagittal plane, causes a 1°-angle of incidence variation along the sample surface. Thus, the required FWHM beam size is smaller than 800 μm (H) × 200 μm (V).

The EUV and soft X-ray beamline BL-11D at PF has been dedicated to photoelectron spectroscopy studies since 1996 [4]. It was fitted with a post-focusing mirror system composed of horizontally and vertically focusing mirrors, HPFM and VPFM, respectively, and a FWHM beam size of 1 mm (H) × 100 μm (V) was achieved [4]. In 2010, BL-12A was closed, and BL-11D was redesigned for the characterization of optical components with the installation of a reflectometer. The requirement for the post-focusing mirror system has become clearer as having a FWHM beam size small enough for the smallest possible acceptance of mirror samples, i.e., 800 μm (H) × 200 μm (V).

In the present paper, the performance of the post-focusing mirror system is evaluated. The beam profile is observed using an EUV microscope, and the beam size is measured by the knife-edge scan test. For long-term reliability of the focusing performance, particularly in the vertical direction, the aberration characteristics are studied using the Ronchi test.

2. Experimental

The optical layout of beamline BL-11D is described here. A schematic side view is shown in figure 1, including the distance of each component from the source. The monochromator is composed of the entrance slit S1, the focusing cylindrical mirror Mf, the deviation-angle-changing plane mirror Mp, the spherical grating G, and the exit slit S2. In the horizontal plane, the cylindrical mirror M0, which is the first deflecting mirror, forms an intermediate focus between S1 and Mf. The elliptic mirror HPFM refocuses the beam to the center of the reflectometer 30.3 m from the source. M0 and HPFM form a 4:1 demagnified image of the source. In the vertical plane, the elliptic mirror VPFM forms a 1:1 image of S2 to the reflectometer. The grazing angles of incidence of M0, M0', Mf, VPFM, and HPFM are all 2°. The upstream and downstream ends of the backside of VPFM are supported by piezo-actuators V1 and V2, respectively; those for HPFM are H1 and H2. During the photoelectron spectroscopy activities over 10 years, the voltages applied to the piezo-actuators have been fixed as V_{V1}, V_{V2}, V_{H1}, V_{H2} = 300 V, 800 V, 500 V, 300 V, respectively.

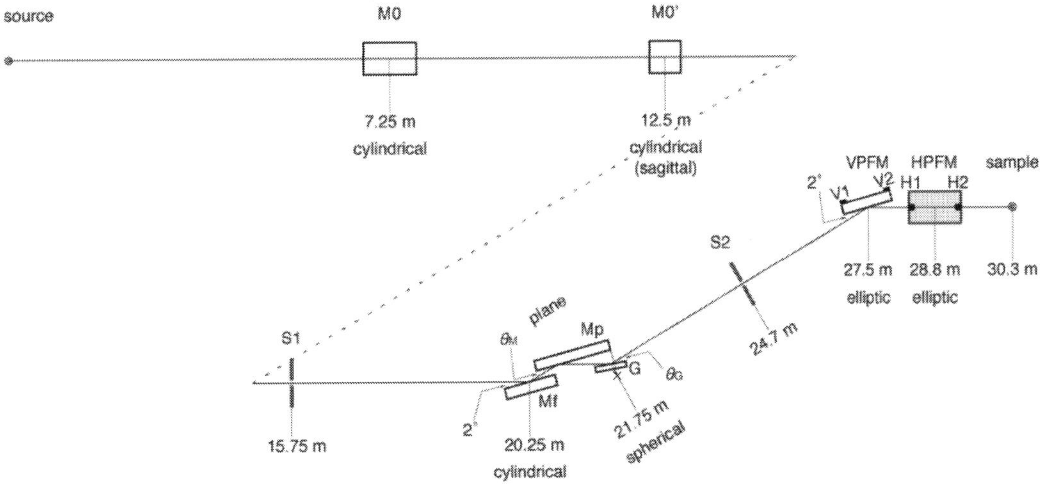

Figure 1. Side view of the beamline optics at BL-11D, PF.

Using an EUV scintillator microscope [5], we observed the beam profile at 92 eV. The reflectometer chamber was moved out of the optical path, and the scintillator plane was set at the sample position. Voltages applied to the piezo-actuators were V_{V1}, V_{V2}, V_{H1}, V_{H2} = 300 V, 800 V, 500 V, 300 V, as before.

A knife-edge scan test was carried out at 275 eV. The reflectometer chamber was put back in the optical path. Figure 2(a) shows the side view of the reflectometer with the knife-edge scan test setup. A photodiode AXUV100Ti/C2 (IRD, Inc.) was used as a detector. The turntable sample stage is shown in gray, and its plan view is inset. A knife labeled "–90" was set with its edge along the vertical axis at a position of z = –90 mm, where z is the coordinate along the beam axis with the origin at the center of the reflectometer. The horizontal beam size at z = –90 mm can be measured by a horizontal scan of the knife "–90." Knives were also set at positions of z = –70 mm, –50 mm, –30 mm, –10 mm, and +10 mm. With the sample stage turned by 180°, the beam size at z = +90 mm can be measured with the knife "–90." A knife labeled "0" was set at the center of the sample stage with its edge along the horizontal axis for the vertical scan. Voltages applied to the piezo-actuators were not changed in the horizontal scan and changed to V_{V1}, V_{V2} = 300 V, 800 V; 100 V, 400 V; 100 V, 200 V; and 0 V, 0 V in the vertical scan.

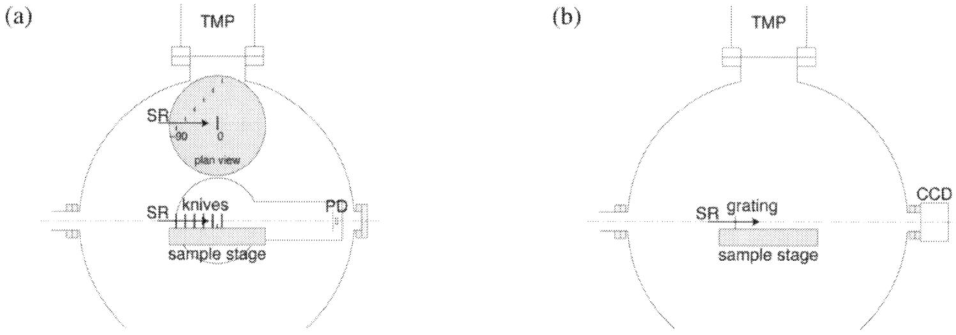

Figure 2. Side view of the reflectometer with the setup for (a) knife-edge scan test and (b) Ronchi test.

The setup for the Ronchi test is illustrated in figure 2(b). A transmission grating with a 50-μm period (NTT-AT Corp.) was set on the sample stage at z = −70 mm with its slits horizontal (with its periodicity vertical). The fringe pattern was observed by a soft X-ray CCD camera BK-501X (BITRAN Corp.) mounted at the downstream end of the reflectometer chamber. The distance between the grating and the camera was 800 mm. Voltages applied to the VPFM were V_{V1}, V_{V2} = 0 V, 0 V.

3. Results and discussions

Figure 3(a) shows the observed EUV microscope image in grayscale on a white background. The S2 opening was 20 μm, and the vertical beam size was 80 μm. Figure 3(b) shows the vertical spot size dependence on the S2 opening supporting the 1:1 imaging of the VPFM. Independent of the S2 opening, the horizontal beam size was 200 μm, which agrees with a quarter of the electron beam size. Thus, basically the performances of VPFM and HPFM agreed with the optical design of the beamline.

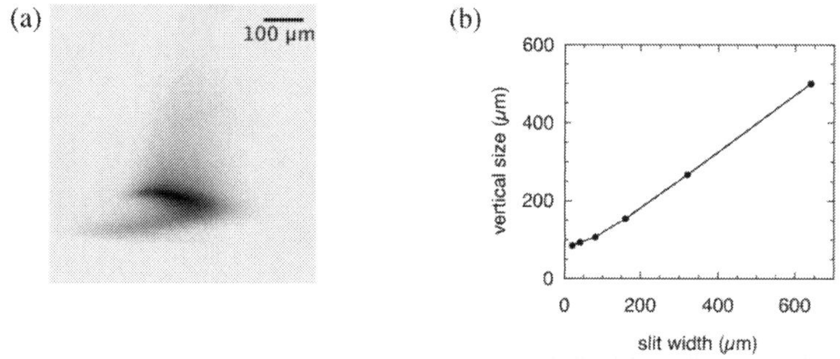

Figure 3. (a) Beam profile and (b) vertical size dependence on the S2 opening.

Figure 4(a) shows the results of horizontal knife-edge scans with the baseline shifted in proportion to z. The beam waist was found just at the center and was 120-μm wide. Figure 4(b) shows the results of vertical knife-edge scans with the baseline shifted by 1. The S2 opening was fixed at 30 μm. As V_{V1} and V_{V2} decreased, the beam size became smaller, which means the focal point is upstream before the center of the reflectometer chamber. At V_{V1}, V_{V2} = 0 V, 0 V, the beam size was the smallest at 30 μm, the same as the S2 opening. Thus, no unexpected broadening was observed, and a sufficiently small FWHM beam size of 120 μm (H) × 30 μm (V) was confirmed.

A Ronchi test was carried out with the best focus condition of V_{V1}, V_{V2} = 0 V, 0 V. Figure 5 shows the observed fringes with the sample stage scanned downward in 8-μm steps. The observed fringes near the center moved downward while fringes near the top and bottom moved upward. A fringe disappeared into line "B," and a new fringe appeared out of line "A." This behavior shows a spherical aberration where the focal length is shorter around the mirror center. If necessary, the focus will be improved by applying a tensile stress on the backside of the VPFM around its center.

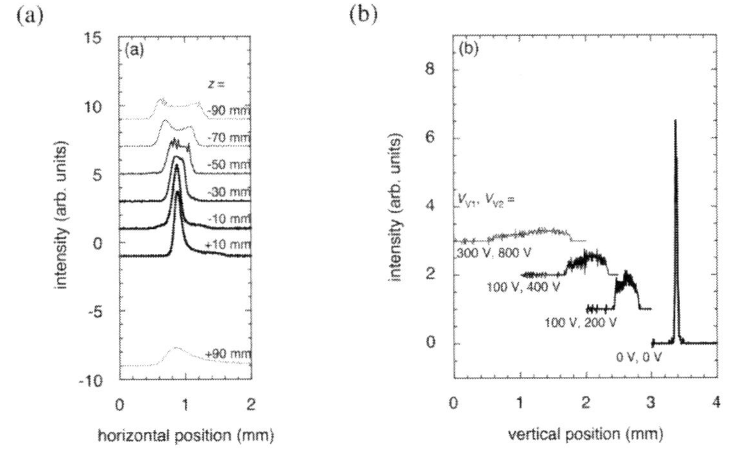

Figure 4. Knife-edge scan along (a) horizontal and (b) vertical axes.

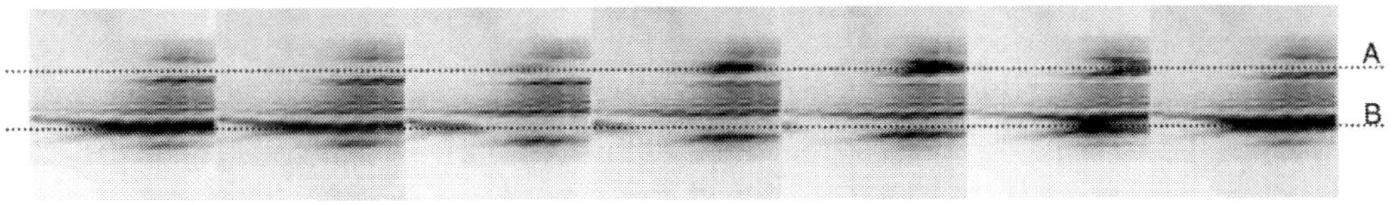

Figure 5. Ronchi fringes with a 50-μm-period grating shifted downward in 8-μm steps.

4. Conclusions

We evaluated the performance of the post-focusing mirror system at BL-11D, PF. Microscopic observations of the beam profile, beam size measurements by the knife-edge scan test, and characterization of the aberration by the Ronchi test were carried out. As a result, the FWHM beam size was confirmed to be 120 μm (H) × 30 μm (V) with an insignificant spherical aberration that is small enough for reflectometry of soft X-ray multilayer microscope optics.

Acknowledgments

The authors thank Prof. K. Ito and Mr. T. Kikuchi of PF for their help at BL-11D. They are also grateful to Prof. T. Ejima and Mr. Y. Neichi of Tohoku Univ. for their help with the EUV microscope observations. This work was partly supported by the Development of Systems and Technology for Advanced Measurement and Analysis, JST. Measurements were performed under the approval of the Photon Factory Program Advisory Committee (Proposal No. 2011G661).

References

[1] Underwood J H, Gullikson E M, Koike M, Batson P J, Denham P E, Franck K D, Tackaberry R E and Steele W F 1996 *Rev. Sci. Instrum.* **67** 1–5
[2] Hatano T, Kubota S, Adachi Y, Tsuru T and Yamamoto M 2004 *AIP Proc.* **705** 839-842
[3] Eriksson F, Johanson G A, Herz H M, Gullikson E M, Kreissig U and Birch J 2003 *Opt. Lett.* **28** 2494–2496
[4] *PF Act. Rep. 2000* A 2001 **#18** 69–71
[5] Ejima T, Ogasawara S, Hatano T, Yanagihara M and Yamamoto M 2010 *AIP Proc.* **CP1234** 811–814

Differential phase microscope and micro-tomography with a Foucault knife-edge scanning filter

N Watanabe, J Hashizume, M Goto, M Yamaguchi, T Tsujimura and S Aoki

University of Tsukuba, Faculty of Pure and Applied Sciences, Division of Applied Physics, Tennoudai 1-1-1, Tsukuba, Ibaraki 305-8573, Japan

E-mail: watanabe@bk.tsukuba.ac.jp

Abstract. An x-ray differential phase microscope with a Foucault knife-edge scanning filter was set up at the bending magnet source BL3C, Photon Factory. A reconstructed phase profile from the differential phase image of an aluminium wire at 5.36 keV was fairly good agreement with the numerical simulation. Phase tomography of a biological specimen, such as an Artemia cyst, could be successfully demonstrated.

1. Introduction

X-ray phase tomography is very attractive for the observation of weakly absorbing material, such as biological samples. In order to realize phase tomography, quantitative phase measurement is required. It was difficult by a conventional method such as a Zernike phase-contrast microscope. Nagayama showed that a quantitative differential phase image was obtained in electron microscopy by introducing a scanning knife-edge filter at a back focal plane of an objective lens [1]. This method can also be applied to x-ray microscopy [2].

In this paper, we report the first results of x-ray differential phase imaging and phase tomography at 5.36 keV by using the synchronous operation between the scanning the knife-edge and the image accumulation.

2. Differential phase microscope with a knife-edge scanning filter

The microscope with a zone plate was constructed using the bending magnet source BL3C at the Photon Factory. The optical system shown in figure 1 is almost the same as the previous report [2]. The magnification ratio was 48. A gold wire of 250 μm in diameter was used as a knife-edge.

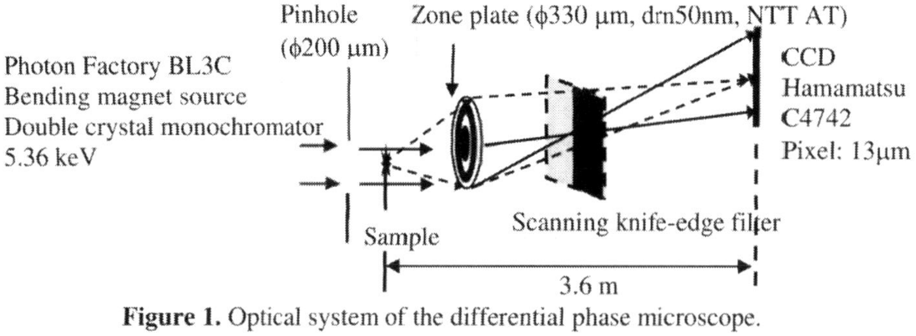

Figure 1. Optical system of the differential phase microscope.

3. Simulation of a differential phase image with a knife-edge scanning filter

Numerical simulations of the differential phase microscope were performed as shown in figure 2. For simplicity, the magnification ratio and the aperture effect of the objective lens are ignored and only the direction of x-axis is considered. A parallel monochromatic beam along the optical axis is incident on the object plane. Let $f(x) \equiv \exp[-g(x)]\exp[i\phi(x)]$ to be the transmitted amplitude through the object at the object plane. Then, the amplitude at the back focal plane of the lens is $F(\mu)$ which denotes the Fourier transform of $f(x)$. Let the filter function at the back focal plane be of the following amplitude transmission formula,

$$T_{\pm}(\mu) = \frac{1}{2}\left(1 \pm \frac{\mu}{S}\right) \qquad |\mu| \le S \tag{1}$$

where S is the half of the filter width. Then, the amplitude $A_{\pm}(x)$ at the image plane is the inverse Fourier transform of $T_{\pm}(\mu)F(\mu)$ assuming that $F(\mu)=0$ when $|\mu| > S$.

$$A_{\pm}(x) = \int_{-\infty}^{\infty} T_{\pm}(\mu)F(\mu)\exp(i2\pi x\mu)d\mu = \frac{f(x)}{2}\left(1 \pm \frac{1}{2\pi S}\frac{d\phi(x)}{dx} \pm \frac{i}{2\pi S}\frac{dg(x)}{dx}\right) \tag{2}$$

Then, the intensity of the image $I_{\pm}(x)$ is the square of the absolute value of $A_{\pm}(x)$.

$$I_{\pm}(x) = |A_{\pm}|^2 = \frac{1}{4}|f(x)|^2\left[1 \pm \frac{1}{\pi S}\frac{d\phi(x)}{dx} + \frac{1}{4\pi^2 S^2}\left\{\left(\frac{dg(x)}{dx}\right)^2 + \left(\frac{d\phi(x)}{dx}\right)^2\right\}\right] \tag{3}$$

Ignoring second order terms of the first derivatives of $g(x)$ and $\phi(x)$, the differential phase can be approximated by the following equations.

$$I_{+}(x) - I_{-}(x) = \frac{1}{2\pi S}|f(x)|^2\frac{d\phi(x)}{dx} \tag{4}$$

$$I_{+}(x) + I_{-}(x) = \frac{1}{2}|f(x)|^2\left[1 + \frac{1}{4\pi^2 S^2}\left\{\left(\frac{dg(x)}{dx}\right)^2 + \left(\frac{d\phi(x)}{dx}\right)^2\right\}\right] \cong \frac{1}{2}|f(x)|^2 \tag{5}$$

$$\frac{d\phi(x)}{dx} = \pi S\left(\frac{I_{+}(x) - I_{-}(x)}{I_{+}(x) + I_{-}(x)}\right) \tag{6}$$

The knife-edge scanning filter is represented by the following equation.

$$T_{\pm}(\mu) = \frac{1}{2}\left(1 \pm \frac{\mu}{S}\right) \cong \frac{1}{4N}\sum_{n=-N}^{N}\left\{1 \pm \mathrm{sgn}\left(\frac{\mu}{S} - \frac{n}{N}\right)\right\} \qquad |\mu| \le S \tag{7}$$

where n is the knife-edge position. When $n = N$, the position is $\mu = S$. Using this equation,

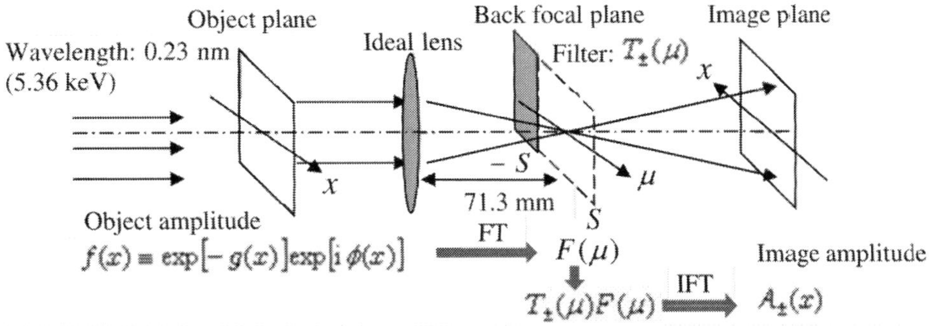

Figure 2. Numerical simulation of the differential phase microscope with a scanning filter.

$$A_{\pm}(x) = \int_{-\infty}^{\infty} T_{\pm}(\mu)F(\mu)\exp(i2\pi x\mu)d\mu = \sum_{n=-N}^{N} A_{n\pm}$$

$$A_{n\pm}(x) = \frac{1}{4N}\int_{-\infty}^{\infty}\left\{1 \pm \mathrm{sgn}\left(\frac{\mu}{S} - \frac{n}{N}\right)\right\}F(\mu)\exp(i2\pi x\mu)d\mu \tag{8}$$

$I_{\pm}(x)$ of equation (3) can be represent as $I_{\pm}(x) = \left|\sum_{n=-N}^{N} A_{n\pm}\right|^2$. On the other hand, using the scanning knife-edge filter, the image intensity $I'_{\pm}(x)$ is $I'_{\pm}(x) = \sum_{n=-N}^{N}\left|A_{n\pm}\right|^2$. Fortunately, the value $I'_+ - I'_-$ is almost the same as $I_+ - I_-$ [2]. The value $I'_+ + I'_-$ is approximately $\left|f(x)\right|^2$ because the half of the incident intensity $\left|f(x)\right|^2$ can be transmitted through the scanning filter. This value is twice of $I_+ + I_-$. Then, using the scanning knife-edge filter, the differential phase can be calculated by the following equations.

$$\frac{d\phi(x)}{dx} = 2\pi S\left(\frac{I'_+(x) - I'_-(x)}{I'_+(x) + I'_-(x)}\right) \tag{9}$$

An aluminum wire of 25 μm in diameter was used for a test sample. The sampling pitch δ was 0.27 μm and the number of samples M was 512. The sampling pitch at the back focal plane was 0.12 μm, which corresponds to the spatial frequency of $1/M\delta$. The phase was calculated by the integration of equation (9).

$$\phi(n) = 2\pi S\delta\sum_{k=0}^{n-1}\left(\frac{I'_+(k) - I'_-(k)}{I'_+(k) + I'_-(k)}\right) \qquad \phi(0) = 0, \quad n = 1, 2, \cdots, M-1 \tag{10}$$

The Nyquist frequency is $1/2\delta$, which corresponds to 30 μm at the back focal plane. Therefore, the value $S\delta$ in the above equation is 0.5 when the scanning width of the filter is ±30 μm.

Figure 3(a) shows the simulation results in various scanning widths of ±5 μm, ±10 μm, and ±30 μm. The phase value in narrow scan range is small compared to that in wide scan range. This is probably because the high-frequency component of the phase is lost in the image in narrow scan range. Figure 3(b) shows the simulation result when the sampling pitch is 0.10 μm and the scan width is 83 μm which corresponds to the Nyquist frequency. The phase profile in figure 3(b) is very close to the dashed line calculated from Henke's table [3]. From these results, accurate phase image can be reconstructed under conditions sufficient scanning range of the filter.

Figure 4 shows the experimental results in scanning widths of ±5 μm, ±10 μm, and ±20 μm. The image contrast in the narrow scanning width is higher than that in the wide scanning width. But the reconstructed phase is lower in the narrow width. This is because $S\delta$ in equation (10) is proportional to the scanning width. A dashed line in figure 4(d) shows the phase profile of figure 4(a) in the scanning width of ±5 μm. The profile has a big slope. Suppose that the center position of the filter is shifted along the x-axis 2 pixels (0.24 μm), the result is fairly good agreement with this profile.

4. Application to phase tomography

Phase tomography of a biological specimen, such as an Artemia cyst, was investigated. Figure 5(a) is the differential phase projection obtained by two scanning steps of opposite directions. The exposure time was 8 s for each scanning step. The three-dimensional images were reconstructed from the 360 phase projections of different angles of view over the range of 360 degrees. The angular step size was 1.0 degrees. A filter function for beam-deflection optical tomography was used for tomographic reconstruction from the differential phase images [4]. Figure 5(b) shows the reconstructed section image and figure 5(c) is a volume rendering phase image. The 3D phase images could be successfully investigated.

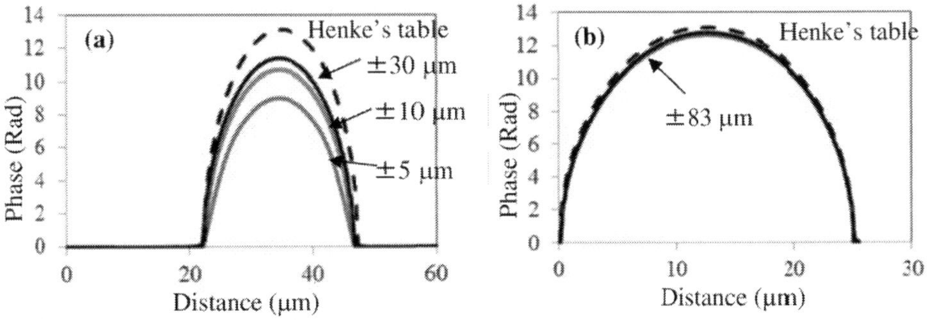

Figure 3. Simulation profiles of an aluminum wire of 25 μm in diameter. (a) Simulations in the scanning width of ±30 μm, ±10 μm, and ±5 μm. (b) Simulation in the sampling pitch of 0.1 μm. The scanning width is ±83 μm. The dashed line is calculated from Henke's table.

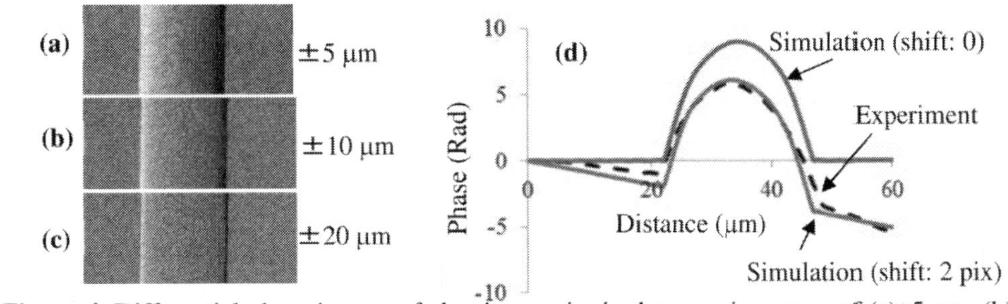

Figure 4. Differential phase images of aluminum wire in the scanning range of (a)±5 μm, (b) ±10 μm, and (c) ±20 μm. (d) Integrated phase profile of (a) is compared to simulation profiles.

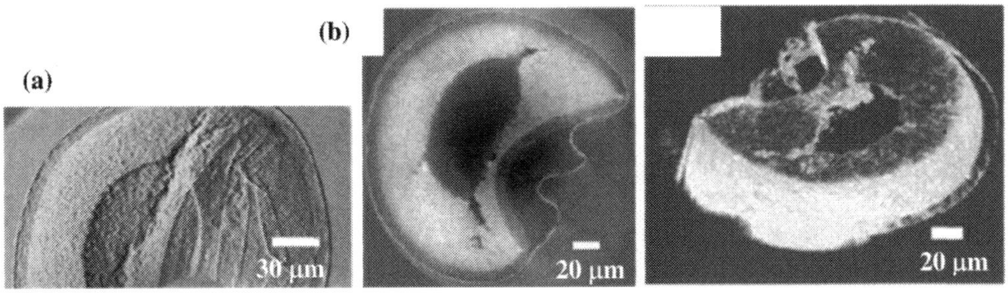

Figure 5. (a) Differential phase image of an Artemia cyst. (b) CT section image. (c) Volume rendering phase image. The x-ray energy is 5.36 keV.

Acknowledgement
This work was supported by a Grant-in-Aid for Scientific Research (C) (22611003) from the Ministry of Education, Culture, Sports, Science and Technology of Japan.

References

[1] Nagayama K 2004 J. Phys. Soc. Jpn. **73** 2725.
[2] Watanabe N, Sasaya T, Imai Y, Iwata S, Zama K and Aoki S 2011 AIP Conf. Proc. **1365** 313.
[3] http://henke.lbl.gov/optical_constants/.
[4] Faris G W and Byer R L 1988 Appl. Optics **27** 5202.

HECTOR: A 240kV micro-CT setup optimized for research

Bert Masschaele[1,2], Manuel Dierick[1], Denis Van Loo[1,2], Matthieu N Boone[1], Loes Brabant[1], Elin Pauwels[1], Veerle Cnudde[3], Luc Van Hoorebeke[1]

[1]UGCT, Dept. Physics and Astronomy, Ghent University, Proeftuinstraat 86, 9000 Gent, Belgium & SIM program SHE
[2]XRE, X-Ray Engineering BVBA, De Pintelaan 111, 9000 Gent, Belgium
[3]UGCT, Dept. Geology and Soil Science, Ghent University, Krijgslaan 281/S8, 9000 Gent, Belgium & SIM program SHE

E-mail: Bert.Masschaele@UGent.be

Abstract. X-ray micro-CT has become a very powerful and common tool for non-destructive three-dimensional (3D) visualization and analysis of objects. Many systems are commercially available, but they are typically limited in terms of operational freedom both from a mechanical point of view as well as for acquisition routines. HECTOR is the latest system developed by the Ghent University Centre for X-ray Tomography (http://www.ugct.ugent.be) in collaboration with X-Ray Engineering (XRE bvba, Ghent, Belgium). It consists of a mechanical setup with nine motorized axes and a modular acquisition software package and combines a microfocus directional target X-ray source up to 240 kV with a large flat-panel detector. Provisions are made to install a line-detector for a maximal operational range. The system can accommodate samples up to 80 kg, 1 m long and 80 cm in diameter while it is also suited for high resolution (down to 4 μm) tomography. The bi-directional detector tiling is suited for large samples while the variable source-detector distance optimizes the signal to noise ratio (SNR) for every type of sample, even with peripheral equipment such as compression stages or climate chambers. The large vertical travel of 1 m can be used for helical scanning and a vertical detector rotation axis allows laminography experiments.
The setup is installed in a large concrete bunker to allow accommodation of peripheral equipment such as pumps, chillers, etc., which can be integrated in the modular acquisition software to obtain a maximal correlation between the environmental control and the CT data taken. The acquisition software does not only allow good coupling with the peripheral equipment but its scripting feature is also particularly interesting for testing new and exotic acquisition routines.

1. Introduction

This paper presents the latest micro-CT scanner developed by UGCT, the Ghent University Centre for X-ray Tomography (http://www.ugct.ugent.be) in collaboration with X-Ray Engineering (XRE bvba, Ghent, Belgium). UGCT specializes in the development of micro-CT scanners which are aimed at pushing the boundaries of lab-based tomography, and at multidisciplinary research applications that require highly customized scanning conditions. To maximize the research possibilities UGCT designs and integrates its own scanners by acquiring the basic components and assembling them in room-size

concrete bunkers. Each scanner is built in a modular way to allow for optimizations, modifications and improvements later on. The available scanners are used for fundamental research on the tomography technique itself as well as for multidisciplinary applications in collaboration with other researchers. CT research involves studying the physics behind tomography (X-ray production, X-ray interactions, phase contrast effects, etc), testing new technologies, optimizing acquisition protocols, programming reconstruction methods and 3D data analysis algorithms. Applied research covers a wide range of disciplines, and consequently a wide range of demands with regard to image resolution, field-of-view, acquisition time etc. The UGCT scanners that were available up to now already cover a wide range of applications. Spatial resolution of these previous scanners can go down to 400 nm, yet samples up to 35 cm in diameter can also be scanned, provided that there is sufficient transmission at the highest previously available tube voltage of 160 kVp.
However, the need was present to extend this range towards larger objects, more absorbing objects, and faster scanning. This resulted in the design of a new scanner, which focuses on this extended range without compromising but even improving the performance for traditional micro-CT.

2. Experimental setup

2.1. Hardware
One of the main components of the system is a XWT 240-SE microfocus source from X-RAY WorX. This high power reflection tube can deliver a target power of up to 280 W for high X-ray flux, while a minimum focal spot size of 4 µm makes the system useful for standard micro-CT scanning. A large 40x40 cm² PerkinElmer 1620 CN3 CS flat panel detector was chosen to allow for a large field-of-view. A rotation stage capable of handling heavy loads was mounted on a vertical stage with 1m travel to allow for helical scanning of long objects such as drill cores and wood logs.

A total of nine motorized stages ensures a high degree of geometrical flexibility. Next to the sample magnification stage to alter the source-to-object distance (SOD) and the sample vertical stage (along the rotation axis) for helical scanning and flat field positioning, the detector is mounted on a sytem of four motorized stages. A vertical and horizontal axis enables tiled radiography and extended tomography up to 4048 x 4048 pixels covering an area of 80 x 80 cm². Using the detector magnification stage, the source-to-detector distance (SDD) can be altered to optimize X-ray flux or cone angle. Furthermore, the detector is equipped with a laminography motor for future research. On top of the high-precision rotation stage, which nevertheless allows for a load of up to 80 kg, is a motorized XY stage to position the sample precisely on the rotation axis.

2.2. Software
To control the scanner, an in-house developed generic platform is used [1]. The platform is developed in LabVIEW and is used at all scanners available at UGCT. The separation between abstract operation of a generic CT scanner and the physical components of each different CT scanner makes the platform flexible and modular. New components can easily be implemented, while the main control software remains unchanged. The command-based operation makes it possible to implement additional devices (e.g. load cell) and include them into the CT scan using the scripting engine, which can be adapted to the specific scanner to include all available acquisition protocols such as step-and-shoot tomography, continuous rotation tomography, helical CT, tiled radiography and tomography, batch scanning, etc.

The tomographic reconstruction is performed using the in-house developed software package Octopus ([2], http://www.octopusreconstruction.com). Any geometrical misalignments of the scanner can automatically be corrected in the reconstruction core, resulting in optimal image quality.

3. Results

Some results, which indicate the possibilities and strengths of the system, are listed in the section below.

3.1. Large object

A concrete drilling core with a diameter of 12 cm was scanned at a tube voltage of 230 kVp. A reconstructed slice is shown in figure 1. Although image contrast is rather low and beam hardening is still present, it must be kept in mind that we are dealing with concrete of 12 cm diameter, which is highly attenuating.

3.2. High resolution and phase-contrast radiography and tomography

Although the setup is designed to scan larger objects, standard micro-CT for smaller objects is still possible. This is illustrated in figure 2, where a reconstructed slice of a Bentheimer sandstone sample is shown. Using a source-to-object distance of 20.4 mm and a source-to-detector distance of 1000 mm, the effective isotropic reconstructed voxel size is 4^3 μm^3. Using the in-house developed image analysis software Morpho+ [3], the average grain size is determined to be 35.6^3 μm^3. Despite the small difference in attenuation coefficients, the distinction between the two phases, quartz and microcline, can be clearly made in the images.

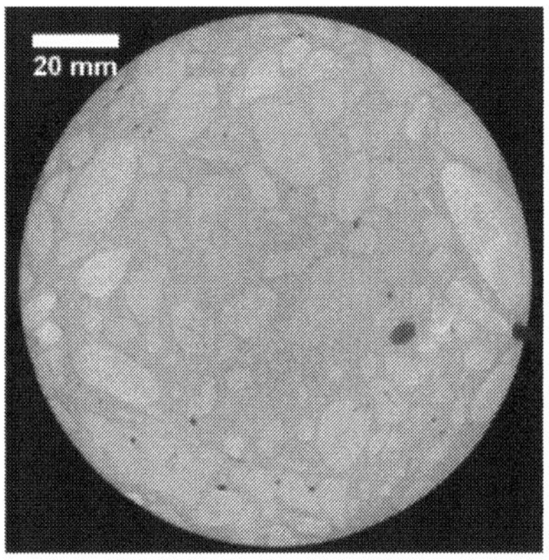

Figure 1. A reconstructed slice of a concrete core with a diameter of 12 cm

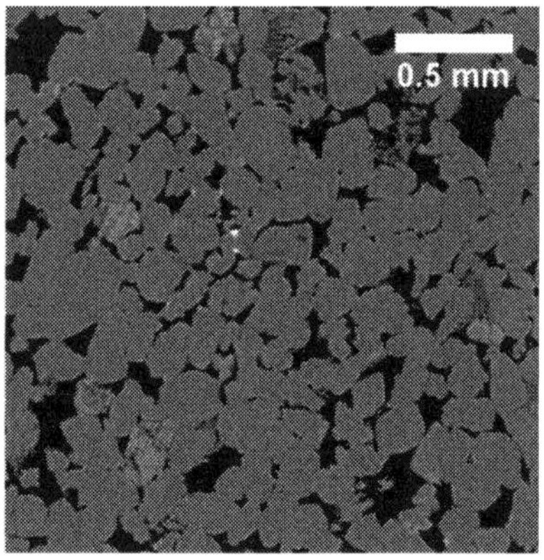

Figure 2. A region of interest from a reconstructed slice of a Bentheimer sandstone sample.

The large spatial coherence of the micro-focus X-ray source makes X-ray phase contrast imaging possible using in-line beam propagation. Since this method requires no additional hardware it can easily be performed, making use of the varying SDD up to 2000 mm. The use of special phase-retrieval algorithms improves significantly the signal-to-noise-ratio of the reconstructed data, making this technique very valuable for low-attenuating samples [4].

3.3. Flexible geometry

The trade-off between cone beam effect and image statistics is shown in figure 3. A bottle of water was scanned twice at an identical voxel size of 100^3 μm^3, but with a different SDD and corresponding SOD. This resulted in a cone angle of 13.24° for an SDD of 850 mm, and 6.01° for the SDD of 1900 mm. All other scanning parameters were kept constant (tube voltage 100 kVp, target power 50 W, 1200 projections at an exposure time of 1000 ms). Figure 3 shows a vertical cross-section through the complete object, where only the upper half is imaged (i.e. the bottom line is the vertical center). The cone effects can be clearly seen on the water surface and the top of figure 3 (left). In figure 3 (right), these effects have been largely reduced. On the other hand, X-ray flux is reduced by a factor 4 which results in lower statistics which can be observed in figure 3 (right). This is confirmed by the SNR measured on the water volume of the central slice, which is calculated to be 26.05 for the small SDD and 12.76 for the large SDD.

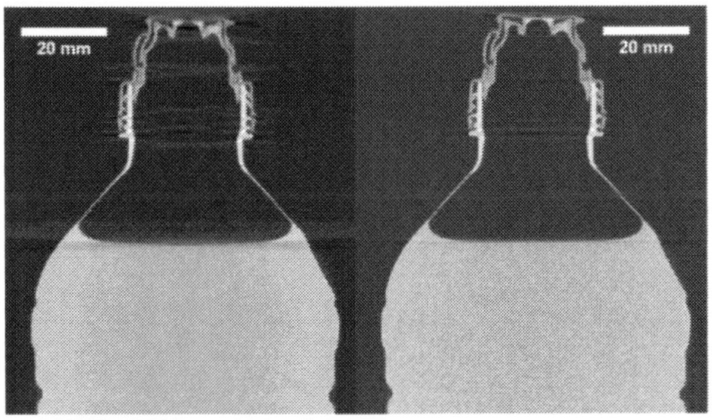

Figure 3. A bottle of water scanned at the same magnification but different source-to-detector distance (SDD).
Left: small SDD hence large cone angle and high flux
Right: large SDD hence small cone angle and low flux

4. Conclusion and outlook

A new 240 kVp setup is installed at UGCT and is now fully operational. The high X-ray energy and flux allow scanning larger samples with a higher content of high-Z material. Additionally, the scanner is capable of performing high-resolution scans down to 4 μm and phase-contrast imaging due to the high spatial coherence of the source.

The scanner is designed to be very flexible using a total of nine motorized stages. At the detector side, these motorized stages can also be used to switch between different detectors. To gain optimal performance from the high energy X-ray tube, a line detector is planned to be installed in the near future.

Acknowledgments

The authors acknowledge the financial support of the Flemish community through the Strategic Initiative Materials in Flanders, the BOF-UGent (doctoral grant to Loes Brabant and GOA 001G01008) and the Research Foundation – Flanders (FWO).

References

[1] Dierick M, Van Loo D, Masschaele B, Boone MN, Van Hoorebeke L. A LabVIEW® based generic CT scanner control software platform. Journal of X-Ray Science and Technology. 2010;18(4):451-61.

[2] Vlassenbroeck J, Dierick M, Masschaele B, Cnudde V, Van Hoorebeke L, Jacobs P. Software tools for quantification of X-ray microtomography. Nuclear Instruments & Methods in Physics Research, Section A: Accelerators, Spectrometers, Detectors, and Associated Equipment. 2007;580(1):442-5.

[3] Brabant L, Vlassenbroeck J, De Witte Y, Cnudde V, Boone MN, Dewanckele J, et al. Three-dimensional analysis of high-resolution X-ray computed tomography data with Morpho+. Microscopy and Microanalysis. 2011;17(2):252-63.

[4] Boone MN, De Witte Y, Dierick M, Almeida A, Van Hoorebeke L. Improved signal-to-noise ratio in laboratory-based phase contrast tomography. Microscopy and Microanalysis. 2012;18(2):399-405.

A compact Laboratory Transmission X-ray Microscope for the water window

H. Legall[1,2], H. Stiel[2], G. Blobel[2], C. Seim[1], J. Baumann[1], S. Yulin[3], D. Esser[4], M. Hoefer[4], U. Wiesemann[5], M. Wirtz[5], G. Schneider[6], S. Rehbein[6] and H. M. Hertz[7]

[1,2]Berlin Laboratory for innovative X-ray technologies (BLiX)
[1]Technische Universität Berlin, Institut für Optik und Atomare Physik, Hardenbergstr. 36, D-10623 Berlin, Germany
[2]Max-Born-Institut, Max-Born-Str. 2A, D-12489 Berlin, Germany
[3]Fraunhofer-Institut für Angewandte Optik und Feinmechanik IOF, Beutenberg Campus Albert-Einstein-Str. 7, D-07745 Jena, Germany
[4]Fraunhofer-Institut für Lasertechnik ILT, Steinbachstr. 15, D-52074 Aachen, Germany
[5]Bruker ASC GmbH, Waltherstrasse 49-51, D-51069 Köln, Germany
[6]Helmholtz-Zentrum Berlin, Institute for Soft Matter and Functional Materials, Albert-Einstein-Str. 15, D-12489 Berlin, Germany
[7]KTH (Royal Institute of Technology), Center Biomedical and x-ray Physics, Roslagstullsbacken 21, SE-106 91 Stockholm, Sweden

E-mail: `legall@mbi-berlin.de`

Abstract. In the water window (2.2 - 4.4 nm) the attenuation of radiation in water is significantly smaller than in organic material. Therefore, intact biological specimen (e.g. cells) can be investigated in their natural environment. In order to make this technique accessible to users in a laboratory environment a Full-Field Laboratory Transmission X-ray Microscope (L-TXM) has been developed. The L-TXM is operated with a nitrogen laser plasma source employing an InnoSlab high power laser system for plasma generation. For microscopy the Ly_α emission of highly ionized nitrogen at 2.48 nm is used. A laser plasma brightness of 5×10^{11} photons/(s $\times$ sr $\times$ μm^2 in line at 2.48 nm) at a laser power of 70 W is demonstrated. In combination with a state-of-the-art Cr/V multilayer condenser mirror the sample is illuminated with 10^6 photons/($\mu m^2 \times$ s). Using objective zone plates 35 - 40 nm lines can be resolved with exposure times < 60 s. The exposure time can be further reduced to 20 s by the use of new multilayer condenser optics and operating the laser at its full power of 130 W. These exposure times enable cryo tomography in a laboratory environment.

1. Introduction

Water window transmission microscopy is suitable for 3-dimensional microscopy of whole cryogenic biological cells [1]. At synchrotron facilities it was shown that by cryo fixation the specimen can be subjected to high radiation doses without showing visible changes in the image [2]. Shock frozen (cryo fixed) samples can be ideally investigated in the water window because of the high penetration depth of radiation in water in this spectral range. Furthermore, the high absorption contrast between the surrounding water and organic structures in the water window enables imaging cells with high resolution. By using zone plates for imaging, resolutions down to 10 nm

are presently achieved at synchrotron facilities [3].
An immediate and permanent access of this technique in the home laboratory would be advantageous for many researchers. In order to make this technique accessible to researchers in a laboratory environment, key components for laboratory microscope setups were developed in the recent decade [4, 5, 6]. In this article we present a further development step of laboratory transmission x-ray microscopy in the water window. The presented laboratory microscope in this article is based on cryo nitrogen jet laser plasma sources driven by a novel high power 130 W InnoSlab laser system and equipped with a Cr/V condenser multilayer mirror and a micro zone plate for imaging. A similar microscope employing a 200 W InnoSlab laser upgrade was taken into operation at the KTH in Stockholm [7]. It was shown that the higher source power decreases the acquisition times for a high quality microscope image to a level which makes cryo tomography feasible in the home laboratory.

2. Arrangement of the laboratory microscope

The arrangement of the microscope is shown in Fig. 1. The setup consists of a laser plasma source for x-ray generation and the x-ray optical arrangement for imaging including a condenser x-ray mirror for sample illumination. A detailed description of the laboratory microscope can be found in [8].

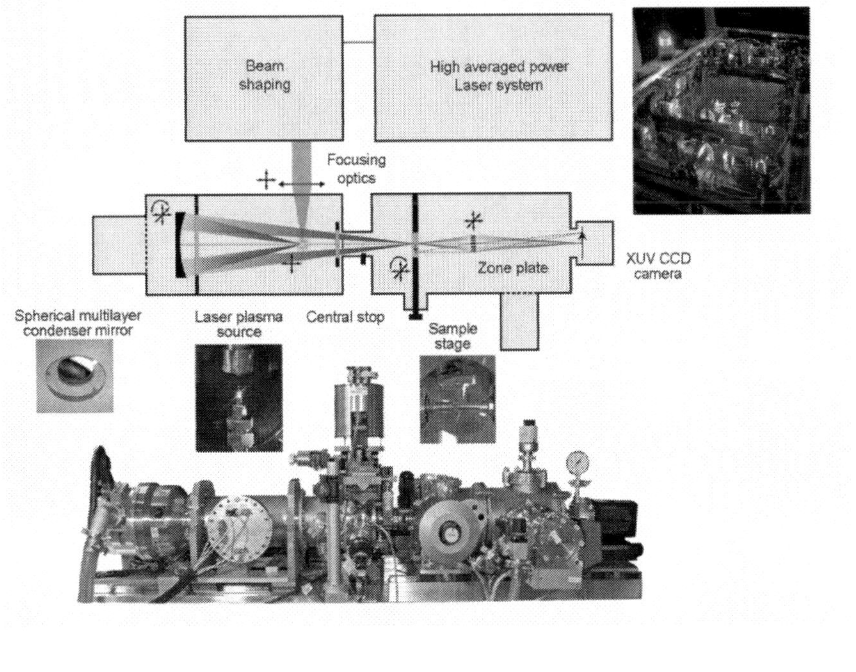

Figure 1. Scheme of the Laboratory Transmission X-ray Microsocpe (L-TXM).

The plasma is generated by focusing the laser beam on a liquid nitrogen cryo jet target. For microscopy the Ly_α emission of highly ionized nitrogen at 2.48 nm is used (Fig. 2). The sample is illuminated by a spherical Cr/V multilayer condenser mirror. The peak reflectivity of the condenser mirror in normal incidence geometry is up to 0.6 % and the bandwidth is 0.008 nm (FWHM). The usage of a multilayer condenser in normal incidence geometry is advantageous for the following reasons: It allows a monochromatization of the plasma light, an easy matching of the zone plate aperture, aberration free imaging of the source into the sample plane and most important, a higher resistance to damages due to debris from the laser plasma source in comparison to condenser zone plates. A central beam stop at the exit of the plasma source unit, placed between source and sample, blocks radiation emitted by the source into the direction of the sample. Scattered laser radiation is eliminated in front of the sample by a 200 nm Al filter behind the laser plasma source. Using an objective zone plate an enlarged image of the specimen is projected onto a CCD camera. Zone plate and sample positioning are integrated in a compact microscopy unit. Samples can be quickly transferred into vacuum with a load lock and are positioned with high accuracy. Cryo tomography of quick-frozen samples can be performed using a cryo sample holder. The characteristic parameters of the L-TXM are summarized in Table 1.

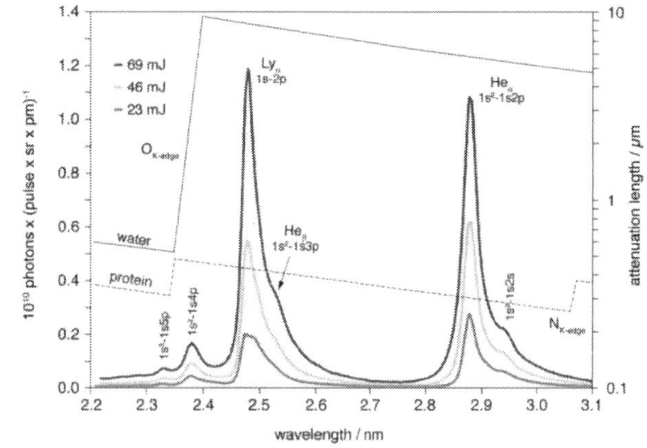

Figure 2. Emission spectra of the liquid-nitrogen plasma. For comparison the attenuation length of radiation within the water window is shown.

Table 1. Characteristics of the L-TXM at a laser power of 70 W. The intensity on the sample is given for 0.1% BW, which is estimated from the measured line width of ≈ 2 pm for the Ly_α emission [9].

Laser system	130 W, 1.3 kHz, 0.4 .. 1 ns, 1.064 μm (InnoSlab)
Source brightness	5×10^{11} photons/(s $\times$ sr $\times$ μm^2) in line at 2.48 nm
Intensity on sample	10^6 photons/ (s $\times$ μm^2 $\times$ 0.1 % BW at 2.48 nm), $R_{ML} = 0.3\%$
Image field	40×40 μm^2
Resolution	35 - 40 nm
Acquisition times	< 60 s (dry samples), $R_{ML} = 0.3\%$, $\eta_{ZP} = 5\%$

3. Results and discussion

In Fig. 3 a long term acquisition of a Siemens star is shown. The resolution (half-pitch) of this image is 41 nm. This high resolution even at longer exposure times emphasizes the stability of the laboratory microscope arrangement. The latter is a precondition for tomography. In Fig. 4 an image of a diatom is shown. This image was collected with 60 W laser power. The acquisition times of this image is 1 min, the pixel size is 26 nm and the resolution is 50 nm (half pitch). Without sample the detector is illuminated at 60 W with averaged 12 ± 2 ph/s per 26 nm pixel. Images at higher laser power were not taken so far. Based on the results at 60 W the recording time can be decreased to < 20 s, if the laser works at full power of 130 W and a multilayer condenser mirror with a reflectivity of 0.6% is used.

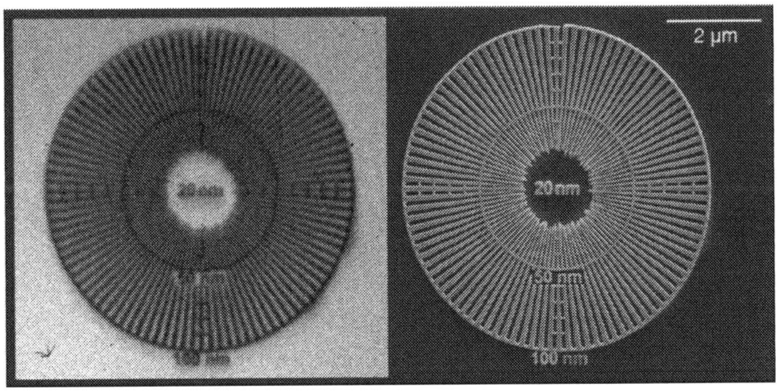

Figure 3. X-ray microscope (left) and SEM (right) image of a Siemens star. The microscope image was taken at a laser power of 10 W and a pulse width of 1 ns with a multilayer reflectivity of 0.45%. The pixel size is 11 nm. The exposure time was 15 min. The efficiency of the used zone plate with an outermost zone width of 25 nm is estimated to be 5%.

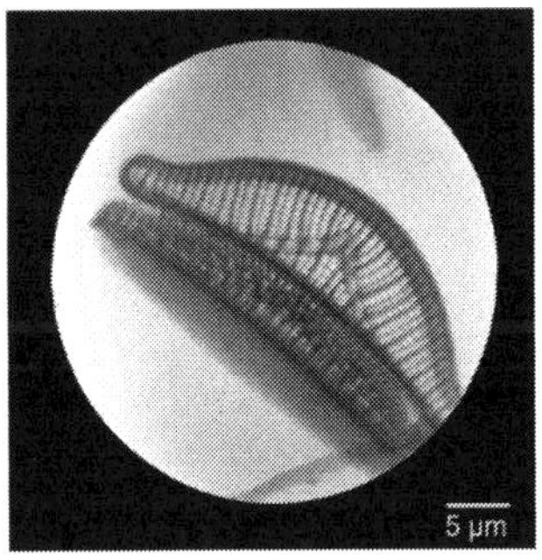

Figure 4. Microscope image of a diatom, taken at a laser power of 60 W, a pulse width of 450 ps and a multilayer reflectivity of 0.3%. The pixel size is 26 nm. The exposure time was 1 min.

4. Summary and perspectives

A novel laboratory x-ray microscope has been developed. Employing a high power InnoSlab laser system, microscopy with exposure times of < 20 s is feasible. This will enable cryo tomography with reduced resolution and sample size in the home laboratory. Further development of the laboratory microscope is in progress, mainly an improvement of the multilayer condenser reflectivity. A reflectivity of 1-2% is envisaged. This would push the performance of the laboratory microscopy to the synchrotron level.
The L-TXM is available for applications at the Berlin Laboratory for innovative X-ray technologies (BLiX).

Acknowledgments

This project was funded by the BMBF (#13N8913) and the BMVBS (WTW #03WWBE106).

References

[1] Kirz J, Jacobsen C and Howells M 1995 *Quarterly Reviews of Biophysics* **28** 33–130
[2] Schneider G 1998 *Ultramicroscopy* **75** 85 – 104
[3] Rehbein S, Guttmann P, Werner S and Schneider G 2012 *Optics Express* **20** 5830–5839
[4] Jansson P, Vogt U and Hertz H 2005 *Review of Scientific Instruments* **76** 043503
[5] Benk M, Bergmann K, Schäfer D and Wilhein T 2008 *Optics Letters* **33** 2359–2361
[6] Bertilson M, von Hofsten O, Vogt U, Holmberg A, Christakou A E and Hertz H M 2011 *Optics Letters* **36** 2728–2730
[7] Martz D H, Selin M, von Hofsten O, Fogelqvist E, Holmberg A, Vogt U, Legall H, Blobel G, Seim C, Stiel H and Hertz H M 2012 *Optics Letters, doc. ID 171680 (posted 20 September 2012, in press)*
[8] Legall H, Blobel G, Stiel H, Sandner W, Seim C, Takman P, Martz D H, Selin M, Vogt U, Hertz H M, Esser D, Sipma H, Luttmann J, Höfer M, Hoffmann H D, Yulin S, Feigl T, Rehbein S, Guttmann P, Schneider G, Wiesemann U, Wirtz M and Diete W 2012 *Optics Express* **20** 18362–18369
[9] Wilhein T, Hambach D, Niemann B, Berglund M, Rymell L and Hertz H M 1997 *Applied Physics Letters* **71** 190–192

The use of zoneplates for point projection imaging

Stefan Baumbach[1] **and Thomas Wilhein**[1]

[1] Institute for X-Optics, RheinAhrCampus Remagen, University of Applied Sciences Koblenz, Suedallee 2, 53424 Remagen, Germany

E-mail: baumbach@rheinahrcampus.de

Abstract. For point projection and phase contrast imaging very small x-ray spot sizes are the key to achieve high resolution images. This article shows a concept of demagnifing the spot of an 8-keV micro focus x-ray tube with a zoneplate. The zoneplate in combination with an order selection aperture acts as a monochromator, which is an enhancement especially for phase contrast imaging. With this concept, the resolution in our point projection microscope has been improved by a factor about two.

1. Introduction

Point projection or projection based phase contrast imaging is very well known. But, due to technical limitations, very small spot sizes are difficult to achieve. In a point projection microscope the resolution is primarily limited by the spot size. So, the idea was to demagnify the x-ray spot with a zoneplate to improve the contrast transfer function (CTF) of the system and therefore to gain higher resolution.

2. Experimental setup

Figure 1 shows a schematical drawing of the experimental setup. A microfocus tube (1) with a copper target to get the Cu-K-α line at 8.05 keV was used as an x-ray source. Zoneplate nr. 1 (3) (for technical information see table 1) forms a demagnified image of the x-ray source into the plane of the order selecting aperture (OSA)(4). This image is now the source of the projection microscope forming an absorption and/or phase contrast image of the object (5) on the detector (8). A beam stop (7) is placed in front of the detector to block the 0th order of the zoneplate.

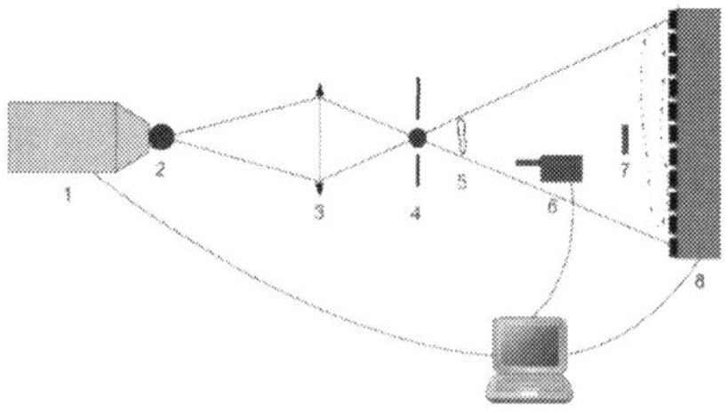

Figure 1. Schematical setup: (1) x-ray tube, (2) tube spot, (3) zoneplate, (4) OSA and demagnified tube spot, (5) object, (6) x-ray spectrometer, (7) beam stop, (8) CCD.

Table 1. Zoneplates technical information.

	Zoneplate nr. 1	Zoneplate nr. 2
Material:	900 nm Tungsten	500 nm Tungsten
Diameter:	180 μm	120 μm
Outer zone width:	80 nm	50 nm
Focal length at 8.05 keV:	93.5 mm	38.96 mm

3. Comparison of the spot sizes and spectral bandwidth

To get quantitative information about the tube spot and the demagnified spot, zoneplate nr. 2 was used as an objective lens forming a magnified image of each x-ray spot on the detector. Figure 2 shows an image of the original tube spot (left) and an image of the demagified spot (right). The tube spot size in diameter is about 3.3 μm (FWHM). The demagnified spot size has a diameter of about 2.10 μm (FWHM). Due to the wave length dependency of the focal

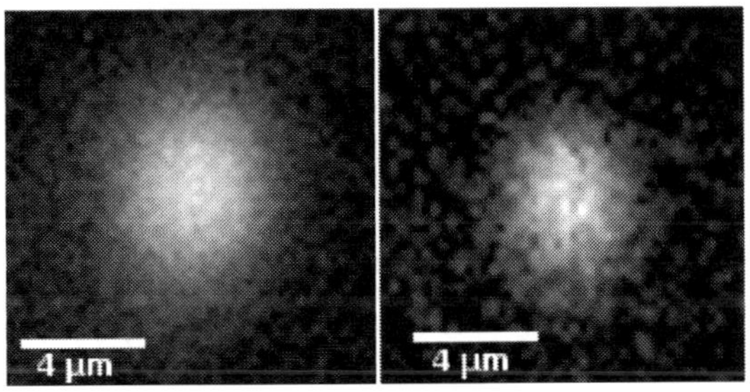

Figure 2. Left: Image of the original tube spot. **Right:** Image of the demagnified tube spot.

length of the zoneplate and the use of an OSA, the spectral bandwidth of the demagnified source is much smaller than the spectral bandwidth of the original tube spot. A comparision of the images in figure 3 shows that the Cu-K-β line emission as well as the continuous spectrum is extenuated. The degree of monochromatisation is proportional to the ratio between zoneplate diameter and two times the diameter of the OSA [1], so a smaller OSA would produce a smaller spectral distribution.

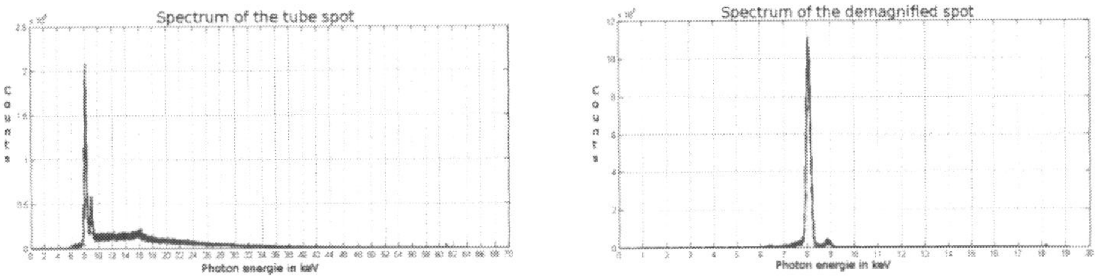

Figure 3. Left: Spectrum of the original tube spot. **Right:** Spectrum of the demagnified spot using a zoneplate and an OSA as monochromator.

4. Contrast Transfer Function (CTF)

The CTF of a projection based phase contrast microscope mainly depends on the x-ray source size (spatial part of the CTF) [2] and on the spectral distribution of the incident radiation (spectral or temporal part of the CTF) [3], assumed that the resolution is not detector limited. The contribution of the source size to the CTF is simply the fourier transform of the spatial distribution of the source. The contribution of the spectral part is given by the following formula [3]:

$$D(u, v) = e^{-(\pi z(f_x^2+f_y^2)\Delta\lambda)^2/8}. \tag{1}$$

Hereby, $\Delta\lambda$ is equal to the gaussian spectral bandwidth of the x-ray source, f_x and f_y are the spatial frequencies in the x and y direction and z is the source to object distance. So, with decreasing z and/or decreasing $\Delta\lambda$, the system is able to visualize smaller object features in propagation based phase contrast images. Figure 4 shows a comparison of the spatial and spectral part of the CTF for the tube spot and the demagnified spot and a combination of both parts in figure 5. Due to the demagnification and the monochromator setup, the resolution has been improved by a factor of about two, assuming 50% contrast as resolution criterion.

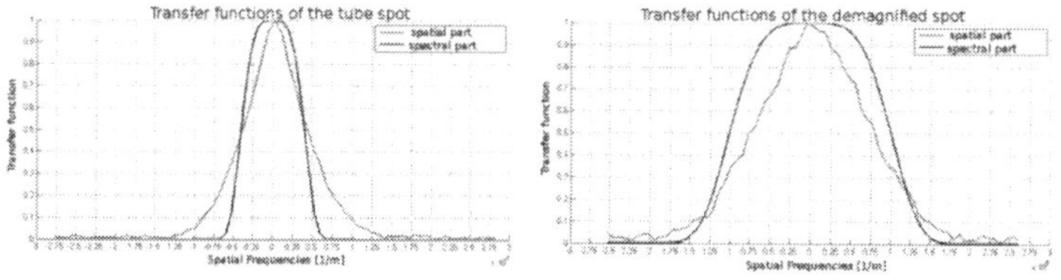

Figure 4. Left: Spatial and temporal CTF of the tube spot **Right:** Spatial and temporal CTF of the demagnified spot.

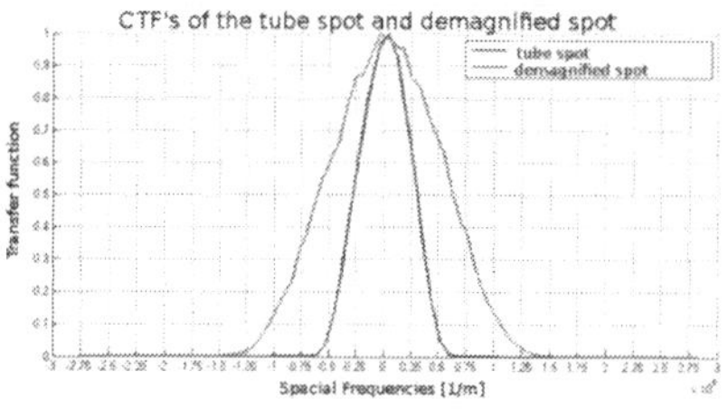

Figure 5. Combined CTF's.

5. Experimental results

An optical fiber core with 125 μm diameter was used as an object and point projection images were taken. Figure 6 shows the images of the core taken with the tube spot (left) and with the demagnified spot (right), respectively.

In figure 6 (right) the more enhanced phase contrast features of dark and bright fringes at the edges of the core are clearly visible.

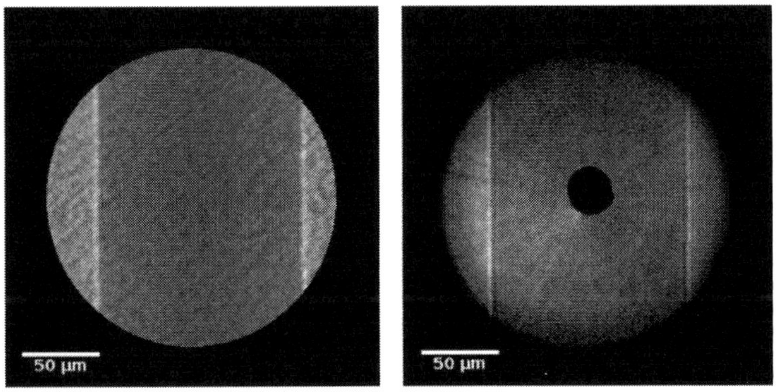

Figure 6. Phase contrast image of a fiber core, illuminated with the tube spot (left) and with the demagnified spot (right).

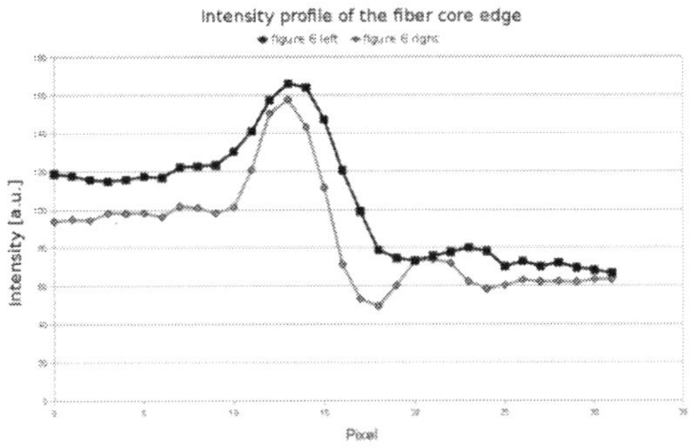

Figure 7. Lineplot through the left edge of figure 6 (left), and through the edge of figure 6 (right), respectively.

6. Conclusion

In this article we have shown that zoneplates can be used to improve the contrast transfer function of a point projection microscope by demagnifying the source of a laboratory x-ray tube. Simultaneously one achieves a smaller spectral bandwidth by using an order selecting aperture. This results in more enhanced dark and bright fringes in projection phase contrast images and smaller object features are visible.

References

[1] Hambach D 2001 *Nanostrukturen mit hohem Aspektverhaeltnis als lichtstarke diffraktive Roentgenoptiken fuer hohe Beugungsordnungen* (Cuvillier Verlag)
[2] Gureyev T and Wilkins S 1998 *Optical Society of America* **15, No.3** 579 – 584
[3] Mayo S and Miller P 2002 *Journal of Microscopy* **207** 79 – 96

Journal of Physics: Conference Series **463** (2013) 012015 doi:10.1088/1742-6596/463/1/012015

Phase-retrieval methods with applications in composite-material tomography

Anna Burvall,[1] Daniel H Larsson,[1] Ulf Lundström,[1] Fredrik Stig,[2] Stefan Hallström,[2] and Hans M Hertz[1]

[1]Biomedical and X-ray Physics, Royal Institute of Technology (KTH), Albanova, SE-106 91 Stockholm, Sweden

[2]Dept. of Aeronautical and Vehicle Engineering, Royal Institute of Technology (KTH), SE-100 44 Stockholm, Sweden

E-mail: anna.burvall@biox.kth.se

Abstract. In-line phase-contrast x-ray imaging is emerging as a method for observing small details when the contrast in absorption x-ray imaging is low. It gives images with strong edge enhancement, and phase retrieval is necessary to obtain quantitative thickness information. In particular for tomography, clarity can be enhanced by phase retrieval, as here demonstrated on a 3D-weave reinforced composite material. Seven suitable phase-retrieval methods are identified and integrated into a single method, where each version is marked by variations in particular steps. The general method and its variations are outlined and a comparison shows which methods are most suitable in different situations.

1. Introduction

In-line phase-contrast imaging [1] is a method for visualizing phase for hard x-rays used, e.g., in material studies [2]. As illustrated in figure 1, the distance between object and detector allows phase variations to develop into intensity differences. The resulting image shows the position of object details very clearly marked by their edges, but to get quantitative information on e.g. object density distribution, phase-retrieval [3] is needed. This post-processing can be of special interest in tomography, as here demonstrated on composite-material samples. We compare phase-retrieval methods intended for use on tomographic data [4].

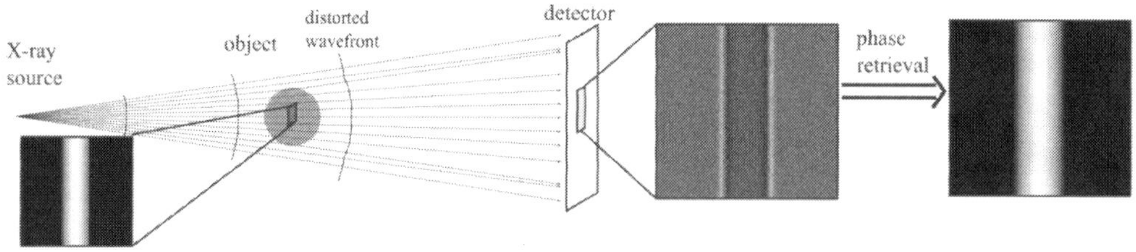

Figure 1. Illustration of in-line phase contrast, and of phase retrieval.

An object, as a function of three-dimensional position $\mathbf{r}$, is characterized by its refractive index $n(\mathbf{r}) = 1-\delta(\mathbf{r})+i\beta(\mathbf{r})$ where $\delta(\mathbf{r})$ describes refractive or phase properties, whereas $\beta(\mathbf{r})$ describes absorption and is related to the absorption coefficient μ as $\mu = 4\pi\beta/\lambda$, where λ is the wavelength. For true phase retrieval $\delta(\mathbf{r})$ and $\beta(\mathbf{r})$ should be recovered independently, but this cannot be done from a single phase-contrast image as it does not contain enough information. It could be done from two images captured at different distances from the object, but this is inconvenient in x-ray tomography. Instead, a relation between $\delta(\mathbf{r})$ and $\beta(\mathbf{r})$ is assumed. Two options are in use: that there is no absorption, i.e. $\beta(\mathbf{r}) = 0$, or that $\delta(\mathbf{r})$ and $\beta(\mathbf{r})$ are proportional to each other. In some cases those assumptions are accurate: for example the second assumption is true if the object is homogeneous, or for light materials at high photon energies. In other cases, such as multi-material samples at lower photon energies, the assumptions are inaccurate but still used for lack of other options.

The scope of this comparison [4] is limited to methods that use a single image. Furthermore, we limit it to analytical methods for speed, and we only consider monochromatic x-rays in order to compare as many methods as possible. Monochromatic theory in this case works surprisingly well for broadband x-rays. We identify seven different phase-retrieval methods in the literature, all summarized in the next section. We use one of them on a 3D-weave reinforced fibre composite sample, and finally outline how phase-retrieval methods are chosen in different situations.

2. Summary of methods

Table 1. Methods of phase retrieval and their properties. The functions $g(I)$, $H_p(\mathbf{w})$ and $f(g_F)$ are introduced in Eq. 1. Table from Burvall et al.[4] .

Method		$g(I)$	$H_p(u,v)$	$f(g_F)$	Assumptions
Bronnikov		$\frac{I}{I_{in}} - 1$	$\left[2\pi\lambda d\lvert\mathbf{w}\rvert^2\right]^{-1}$	g_F	$\beta = 0$, $F \gg 1$
Modified Bronnikov		$\frac{I}{I_{in}} - 1$	$\left[2\pi\lambda d\lvert\mathbf{w}\rvert^2 + \alpha\right]^{-1}$	g_F	$\beta \approx 0$, $F \gg 1$
Phase-att. duality		$\frac{I}{I_{in}}$	$\left[2\pi\frac{r_e\lambda^2 d}{\sigma_{KN}}\lvert\mathbf{w}\rvert^2 + 1\right]^{-1}$	$\frac{\lambda r_e}{\sigma_{KN}}\ln g_F$	$\beta \propto \delta$, $F \gg 1$, SVP
Single material		$\frac{I}{I_{in}}$	$\left[4\pi^2 d\frac{\delta}{\mu}\lvert\mathbf{w}\rvert^2 + 1\right]^{-1}$	$-\frac{1}{\mu}\ln g_F$	$\beta \propto \delta$, $F \gg 1$
Two materials		$\frac{I\exp[\mu_2 A(\mathbf{r}_\perp)]}{I_{in}}$	$\left[4\pi^2 d\frac{\delta_1-\delta_2}{\mu_1-\mu_2}\lvert\mathbf{w}\rvert^2 + 1\right]^{-1}$	$-\frac{1}{\mu_1-\mu_2}\ln g_F$	$\beta \propto \delta$, $F \gg 1$, SVP
Fourier method, Born type	a	$\frac{1}{2}\left(\frac{I}{I_{in}} - 1\right)$	$\left[\sin\left(\pi\lambda d\lvert\mathbf{w}\rvert^2\right)\right]^{-1}$	g_F	$\beta = 0$, Born
	b		$\left[\gamma\cos\left(\pi\lambda d\lvert\mathbf{w}\rvert^2\right) + \sin\left(\pi\lambda d\lvert\mathbf{w}\rvert^2\right)\right]^{-1}$		$\beta \propto \delta$, Born
Fourier method, Rytov type	a	$\frac{1}{2}\ln\frac{I}{I_{in}}$	$\left[\sin\left(\pi\lambda d\lvert\mathbf{w}\rvert^2\right)\right]^{-1}$	g_F	$\beta = 0$, Rytov
	b		$\left[\gamma\cos\left(\pi\lambda d\lvert\mathbf{w}\rvert^2\right) + \sin\left(\pi\lambda d\lvert\mathbf{w}\rvert^2\right)\right]^{-1}$		$\beta \propto \delta$, Rytov

The seven methods included in this comparison are 1) the Bronnikov method [5], 2) the modified Bronnikov algorithm by Groso et al. [6], 3) the phase-attenuation duality algorithm by Wu et al. [7], 4) the method for homogeneous object or single material by Paganin et al. [8], 5) the two-material method by Beltran et al. [9], and the Fourier method in 6) the Born or 7) the

Rytov approximation derived by Gureyev et al. [10]. Number 6, the Fourier method in Born approximation, has also been derived by Zabler [11], Guigay [12], and Turner [13] under slightly different approximations. Modifications to this method have been suggested by Moosmann [14].

All these methods follow the same pattern – the phase $\varphi(\mathbf{r}_\perp)$, where $\mathbf{r}_\perp$ is the two-dimensional space coordinate in the plane transverse to the direction of propagation, is retrieved from registered intensity $I(\mathbf{r}_\perp)$ as

$$\varphi(\mathbf{r}_\perp) = f\left(\mathcal{F}^{-1}\left\{H(\mathbf{w})\cdot\mathcal{F}\left\{g\left[I(\mathbf{r}_\perp)\right]\right\}[\mathbf{w}]\right\}[\mathbf{r}_\perp]\right) \tag{1}$$

where $\mathcal{F}$ denotes Fourier transform and $\mathbf{w}$ spatial frequency. The three functions $g(I)$, $H(\mathbf{w})$, and $f(g_F)$ will differ between the different methods, and are summarized in Table 1. For derivation and details on method parameters, see Ref. [4]. All methods use the Fresnel approximation; the other assumptions listed in Table 1 are large Fresnel numbers ($F \gg 1$), slowly varying phase (SVP), a Born-type approximation (Born), and a Rytov-type approximation (Rytov).

3. Examples of phase retrieval

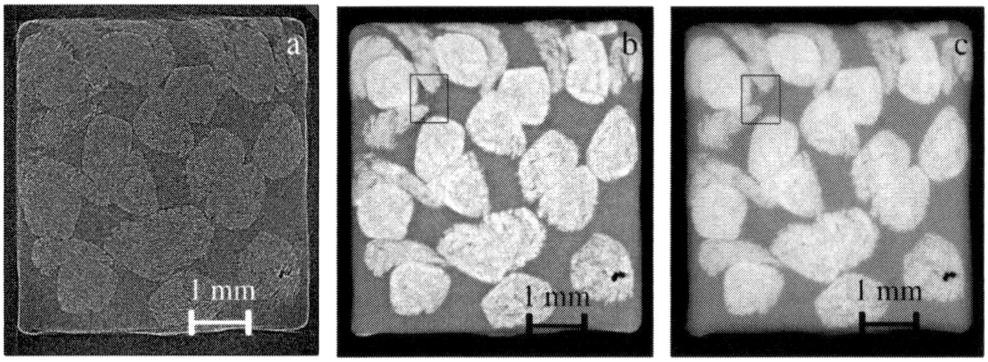

Figure 2. Reconstructed slices of a 3D-weave reinforced composite sample a) without phase retrieval, b) using the single-material method for $\delta/\mu = 6.8\cdot10^{-9}$ m and c) $\delta/\mu = 1.5\cdot10^{-8}$ m.

A 3D-weave reinforced composite sample, consisting of vinyl ester resin surrounding carbon fibre bundles and some air voids, has been imaged using phase-contrast x-ray tomography. This is a multi-material sample, but the single-material phase-retrieval method has still been applied to demonstrate how the parameter δ/μ can be adjusted to reconstruct interfaces between different materials. A liquid-metal-jet x-ray source [15] with Galinstan as anode material, operated at 40 W with a 50 kV acceleration voltage, was used. The source-to-object distance was 625 mm and the object-to detector distance 2375 mm. The detector is a Gadox scintillator-based fiber-coupled CCD detector from Photonic Science with 9×9 μm^2 pixels and a total area of $36\times24\,\text{mm}^2$, and images are binned 4×4 times before phase retrieval and the subsequent tomographic reconstruction using filtered back-projection (FBP). A slice through the tomographic reconstruction is shown in figures 2 and 3 with and without phase retrieval. For lower values of δ/μ, as in figures 2b or 3a, the air-vinyl interface is properly reconstructed, while edge amplification is still visible for vinyl-carbon. For higher values of δ/μ, as in figures 2c or 3b, the vinyl-carbon interface is well reconstructed but the air-vinyl interface smeared. This shows particularly well around the voids in figure 3, which are quite well reconstructed for a low δ/μ but less visible for a higher value.

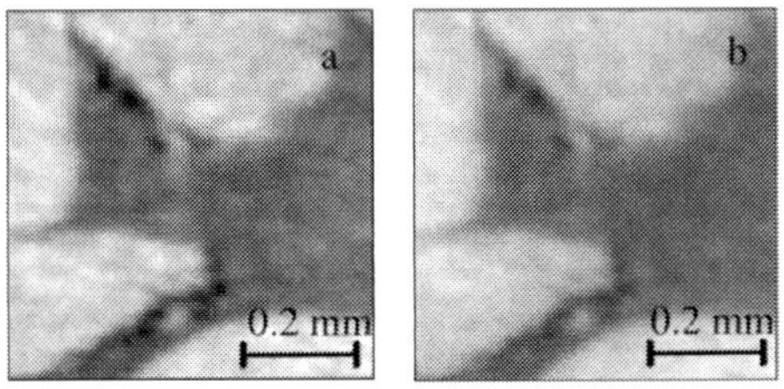

Figure 3. Zoom-in of the regions indicated in figure 2 for a) $\delta/\mu = 6.8 \cdot 10^{-9}$ m and b) $\delta/\mu = 1.5 \cdot 10^{-8}$ m.

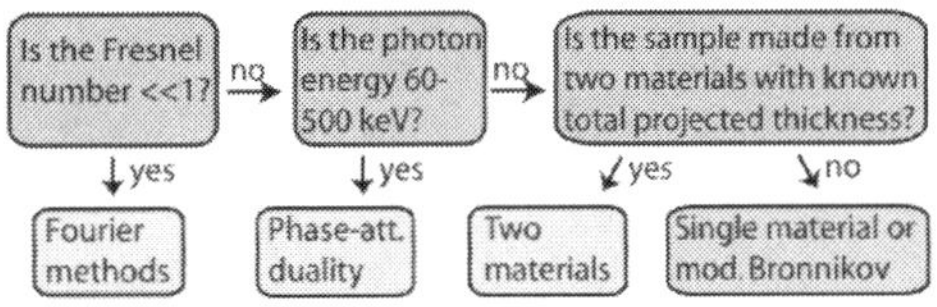

Figure 4. Scheme for choosing the appropriate phase-retrieval method.

4. Conclusions: choosing your phase-retrieval method

Comparison of the different methods, based mainly on the approximations and assumptions of their derivations but also on testing their noise sensitivity, lead to a simple scheme for choosing a reconstruction method [4], as illustrated in figure 4.

References

[1] Wilkins S W, Gureyev T E, Gao D, Pogany A, and Stevenson A W 1996, Phase-constrast imaging using polychromatic hard X-rays, *Nature* **384** 335
[2] Mayo S C, Stevenson A W, and Wilkins S W 2012, In-line phase-contrast x-ray imaging and tomography for materials science, *Materials* **5** 937
[3] Nugent K A 2007, X-ray noninterferometric phase imaging: a unified picture, *J. Opt. Soc. Am. A* **24** 536
[4] Burvall A, Lundström U, Takman P A C, Larsson D H, and Hertz H M 2011, Phase-retrieval in X-ray phase-contrast imaging suitable for tomography, *Optics Express* **19** 10359
[5] Bronnikov A V 1999, Reconstruction formulas for phase-contrast imaging, *Opt. Commun.* **171** 239
[6] Groso A, Abela R, and Stampanoni M 2006, Implementation of a fast method for high resolution phase contrast tomography, *Optics Express* **14**, 8103
[7] Wu X and Liu H 2005, X-ray cone-beam phase tomography formulas based on phase-attenuation duality, *Optics Express* **13** 6000
[8] Paganin D, Mayo S C, Gureyev T E, Wilkins P R, and Wilkins S W 2002, Simultaneous phase and amplitude extraction from a single defocused image of a homogeneous object, *J. Microsc.* **206** 33
[9] Beltran M A, Paganin D M, Uesugi K, and Kitchen M J 2010, 2D and 3D X-ray phase retrieval of multi-material objects using a single defocus distance, *Optics Express* **18** 6423
[10] Gureyev T E, Davis T J, Pogany A, Mayo S C, and Wilkins S W 2004, Optical phase retrieval by use of first Born- and Rytov-type approximations, *Appl. Opt.* **43** 2418
[11] Zabler S, Cloetens P, Guigay J P, and Baruchel J 2005, Optimization of phase contrast imaging using hard x rays, *Rev. Sci. Intstr.* **76** 073705
[12] Guigay J P 1977, Fourier transform analysis of Fresnel diffraction patterns and in-line holograms, *Optik* **49** 121
[13] Turner L D, Dhal B B, Hayes J P, Mancuso A P, Nugent K A, Paterson D, Scholten R E, Tran C Q, and Peele A G 2004, X-ray phase imaging: Demonstration of extended conditions with homogeneous objects, *Optics Express* **12** 2960
[14] Moosmann J, Hofmann R, and Baumbach T 2011, Single-distance phase retrieval at large phase shifts, *Optics Express* **19** 12066
[15] Larsson D H, Lundström U, Takman P A C, Burvall A, and Hertz H M 2011, A 24 keV liquid-metal-jet x-ray source for biomedical applications, *Rev. Sci. Instr.* **82** 123701

Journal of Physics: Conference Series **463** (2013) 012016 doi:10.1088/1742-6596/463/1/012016

Moonlight receptor of the "1-h-midge" *Clunio marinus* studied by micro-XRF

G Falkenberg[1], Ge Fleissner[2], D Neumann[3], G Wellenreuther[1], P Alraun[1], and Gue Fleissner[2]

[1] Deutsches Elektronen-Synchrotron DESY, PETRA III beamline P06, Hamburg, Germany
[2] Institute for Cell Biology and Neurosciences, Goethe University Frankfurt a. M., Germany
[3] Zoological Institute, University of Cologne, Germany

Email: Gerald.Falkenberg@desy.de

Abstract. Melanin is a pigment widely occurring in animals, plants, fungi and algae. It does not only colour skin, hair and eyes but serves mainly as photoprotectant and prevents overload with minerals induced by inflammations, infections and degenerative diseases. Therefore, the mechanisms underlying melanisation gained increasing interest in the field of biomedical research and clinic. So far, the processes of melanogenesis are only partly analysed, nearly nothing is known on a putative switch between melanins of different types. Here we offer a model organism to study these mechanisms as part of a naturally cycling change of transparency of the retinal shielding pigment. A marine midge, *Clunio marinus*, living in coastal regions, underlies a complex timing of its development by solar and lunar climatic periodicities, which synchronise biological clocks. The question was how the animals can discriminate changing sunlight from moonlight intensities. For the first time, we could show a "moonlight window" in the larval ocelli of this midge, and propose a hypothesis on the underlying mechanisms. Driven by a lunar clock the image forming ocelli become transparent and convert during moonlit nights to a sensitive photometer, which can record the dynamics of environmental light. High resolution X-ray fluorescence (XRF) measurements of the distribution of trace minerals in single melanosomes combined with their fine structural details in various states of the lunar cycle provide a first insight into the enzymatic pathways for the generation of a dark melanin (like eumelanin) and a light coloured melanin (like phaeomelanin). Essential advantage of this approach is the spatial and temporal resolution of the metals associated with melanisation processes, which could never before be demonstrated in these details. The data may stimulate further research projects in biomedicine.

1. Introduction

In organisms melanin is a ubiquitous pigment molecule with a far reaching functional meaning, like pattern of skin or fur, colour of hair and eyes; nevertheless it also serves as photoprotectant, and storage of excess minerals, which might occur along with inflammation, injury and infection. The molecule is highly concentrated intracellularly within membrane vesicles, so called "melanosomes". Understanding genetics and biochemistry of melanogenesis and also hormonal or neuronal controlling pathways has nowadays gained a fast growing relevance in medicine for the early diagnosis and treatment: cancer research, degenerative diseases of the central nervous system (e.g., Alzheimer or Parkinson disease) and other age dependent illnesses; also some genetic defects like albinism and special varieties of blindness can be relayed to mutations in the melanisation processes. Nevertheless, a concise theory about all these complex interrelations is still missing. The tiny midge *Clunio* which lives in Atlantic and Pacific coastal regions [1, 2] offers a surprising model for solving some of the open questions of melanisation. Different from most insects, where ommochromes fulfil this function, *Clunio* has melanin as screening pigment in its larval ocelli [3], and here a clear rhythmic change of the screening pigment transparency can be observed [4, 5]. Endogenously controlled by a lunar clock [5, 6], this mechanism alters the function of the ocelli from image forming eyes - with light access only from a distinct direction like in a pinhole camera (Fig. 1) - to a non-visual photoreceptor for a temporal and spatial integration of environmental light during moonlit nights [6].

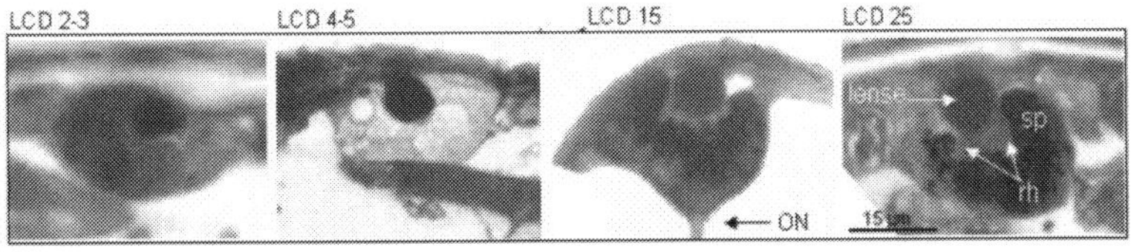

Fig. 1 Semithin sections of larval ocelli in various lunar states: LCD 2-3 begin of Full moon; LCD 4-5 Midst of Full moon; LCD 15 New moon, about a week past nocturnal moonlight; LCD 25 New moon about one week before nocturnal moonlight restarts. [Lunar cycle day (LCD), on LCD 1 "moonlight" is switched on for four nights, the entire lunar cycle lasts 30 days; ON = optic nerve, rh = rhabdom, sp = screening pigment]

This moonlight window enables the organism to use the dynamics of nocturnal moonshine as a precise Zeitgeber signal for the lunar clock. The complex timing of metamorphosis and reproduction by solar and lunar changes of light intensity is essential for the survival of organisms like *Clunio* which live in an intertidal environment [1, 2]. Here we try to develop a hypothesis how this change of transparency in the larval ocelli could be achieved - possibly by a switch between two different forms of the melanin protein. However, a detailed knowledge of the decisive steps of melanisation is missing mainly caused by the lack of a general concept, which comprises the generation of all the various melanin forms - eumelanins (black-brown colour of e.g., skin, hair and eyes), phaeomelanins (yellow-light red pigment in similar sites of the organism), neuromelanins (in the nervous system) and all "allomelanins", which occur in plants, fungi and algae. In insects melanisation has not been analyzed in detail. We hypothesize that in *Clunio* melanogenesis, at least partially, may take place similarly to the mammalian processes. Enzymes involved in melanogenesis depend on trace metals, which can be associated with certain steps of the pigment formation and are also indicative for the different forms of melanin [7]. Thus, we investigate the elements occurring in the ocellar screening pigment layer of the midge along with the changing transparency in different lunar states.

2. Material and Methods

The midges *Clunio marinus* were raised as synchronized laboratory cultures in the animal houses of the Zoological institutes of the Universities Cologne and the Max F. Perutz Laboratories, Vienna, under a daily 14:10 hrs light/dark regime. Every 30 days an artificial "moon" was switched on for four nights, simulating a nocturnal moonlight around "Full moon" [for more details see 5]. For convenience we refer to the time span without nocturnal moonlight as "New moon". Lunar Cycle Day (LCD) 1 is the day with the first of four moonlit nights, the entire lunar cycle lasts 30 days, ending with LCD 30. Histological processing was performed with the strict avoidance of metal tools for dissection, cutting and further handling, and of metal containing chemicals during fixation and embedding [see 5]. For micro-XRF measurements, semithin sections (2-3 µm) were mounted on silicon nitride membranes windows (Silson Ltd., Northampton, England). In order to control the fine structure of single melanosomes, parallel ultrathin sections were collected on Formvar coated copper grids for the inspection in the transmission electron microscope (ZEISS C10R and C12R, Oberkochen). The micro-XRF measurements were carried out at the PETRA III beamline P06 at the Hard X-ray Microprobe experiment. A monochromatic beam of 12 keV was focused by KB-mirrors to a size of 300×300 nm^2 FWHM (10^{10} ph/s). The sample was raster-scanned across the beam at normal incidence and under visual inspection by an in-line video microscope. The micro-XRF signals were collected in the horizontal plane by means of a VORTEX EM silicon drift detector oriented at an angle of 80 ° with respect to the sample surface normal. Sample dwell times are 1-2 s per scan point.

3. Results and Discussion

Our pilot studies at the P06 microprobe on the cyclic change of transparency of the screening pigment in the larval ocelli of *Clunio marinus* provided surprising results, thanks to 50 times higher spatial resolution of this new experiment compared to the previously applied XRF-setup at DORIS III beam line L [4, 5]. Already the beamline L data have made us assume that the screening pigment in *Clunio* larval ocelli is a melanin, not an ommochrome, which generally occurs in insect eyes, but does not contain metal ions. Now, at P06, we could clearly identify different elements in the retinal shielding pigment layer, and even in individual pigment granules, which are changed according to the time of fixation at different lunar phases. The great advantage of XRF element mapping with sub-micrometer resolution is the topological association of metal content with fine structural details as shown in the transmission electron microscope - in contrast to biochemical analyses of tissue extracts (Fig. 2).

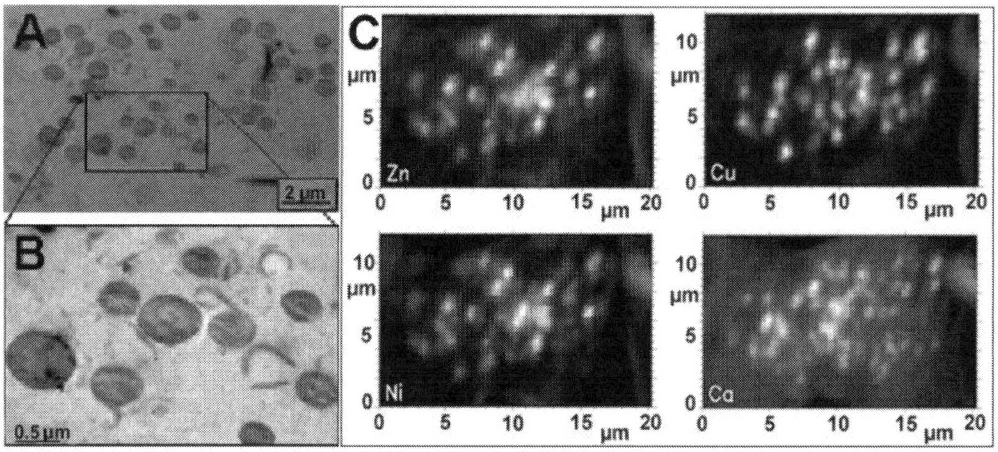

Fig. 2 Melanosomes inside one of the retinula cells of an ocellus at the beginning of New moon (LCD 10). A, B electron microscopic view shows partly filled pigment granules (B higher magnification of A); C µ-XRF element distribution in a section parallel to A, B. The elements are indicative for ongoing eumelanisation.

We can easily discriminate the individual receptor cells (Fig. 3) as well as single pigment granules inside the ocelli (Fig. 2, 3). This is relevant, as the lunar clock signal is most likely transferred to the eye via the optic nerve. The receptor cells next to the entrance of the eye nerve into the retina are the fastest to change their transparency (see Fig. 1). So far, we could analyze ocelli fixated during "Full moon" (LCD 5), the time with nocturnal moonlight, and ocelli fixated during the first week of "New moon" (LCD 10 and LCD 14) and in its middle phase (LCD 19) (Fig. 2, 3).

As the details of different preconditions (animal age, precision of lunar phase, also section plane, section thickness) and their correlation with the XRF data are not yet systematically investigated, we can here only offer a first hypothesis. Corresponding to the different states of the pigment granules we show that the amount of various trace elements like Zn, Ni, Cu, and Ca, as well as their concentration within the pigment granules is changing. Thus, we do not find melanisation versus demelanisation, but rather a conversion between two molecular configurations of melanin like the dark eumelanin and the transparent phaeomelanin [cf. 7]. This conversion seems to be controlled by trace elements (with Cu in the first place) and is likely to occur as the lunar cycle proceeds.

The transport of trace metals from the melanosomes into the surrounding cytoplasm seems to be enhanced as indicated by the increased Ca content, which is essential for membrane transporter enzymes. These pilot analyses must be supplemented and critically verified by further data from eyes fixated during different phases of the experimental lunar light programme in order to resolve the dynamics of the change of transparency in greater details. But already now, they have lead us to the hypothesis that melanogenesis in *Clunio* ocelli has great similarity to the same process in mammalian tissues [7].

Acknowledgements

The project was funded by several grants from HASYLAB (at DESY), from DFG (Deutsche Forschungsgemeinschaft), from the Polytechnische Gesellschaft Frankfurt a. M. and the Alumni of the Goethe University Frankfurt a. M. The authors thank T Kaiser (Max Perutz Institute, Vienna) for providing *Clunio* specimen from synchronized cultures.

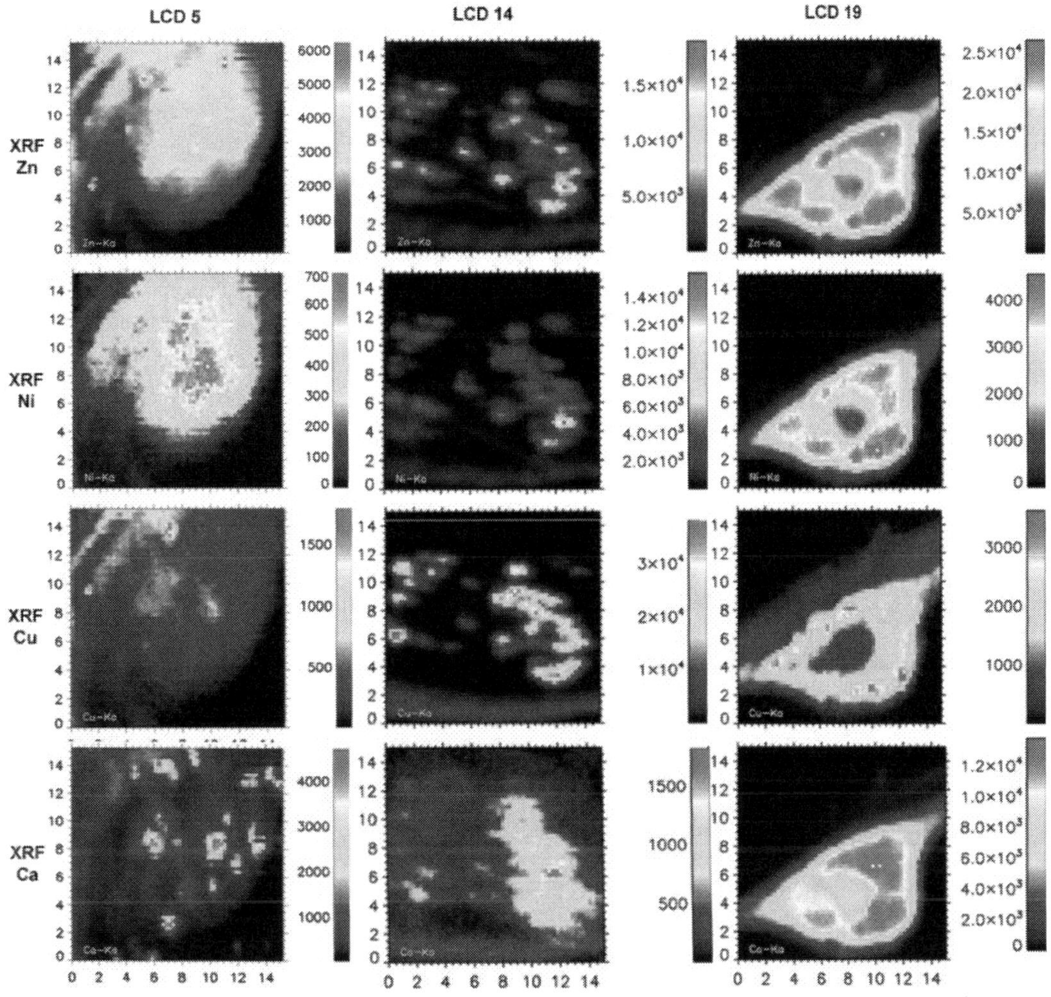

Fig. 3 μ-XRF maps of ocelli in different lunar states. Left row: ocellus during Full moon putatively filled with phaeomelanin; Middle row: Dramatic change of element distribution at the beginning of melanisation during New moon; Right row: Midst of new moon phase, element distribution matches eumelanin. [The x, y axes of the XRF-maps are calibrated in μm; notice that the colour coded intensity scale bars (counts per sec) differ in the different plots.]

References

[1] Neumann D (1985) *Int. J. Biometeorology* **29**, Suppl. 2, 165-177
[2] Neumann D (1986) In: FR Taylor & R Karban (eds.) *Proc. Life Sciences*, Springer Verlag, 3-19
[3] Fleissner Ge *et al* (2005) *Proc. Europ. Biol. Soc.* **III** 131
[4] Fleissner Ge *et al* (2007) *HASYLAB Annual Report (DESY Hamburg)* **2006** 1337
[5] Fleissner Ge *et al* (2008) *Chronobiol. Int.* **25** 17-30
[6] Fleissner Gue *et al* (2008) In: E J Warrant and D E Nilsson (eds.) *2nd International Conference On Invertebrate Vision.* Lund University Press, pp 174 (08/01/2008-08/07/2008)
[7] Slominski AD *et al* (2004) *Phys. Rev.* **84 (4)** 1155-1228

Development of achromatic full-field hard X-ray microscopy using four total-reflection mirrors

S Matsuyama[1], Y Emi[1], Y Kohmura[2], K Tamasaku[2], M Yabashi[2], T Ishikawa[2] and K Yamauchi[1,3]
[1] Department of Precision Science and Technology, Graduate School of Engineering, Osaka University, 2-1 Yamada-oka, Suita, Osaka 565-0871, Japan
[2] SPring-8/RIKEN, 1-1-1 Kouto, Sayo-cho, Sayo-gun, Hyogo 679-5148, Japan
[3] Research Center for Ultra-Precision Science and Technology, Graduate School of Engineering, Osaka University, 2-1 Yamada-oka, Suita, Osaka 565-0871, Japan

E-mail: matsuyama@prec.eng.osaka-u.ac.jp

Abstract. We developed achromatic full-field hard X-ray microscopy on the basis of advanced Kirkpatrick-Baez mirrors consisting of four total-reflection mirrors. The microscope system consists of advanced Kirkpatrick-Baez mirrors as an objective, Kirkpatrick-Baez mirrors as a condenser and an X-ray CCD camera. The performance of the system was investigated using a Siemens star chart with a minimum resolution of 50 nm at BL29XUL of SPring-8 at 10 keV. As a result, a spatial resolution of ~150 nm was achieved. Also, no chromatic aberration was confirmed by taking images at an X-ray energy from 8 to 11.5 keV.

1. Introduction

To realize achromatic full-field hard X-ray microscopy, it is essential to employ imaging optics with total-reflection mirrors. Kirkpatrick–Baez (KB) mirrors, which consist of two elliptical mirrors aligned perpendicular to each other, are often used to focus X-rays. However, they are not suitable for full-field X-ray microscopy because they suffer from comatic aberration. To overcome this aberration, advanced Kirkpatrick–Baez mirrors have been proposed [1]. They consist of two elliptical mirrors and two hyperbolic mirrors oriented perpendicular to each other, as in KB mirrors (Fig. 1).

To develop advanced KB mirrors for microscopy with sub-50 nm resolution, ultraprecise aspherical mirrors with a figure accuracy on the order of 1 nm and precise four-mirror alignment are needed [2]. We have successfully fabricated four mirrors with figure errors of 2 nm by EEM [3]. We also have developed an alignment procedure for aligning the mirrors with the required accuracies [2, 4].

In this paper, we report the results of our first experiment, in which a magnified X-ray image is formed using advanced KB mirrors. The performance test was carried out at BL29XUL of SPring-8 at an X-ray energy of 10 keV. Also, the X-ray energy dependence of image formation was investigated by varying the X-ray energy.

2. Experiment

We constructed an imaging system for magnifying transmission X-rays through a sample between the first and second experimental hutches of BL29XUL (Fig. 2). A monochromatized X-ray is focused at the object plane using KB mirrors as a condenser. The transmission X-ray through the sample is collected by advanced KB mirrors. The sample can also be observed using a light microscope to find the target sample. Finally, a magnified image is formed at the X-ray CCD system placed ~45 m downstream of the objective, which consists of a scintillator (P43), lenses (×2) and a CCD (12.4 μm/pixel, Hamamatsu Photonics). The sample used is a Siemens star chart with a minimum resolution of 50 nm (XRESO-50, NTT AT). The specifications of the advanced KB mirrors are shown in Table 1 (For more details, see references [2, 5]). Also, the parameters of the KB mirrors used as a condenser are shown in Table 2.

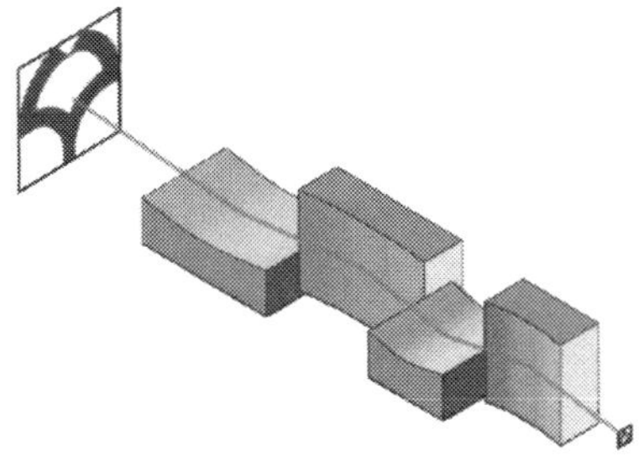

Figure 1. Advanced Kirkpatrick-Baez mirrors

Table 1. Specifications of the advanced KB mirrors used as objective

Numerical aperture (Both directions)	1.2×10^{-3}
Working distance (mm)	50
Magnification factor (H × V)	385 × 210
Acceptable X-ray energy (keV)	< 11.5

Table 2. Parameters of the KB mirrors used as condenser

	Vertical focusing	**Horizontal focusing**
Mirror length (mm)	200	200
Glazing incidence angle[a] (mrad)	7.8	7.4
Focal length[a] (mm)	555	350
Working distance (mm)	150	
Coating	Pt	

[a] At the center of the mirror

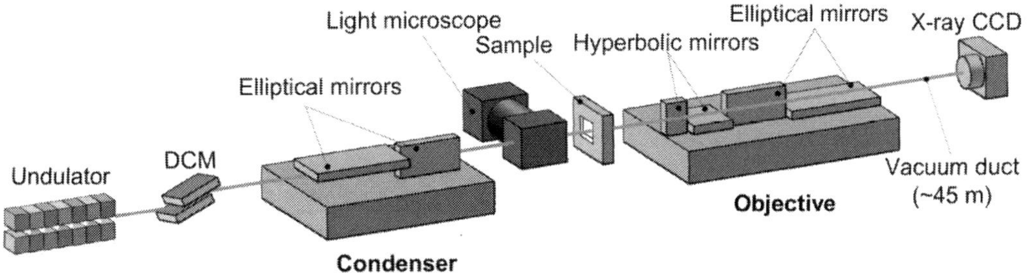

Figure 2. Experimental setup for magnification imaging

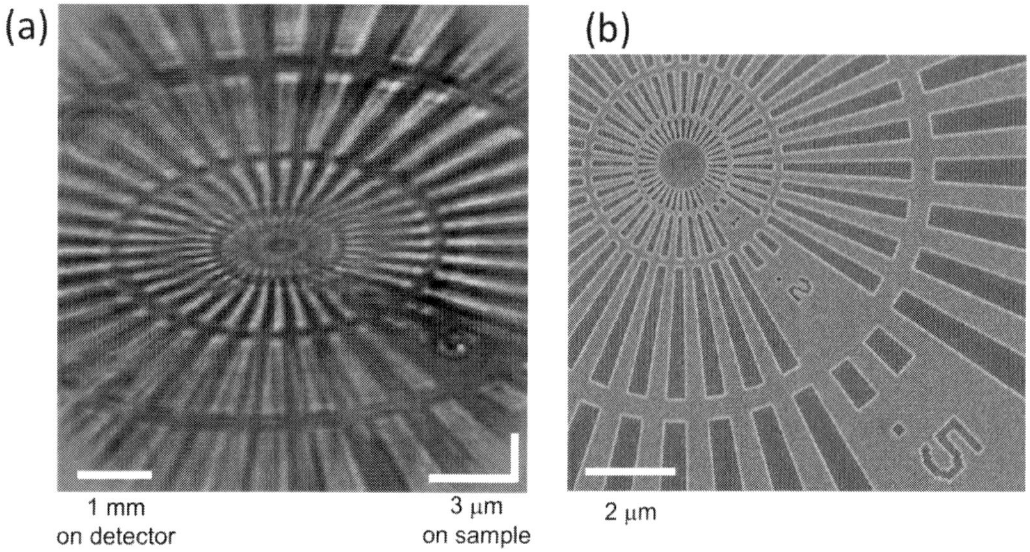

Figure 3. (a) X-ray image and (b) SEM image. Experimental conditions for X-ray imaging: X-ray energy = 10 keV, exposure time = 20 sec. The image was corrected by subtracting the background.

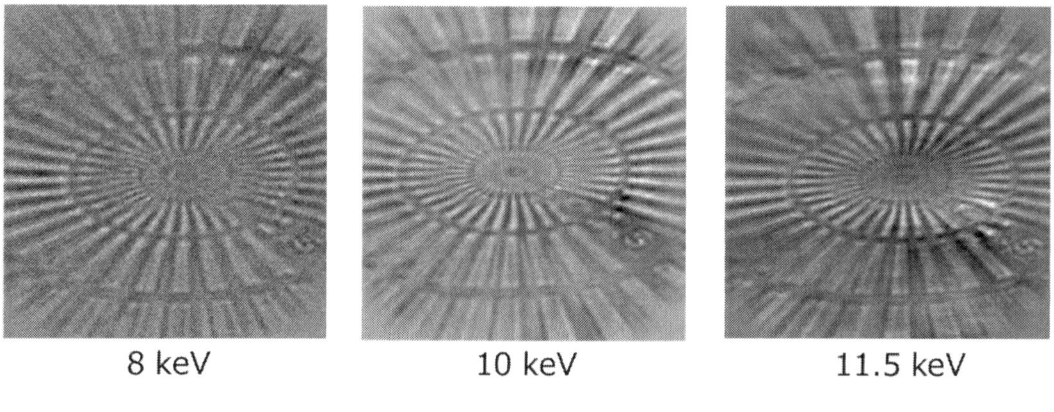

Figure 4. X-ray energy dependence. The exposure time is 20 sec. The image was corrected by subtracting the background.

3. Results and discussion

Figure 3(a) shows the acquired X-ray image. By comparing the X-ray and SEM images, we can conclude that a spatial resolution of ~150 nm was obtained. The observed magnifications are ×393 and ×213 in the horizontal and vertical directions, respectively. The result is in good agreement with that expected from the design. However, speckle noises from a Kapton film used as a vacuum window were observed. Also, the contrast of the line with a certain spatial frequency region was reversed. This is because the MTF (modulation transfer function) has a negative value. That is, the experimental PSF (point spread function) seems to have multiple peaks. This may result from the misalignment of the focal length. Because the experiment was preliminary and we had insufficient time, we did not make sufficient adjustment of the focal length.

Figure 4 shows the X-ray energy dependence of image formation. We varied the X-ray energy from 8 to 11.5 keV, keeping other conditions fixed. As can be seen, the image quality was not changed. The noise in the image at 8 keV resulted from the absorption of X-rays by air. Because the light path in the atmosphere was ~3 m, the transmissions at 8, 10 and 11.5 keV are 3, 18 and 32%, respectively.

4. Summary and outlook

We have successfully developed a full-field hard X-ray microscope based on advanced KB mirrors. So far, we have achieved ~150 nm resolution at 10 keV. Also, we confirmed that image formation is independent of X-ray energy in the range from 8 to 11.5 keV. In our next experiment, we will take clear images with a spatial resolution of ~ 50 nm by finely adjusting the whole system and eliminating speckle noises. We aim to realize a full-field X-ray fluorescence microscope with sub-50 nm resolution because its achromatic property will be effective for detecting polychromatic X-ray emissions with a wide field of view. Such a fluorescence microscope will become a powerful tool for observing elemental distributions. Also, in the near future, we will apply the system in X-ray free electron lasers. The combination will enable us to image X-ray emissions on an ultrashort time scale.

Acknowledgments

This research was mainly supported by Sentan from the Japan Science and Technology Agency (JST). It was also partially supported by a Grant-in-Aid for Specially Promoted Research (18002009) and the Global COE Program "Center of Excellence for Atomically Controlled Fabrication Technology" from the Ministry of Education, Culture, Sports, Science and Technology, Japan, and the Iketani Science and Technology Foundation. The use of BL29XUL at SPring-8 was supported by RIKEN. We are grateful to Mr. Masaki Fujii, Mr. Toshiyuki Wakioka and Mr. Naotaka Kidani for their support.

References

[1] Kodama R, Ikeda N, Kato Y, Katori Y, Iwai T and Takeshi T 1996 *Opt. Lett.* **21** 1321-1323
[2] Matsuyama S, Kidani N, Mimura H, Sano Y, Kohmura Y, Tamasaku K, Yabashi M, Ishikawa T and Yamauchi K 2012 *Opt. Express* **20** 10310-10319
[3] Yamauchi K, Mimura H, Inagaki K and Mori Y 2002 *Rev. Sci. Instrum.* **73** 4028
[4] Matsuyama S, Fujii M and Yamauchi K 2010 *Nucl. Instrum. Meth. Phys. Res. A* **616** 241-245
[5] Matsuyama S, Wakioka T, Kidani N, Kimura T, Mimura H, Sano Y, Nishino Y, Yabashi M, Tamasaku K, Ishikawa T and Yamauchi K 2010 *Opt. Lett.* **35** 3583-3585

New Energy Sources: in-situ Characterisation of Fuel Cell and Supercapacitor Components. Complementary Studies using Transmission, Fluorescence and Photoelectron Microscopy and Imaging

B Bozzini[1*], M Amati[2], A Gianoncelli[2], L Gregoratti[2], B Kaulich[3], and M Kiskinova[2]

[1] Dipartimento di Ingegneria dell'Innovazione, Università di Lecce, via Monteroni, 73100 Lecce, Italy
[2] Elettra - Sinctrotrone Trieste S.C.p.A. S.S. 14, km 163.5 in Area Science Park, 34149 Trieste-Basovizza, Italy
[3] Diamond Light Source Ltd, Diamond House, Harwell Science and Innovation Campus, Didcot, Oxfordshire OX11 0De, UK

[*] corresponding author, E-mail: benedetto.bozzini@unisalento.it

Abstract. Fuel cells and supercapacitors are electrochemical devices providing efficient and pollution-free production and transformation of electricity. Notwithstanding their environmental appeal, a host of materials-science problems – chiefly related to the limited durability of crucial functional components – are hindering their widespread application. The present knowledge of the relevant materials-science notion is mostly at the macroscopic and empirical trial-and-error level and the answers to many questions require much deeper scientific understanding of the origin of degradation processes. In this regard, the development and the implementation of appropriate methods for in-situ characterization of cell components at the functionally relevant length scales is highly required. Soft X-ray spectroscopies, such as X-ray absorption spectroscopy, X-ray emission (fluorescence) spectroscopy, resonant inelastic X-ray spectroscopy and X-ray photoelectron spectroscopy have been extensively employed for ex-situ characterization of materials used in electrochemical systems. Furthermore, adding spatial resolution capabilities by implementing proper optical solutions has opened unique opportunities for monitoring material changes and mass transport events occurring at submicron length scales. The input from these methods is providing correlative information about the status of the electrode surface and of the electrode/electrolyte interface and also of the processes occurring under operation conditions.

1. Introduction

This paper is focussed on "in operando" X-ray microspectroscopy investigations of materials-science aspects of devices for the electrochemical generation and storage of electrical energy - also denominated "electrochemical energetics" - such as fuel cells and supercapacitors. This work summarises some achivements of the collaboration between the TwinMic and ESCAmicroscopy beamlines of the Elettra synchrotron, Trieste (Italy) and the electrochemistry group of University of Salento, Lecce (Italy) [1-20]. The unicity of electrochemical energetics is twofold: (i) it enables the direct conversion of chemical energy to electricity, allowing the so called "thermodynamic advantage" of getting rid of the Carnot efficiency, as well as the virtual elimination of emissions; (ii) it renders electrical energy storage feasible, impacting a major concern for the rational exploitation of the intrinsically localised and intermittent renewable energies. Electrochemical energetics devices rely on: (i) reversible chemical/electrical conversion processes - open (FC) or closed (batteries) to chemical fluxes – and (ii) electrostatic or faradaic charge storage (SC). In terms of the well-known trade-off between power density and energy density - often represented in Ragone-type plots - FCs cover the high energy density zone, while SCs allow to jointly optimise power and energy densities. It is worth noting that batteries lie between FCs and SCs and that conventional capacitors exhibit very high power densities, but dramatically low energy densities. In this paper we shall set forth three points: (i) offer some introductory concepts on FCs and SCs and the relevance of XRM methods for their investigation; (ii) provide details on the design and fabrication of the electrochemical cells developed for in situ XRM work; (iii) discuss a selection of examples.

2. Fuel cell (FC) and supercapacitor (SC) concepts, operating principles and typical problems; prospects of XRM tools for their investigation

An FC is a device directly converting the electrons exchanged in a redox reaction (such as a combustion, most commonly of H_2 and O_2) into a usable electric current. In order to implement this functionality, FCs feature – in an essentially symmetric way –: (i) two gas distribution systems, (ii) individual catalysts for the red and ox semi-reactions and (iii) an ionic-conduction pathway (electrolyte) allowing simultaneously electronic insulation of the red- and ox-catalysts and electrical continuity through an ionic pathway. This general concept can materialise in several ways, among which the principal ones are a room- or low temperature (<120°C) version using a polymeric electrolyte (typically Nafion, conducting protons) and a high-T one (generally in the range 650-1000°C) using a ceramic electrolyte (typically YSZ, conducting O^{2-}). Since a single cell yields a typical voltage of less than 1 V, for practical applications several single cells have to be stacked and connected in series. Of course, the electronic pathways connecting the cells, distributing the gases, supporting the catalysts and in chemical contact with both catalyst and electrolyte – denominated interconnects or bipolar plates - play a key role, as detailed in Section 4.

An SC is a peculiar capacitor with ultra-high capacitance values, achieved by the implementation of the peculiar charge-separation modes that are allowed by electrochemical processes, chiefly: (i) electrostatic charge storage in ultra-high surface area electrodes; (ii) charge storage through faradaic reactions of immobilised species, such as multivalent transition metal oxides; (iii) ultra-thin dielectrics and electrochemical double layers.

From the previous descriptions of the working principles of FCs and SCs, it is obvious that electrochemical devices – thriving on subtle combinations of surface processes coupled with ionic and electronic conductance pathways – are complex systems with challenging materials-science problems on several lengthscales. In particular, special material combinations are required – not found in other technologies –, practically relevant processes are out of equilibrium and transient changes in material properties occur: these characteristics lead to the fact that these devices are daunted by durability problems, that typically impact the practical operation of these systems. For these reasons, the availability of analytical tools exhibiting chemical and structural sensitivity with space and time resolution are obviously highly desirable: hence tackling the painstacking job of setting up in situ microspectroscopy studies of these devices is expected to be highly rewarding. In the recent literature several papers have appeared, reporting mainly hard X-ray based studies of electrochemical energetics systems, centred on XANES, XRD, SAXS, radiography and tomography. The chief topics addressed have been: (i) fine chemical analysis of catalysts, generally ex situ, and (ii) imaging of water and gas flow; space constraints do not permit to review this work more in detail. The authors, instead, have concentrated on the development of soft X-ray based methods aimed at the detailed, space dependent analysis of the single, operating components of FCs and SCs. Notwithstanding the operative drawbacks due to absorption, soft X-ray offer the unique advantage of utmost space resolution with minimal bulk effects.

3. Cell fabrication and experimental details

Our microspectroscopy investigations have been carried out the TwinMic (STXM) and ESCAmicroscopy (SPEM) beamlines at Elettra: instrumental insight can be found in the beamline websites (www.elettra.trieste.it/experiments/beamlines). The main interest here is to emphasise in situ electrochemical work and to this aim instrumental requirements are crucial. The TwinMic sample stage has the following constraints: (i) limited space, (ii) vacuum and (iii) work in transmission. For this reason, we have developed a three-electrode transmission electrochemical cell, fabricated along the following conceptual lines (for details, see [1-9,17,18]): we have used a Si_3N_4 film as the cell support, onto which we have grown the electrode system by lithography. On top of the electrode system, an electrolyte reservoir is built by lithographic deposition of a resist frame. After the electrolyte is applied by drop-casting or dipping, a second Si-supported Si_3N_4 frame is overlaid in order to close the cell that is eventually vacuum-sealed. We have developed two versions of the cell: (i) sealed for work with aqueous solution [1,4-6,8,9] and (ii) open, for studies using as the electrolyte a vacuum-stable ionic liquid [2,3,7,17]. As far as SPEM experiments are concerned, the electrochemical cells consist in electrolyte-supported, planar systems with metallic electrodes grown by lithography: the cell geometry allows access of both electrodes as well as of the electrolyte to analysis [10-16,18,19].

4. Results and discussion

As far as STXM is concerned, our recent work included: (i) cell development and demonstrative work [1,5,17]; (ii) studies of catalyst electrodeposition [8,9]; (iii) corrosion of PEM interconnects [2-6] and (iv) electrochemical growth of materials for supercapacitors [7]. Our SPEM-based studies encompass: (i) corrosion of SOFC interconnects [11,12]; (ii) electrocatalyst degradation [13,14,20]; (iii) electrodic deposition of C in hydrocarbon-fuelled SOFCs [10], (iv) electrodeposition of catalyst-support alloys [15], (v) study of full SOFCs in single-chamber configuration [19]. As representative examples here we have chosen: (i) the corrosion of interconnects in low- and high- temperature fuel cells and (ii) the electrochemical growth of nanostructured electrode materials for supercapacitors. FC interconnects are critical components because, in the quest for durable, low cost devices, metals have proved the material of choice. Unfortunately, they exhibit corrosion problems, causing a series of problems: (i) increase of ohmic drops due to the build-up of corrosion products; (ii) clogging of the porous, gas-transfer zones; (iii) catalyst poisoning due to the deposition of corrosion products; (iv) electrolyte poisoning, again due to the fixation of corrosion products. In one of our STXM-based studies, we have investigated the release of Ni and Fe from PEM interconnects into aqueous solutions in contact with a Nafion membrane [6]. In this work, we have run single-cell experiments by driving the oxidising current through an auxiliary electrode. In Figure 1 we report absorption images of the interfaces between the Fe (a) electrode and the electrolyte, local XAS spectra (b) and XRF maps (c). From panels (a, c) it is straightforward to note that the Fe electrode develops a complex morphology related to the progress of the corrosion reaction. Local XAS spectra (b) and XRF maps (c) pinpoint the Fe space distribution. Moreover, different oxidation states are found in different positions within the area originally occupied by the electrolyte, that can be correlated with the local electrochemical potential. Furthermore, Fe corrosion products are injected into the liquid phase and end up fixed in the Nafion membrane (c), a mechanism related to the PEM electrolyte deterioration.

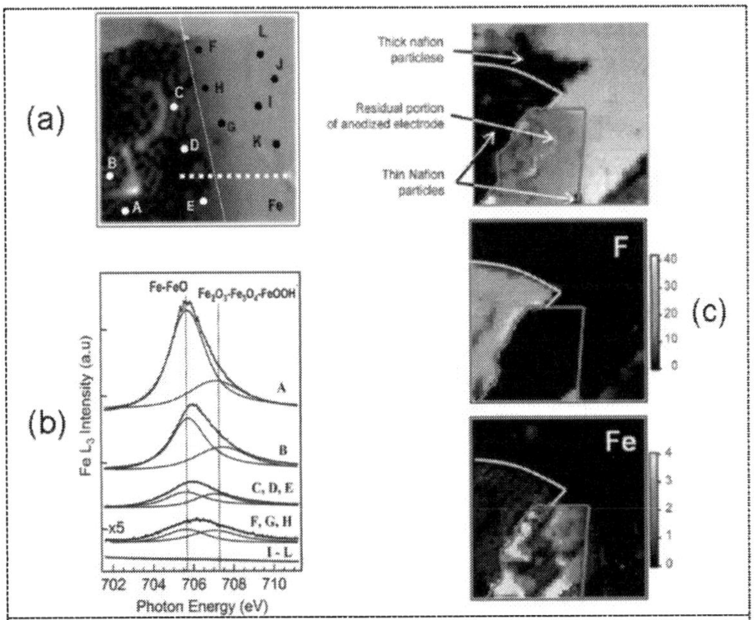

Figure 1. Ni and Fe electrodes corroded electrochemically in a Nafion-containing aqueous solution. (a) STXM image of Fe (22×22 μm^2, 707 eV); (b) XAS spectra measured in the points indicated in (a); (c) XRF map of Fe.

While STXM is suited for room-temperature studies, SPEM lends itself ideally to high-temperature work. We used this method for the investigation of interconnect stability in SOFC environment in O_2 10^{-5} mbar at 650°C [12]. In particular, we have concentrated on the interaction between the Ni electrocatalyst and the Cr-based interconnects. Figure 2 shows that the initially well defined electrode geometry and flat electrode morphology (a) - where Ni is present in the elemental state, while Cr is environmentally oxidised -, is deeply altered at high temperature in O_2 ambient (b). In addition to the expected oxidation of Ni, this element diffuses onto the electrolyte patch, forming islands, while oxidised Cr spreads over the Ni electrode. Upon applying a cathodic polarisation (panel c, upper electrode), notwithstanding the oxidising gas environment, Ni can be easily reduced, while Cr requires a cathodic potential exceeding -3 V. Moreover, we found that oxidised Cr retracts from the reduced Ni patch. This type of behaviour offers mechanistic insight into the empirically observed Cr contamination problems in SOFCs implementing Cr-based interconnects.

Figure 3 reports STXM and XAS results regarding the electrodeposition of Mn from a choline-chloride/urea ionic liquid spun on top of an open electrochemical cell. This metal electrodeposition system is important for the fabrication of advanced supercapacitor electrodes, in view of the formation of MnO_2-decorated Au NPG, yielding exceptionally high capacitance values, due to a pseudocapacitance effect [7,21]. STXM allowed to analyse simultaneously the morpology development and the localisation of the Mn valence types generated by the applied electrochemical conditions.

Typical globular electrodeposition morphologies, controlled by the local current density distribution, were found on the cathode, while fibrous structures grow at the anode. Localised XAS demonstrates that elemental Mn is obtained at the cathode, Mn(III) dominates at the anode, while outside the electrode patches the Mn(II) spectra corresponding to the electrodeposition bath are found.

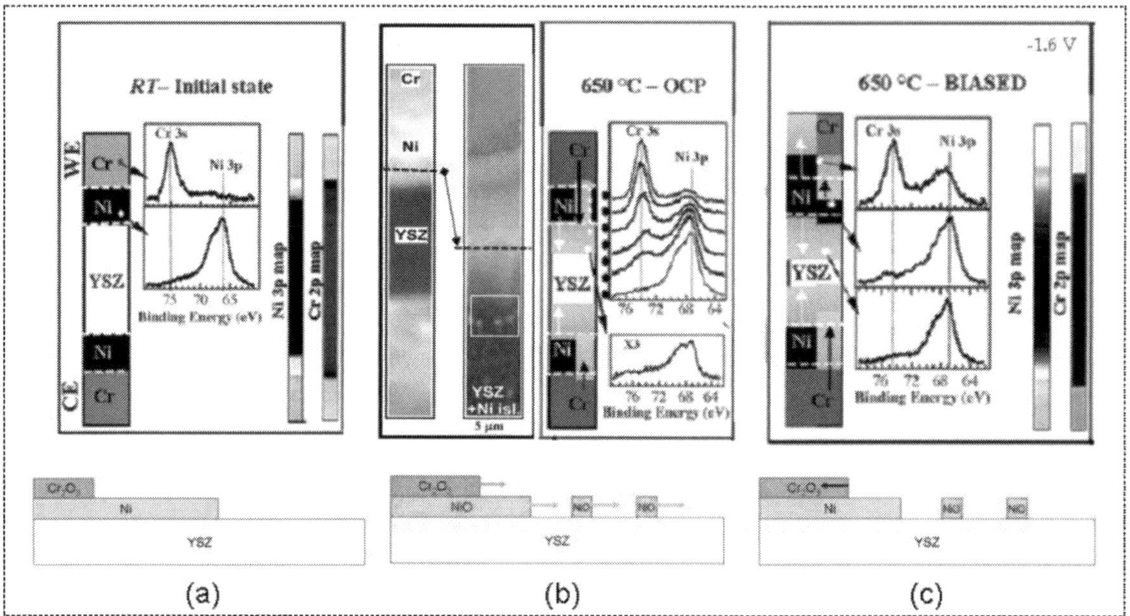

Figure 2. SPEM maps (750×50 μm^2) and PES spectra measured in the indicated locations. (a) Pristine conditions, room temperature; (b) At 650°C in 10^{-6} mbar O_2, open circuit potential; (c) same as (b), but with applied cathodic polarisation of -1.6 V at the upper electrode.

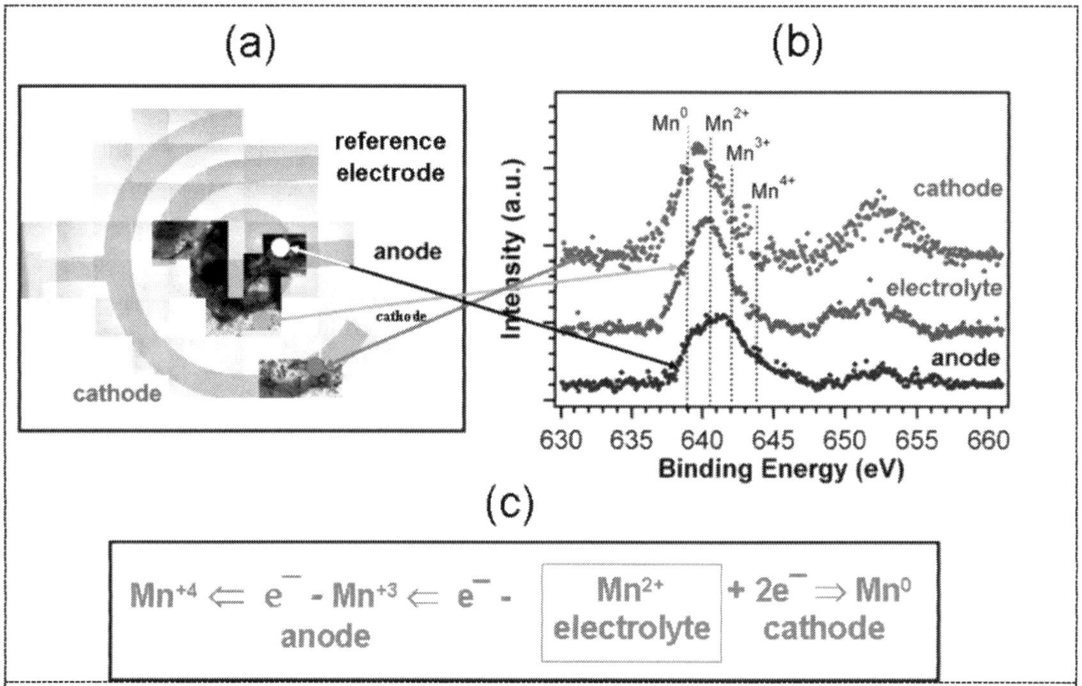

Figure 3. (a) STXM images (each square 80×80 μm^2, 1189 eV) of the cell after application of -1.7 V. (b) XAS spectra measured in the points indicated in (a). (c) Scheme of the cathodic and anodic reactions.

5. Conclusions

In situ electrochemical microspectroscopy has been proved capable of following morphological and chemical effects of electrochemistry. In the studies reviewed in this paper, we have developed three types of microfabricated cells with nm-thick electrodes: (i) sealed cells for SXTM with aqueous electrolytes; (ii) open cells for STXM with spun IL-based electrolytes; (iii) solid electrolyte-supported cells for high-temperature SPEM analyses. Moreover, we have shown the possibility of dosing electroactive gases in the analysis chamber: a first step towards environmental in situ X-ray microspectroscopies. We have thus demonstrated that SXTM and SPEM can contribute unique information to electrochemical materials science, in particular for energy-conversion devices, in terms of: (i) assessment of localisation processes (chemical and overvoltage); (ii) rationalisation of growth/damaging mechanisms; (iii) acquisition of mechanistic information on coupling of materials.

References

[1] Bozzini B, D'Urzo L, Gianoncelli A, Kaulich B, Kiskinova M, Prasciolu M, Tadjeddine A 2008 *Electrochem. Comm.* **10** 1680
[2] Bozzini B, Gianoncelli A, Kaulich B, Mele C, Prasciolu M and Kiskinova M *Fuel Cells* in press
[3] Bozzini B, Gianoncelli A, Kaulich B, Kiskinova M, Mele C, Prasciolu M 2011 *PhysChemChemPhys* **13** 7968
[4] Bozzini B, Mele C, Gianoncelli A, Kaulich B, Kiskinova M, Prasciolu M 2011 *Microelectronic Engineering* **88** 2456
[5] Gianoncelli A, Kaulich B, Kiskinova M, Prasciolu M, D.Urzo L, Bozzini B 2011 *Micron* **42** 342
[6] Bozzini B, Gianoncelli A, Kaulich B, Kiskinova M, Prasciolu M, Sgura I 2010 *ChemSusChem* **7** 846
[7] Bozzini B, Gianoncelli A, Kaulich B, Mele C, Prasciolu M and Kiskinova M 2012 *J. Power Sources* **211** 71
[8] Bozzini B, D'Urzo L, Gianoncelli A, Kaulich B, Prasciolu M, Sgura I, Tondo E and Kiskinova M 2009 *J. Phys. Chem. C* **113** 9783
[9] Bozzini B, D'Urzo L, Gianoncelli A, Kaulich B, Kiskinova M, Prasciolu M, Tadjeddine A 2009 *J. of Physics: Conf. Series* **186** 012103
[10] Bozzini B, Amati M, Gregoratti L, Kazemian Abyaneh M, Prasciolu M, Trygub AL and Kiskinova M *J. Phys. Chem. C* in press
[11] Bozzini B, Amati M, Gregoratti L, Kazemian M, Prasciolu M, Tondo E, Trygub AL and Kiskinova M 2012 *J.Phys. Chem. C* 116 7243
[12] Bozzini B, Tondo E, Amati M, Kazemian M, Gregoratti L and Kiskinova MM 2011 *ChemSusChem* **4** 1099
[13] Bozzini B, Amati M, Kazemian Abyaneh M, Gregoratti L and Kiskinova M 2011 *J. Electroanal. Chem.* **657** 113
[14] Bozzini B, Amati M, Boniardi M, Kazemian Abyaneh M, Gregoratti L and Kiskinova M 2011 *J. Power Sources* **196** 2513
[15] Bozzini B, Dalmiglio M, De Gaudenzi GP, D'Urzo L, Gregoratti L 2009 *Corros. Sci.* **51** 1675
[16] Bozzini B, D'Urzo L, Gregoratti L, Tadjeddine A 2008 *J. Electrochem. Soc.* **155** F165
[17] Gianoncelli A, Kaulich B, Kiskinova M, Mele C, Prasciolu M, Sgura I and Bozzini B *J. of Physics: Conf. Series* in press
[18] Bozzini B, Kazemian Abyaneh M, Amati M, Gianoncelli A, Gregoratti L, Kaulich B, Kiskinova M 2012 *Chemistry - A European Journal* **18** 10196
[19] Bozzini B, Amati M, Gregoratti L, Mele C, Kazemian Abyaneh M, Prasciolu M and Kiskinova M 2012 *Electrochem. Comm.* **24** 104.
[20] Bozzini B, Kazemian Abyaneh M, Busson B, De Gaudenzi GP, Gregoratti L, Humbert C, Amati M, Mele C, Tadjeddine A *J. Power Sources* in press
[21] Lang X, Hirata A, Fuhjita T and Chen M 2011 *Nature Nanotechnology* **6** 232

Journal of Physics: Conference Series **463** (2013) 012019 doi:10.1088/1742-6596/463/1/012019

Hard X-ray Nanoprobe with Spherical-aberration-corrected Quad-spherical-mirror Optics

Yoshio Suzuki, Akihisa Takeuchi

JASRI/Spring-8, Koto 1-1-1 Sayo, Hyogo 679-5198, Japan

yoshio@spring8.or.jp

Abstract. A quad-mirror system is developed for nanofocusing of hard x-ray beam. The optical system consists of four spherical-concave total-reflection mirrors. Each two mirrors are combined in a tandem-mirror configuration to compensate spherical aberrations in grazing-incidence mirror optics, and a pair of the tandem-spherical mirrors is configured in a crossed mirror system (Kirkpatrick-Baez optics) to eliminate astigmatism in grazing incidence optics. Focusing properties are tested at the beam line 20XU of SPring-8. The focused beam size measured by knife-edge scan method is 170 nm x 190 nm at 10 keV. Preliminary experiments on scanning x-ray microscopy are also carried out, and a fine structure of 100 nm-line/100 nm-space is resolved in the measured image.

1. Introduction

The x-ray microprobe is now widely used in many application fields at synchrotron radiation facilities. Several types of optical systems are developed for microfocusing of x-ray beam: Fresnel zone plates, multilayer Laue lens, grazing-incidence total-reflection mirrors, multilayer mirrors, and refractive lens systems. The most important advantage of total-reflection mirror system is its achromatic feature. Wide spectral range is available at a fixed focal length. The most serious problem of total-reflection mirrors in the hard x-ray region is spherical aberration and astigmatism in grazing incident optics.

It is well known that the astigmatism in grazing incidence mirror system is perfectly eliminated by the crossed-mirror optics [1]. The simple way of spherical aberration-correction is use of aspherical mirrors [2]. Then, the most widely used total–reflection mirror system for x-ray microfocusing is the crossed-mirror system with aspherical total-reflection mirrors (Kirkpatrick-Baez configuration with elliptic or parabolic mirrors) [3-5]. The focused beam size less than 50 nm is already achieved in the hard x-ray region. However, the fabrication of high-precision aspherical surface is not easy at present. On the contrary, the fabrication of spherical surface is much easier than that of aspherical figure. Therefore, it is still useful to find a way of microfocusing using spherical mirrors. The method of spherical-aberration-correction by a combination of spherical mirrors was firstly suggested by Sakayanagi in tandem-toroidal mirror system for soft x-ray imaging microscopy [6]. We have recently demonstrated spherical-aberration-correction in tandem-spherical-mirror system [7], and applied it to microfocusing of hard x-ray beam. However, in our previous experiment, the achieved beam size was about a half micrometer [8], probably owing to the surface figure errors and limitation on optical configuration, while a focused beam width of 170 nm has been achieved in the one-dimensional focusing experiment.

In this report, newly developed quad-spherical-mirror system that consists of a pair of tandem-spherical-mirror systems is described, and results on x-ray nanofocusing and its application to scanning microscopy are also described.

2. Optical System and Experimental Setup

The concept of spherical aberration correction is shown in Fig. 1. Two spherical (or cylindrical) concave mirrors are aligned in a tandem mirror configuration. The x-ray beam firstly incident on the M2 mirror is reflected again by the next mirror M1 placed just after the M2 mirror. The spherical aberration of the second mirror has opposite sign of that of the first mirror. Therefore, by choosing appropriate parameters, radii of concave surface, angles of incidence, and a distance between two mirrors, it is possible to compensate the spherical aberration of the total optical system. The idea of the spherical-aberration-correction (Cs-correction) was suggested by Sakayanagi in the tandem-toroidal-mirror system for soft x-ray imaging microscopy [6]. The sequential reflection by grazing-incidence mirrors was also important for image forming optics, because the Abbe's sine condition is approximately satisfied by the sequential reflection. The quad-mirror system with similar configuration was already developed and is called "advanced Kirkpatrick-Baez system" [9]. However, in the previous advanced-KB system, the compensation of spherical aberration was not considered. Only the coma and field distortion (field obliquity) were discussed in the previous advanced-KB optics.

The general theory of spherical-aberration-correction is not a simple equation. However, the formula can be simplified at a typical condition. When the two concave reflectors have the same curvature, the Cs-correction is simply expressed by the following formula [10].

$$(1 + \sqrt{5})/2 \sim R\theta_1/(R\theta_2 - 2D),$$

where R is the radius of curvature of reflector surfaces, θ_1 and θ_2 is a glancing angle to each mirror, and D is a distance between two mirrors. The tandem-spherical mirror system is not a stigmatic optics as a single spherical mirror system. Therefore, for the purpose of two-dimensional focusing, two tandem-spherical-mirror systems are combined in the crossed mirror geometry (an analogue of Kirkpatrick-Baez configuration) in order to eliminate astigmatism, as shown in Fig. 2. The size of beam incident on the mirror is limited with a cross slit placed in front of the mirror system, and the beam dimension is slightly smaller than the aperture of the mirror (~80% of full aperture), because, usually, the marginal region of mirror surface is not precisely formed.

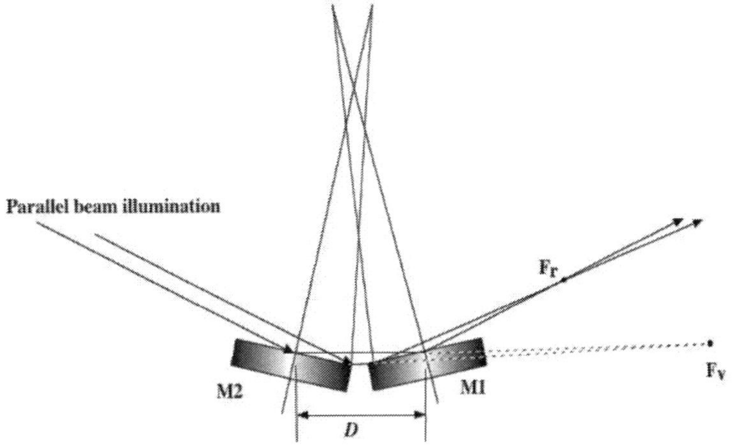

Fig. 1. Concept of spherical aberration correction in grazing-incidence total-reflection-mirror optics by tandem-spherical-mirror system. M1 and M2 are spherical-concave mirrors. Fv is a virtual focal point produced by the M2 and Fr is a real focus generated by the tandem mirror system.

The experiment has been carried out at the beam line 20XU of SPring-8. The four concave mirrors are general-purpose products supplied from Sigma Koki (Tokyo, Japan). The design radius of curvature of the two concave mirrors (M4 and M3) is 90 m, and the other two concave mirrors (M2 and M1) have radius of 45 m. The length of each mirror is 30 mm. The substrate material of the concave mirrors is borosilicate crown glass (BK7), and the reflection surface is coated by platinum. The concave surface is fabricated by conventional polishing technique. The precise alignment of mirrors was done by monitoring the spherical-aberrations and the astigmatism by means of Foucault's knife-edge test. The glancing angles of the each mirror experimentally determined to compensate spherical aberration and astigmatism are 4.1 mrad, 3.8 mrad, 5.0 mrad, and 4.6 mrad for M1, M2, M3 and M4, respectively. The focal point is about 23 mm from the edge of M1 mirror.

The beam focusing properties are characterized by conventional edge scan method using a gold wire of 200 μm-diameter as a knife-edge. The focused beam shapes are derived by differentiation of the measured knife-edge scan profiles. Preliminary experiments on scanning microscopy were also carried out using an x-ray resolution test chart as a specimen. The resolution test chart is fabricated at NTT Advanced Technology using electron beam lithography technique and reactive ion etching. The pattern material is tantalum with a thickness of 0.5 μm deposited on a SiN membrane.

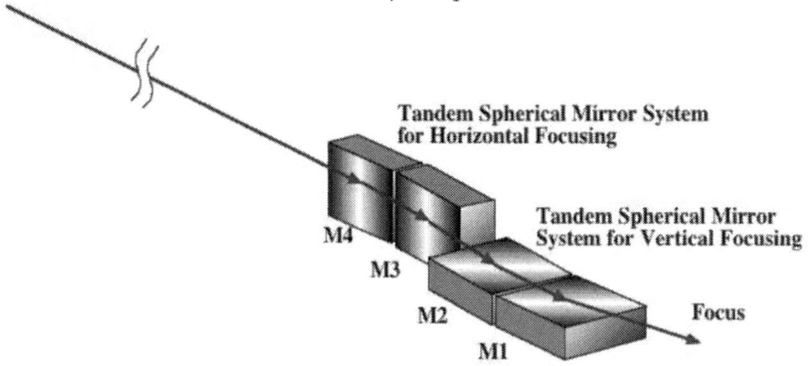

Fig. 2. Schematic diagram of optical system with quad spherical mirrors.

3. Results and Discussion

Fig. 3 shows the focused beam profiles measured by knife-edge scan method at an x-ray energy of 10 keV. The full-width at half-maximum (FWHM) of focused beam profile is 170 nm in the horizontal direction and 190 nm in the vertical direction. The measured FWHM of the focus spot is less than 200 nm in the horizontal direction, and less than 230 nm in the vertical direction within the x-ray energy range from 8 keV to 12 keV. The example of scanning microscopy experiments is shown in Fig. 4. The scanning microscopy image of resolution test patterns with a 100 nm-line/100 nm-space and a 50 nm-line/50 nm-space is measured with the focused probe. Here, the x-ray energy of 9.90 keV, just above the Ta L3-absortion edge (9.88 keV), was chosen to achieve the highest absorption contrast for tantalum test pattern. Although the 50 nm-pattern cannot be resolved, the 100 nm line/100 nm space patterns are clearly resolved in the both directions.

The effective numerical aperture of the quad-mirror system is experimentally estimated to be 4.3×10^{-4} in the horizontal direction and 7.4×10^{-4} in the vertical direction. Therefore, the diffraction-limited beam size for 10 keV x-rays defined by the Rayleigh's criterion is calculated to be 144 nm in the horizontal direction, and 84 nm in the vertical direction. The measured spot size in the horizontal direction (170 nm in FWHM) is very near to the diffraction-limited resolution. Although the vertical focus (190 nm inFWHM) is about twice of the diffraction-limited spot size, this blurring may come from imperfection of the mirror figure and surface roughness. We consider that the spatial resolution of about 100 nm would be possible using the Cs-corrected tandem-spherical-mirror system.

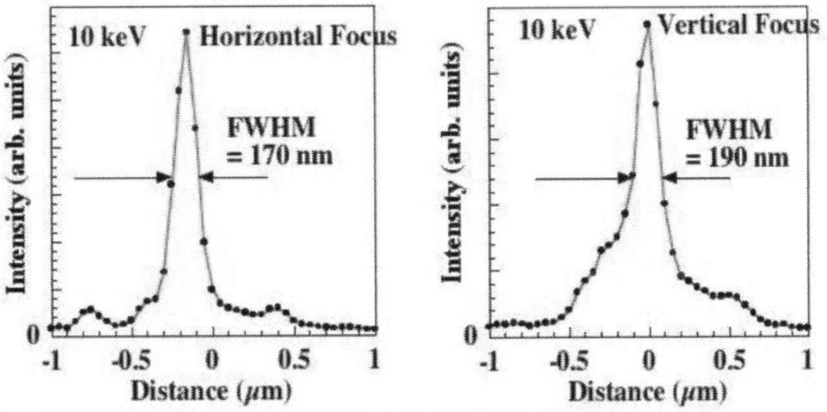

Fig. 3. Focused beam profiles measured by knife-edge scan method. X-ray energy is 10 keV.

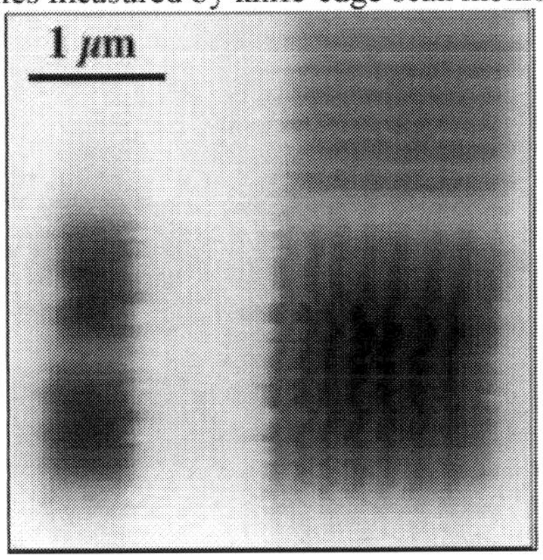

Fig. 4. Scanning XRM image of resolution test chart. X-ray energy: 9.9 keV. Raster scan image of 81 x 81 pixels is acquired with a 50 nm step and 100 ms dwell time.

4. Acknowledgements
The experiment has been done with an approval of SPring-8 Proposal Review Committee (Proposal No. 2011B1190 and No. 2010B1136)

References
[1] P. Kirkpatrick and A. V. Baez: J. Opt. Soc. Am. **38** (1948) 766.
[2] Y. Suzuki and F. Uchida: Rev. Sci. Instrum. **63** (1992) 578.
[3] H. Mimura, S. Matsuyama, H. Yumoto, H. Hara, K. Yamauchi, Y. Sano, M. Shibahara, K. Endo, Y. Mori, Y. Nishino, K. Tamasaku, M. Yabashi, T. Ishikawa, and K. Yamauchi: Jpn. J. Appl. Phys. **44** (2005) L539.
[4] W. Liu, G. E. Ice, J. Z. Tishler, A. Khounsary, C. Liu, L. Assoufid, and A. Macrander, Rev. Sci. Instrum. **76** (2005) 113701.
[5] Y. Suzuki, A. Takeuchi, and Y. Terada: Rev. Sci. Instrum. **78** (2007) 053713.
[6] Y. Sakayanagi: Opt. Acta **23** (1976) 217.
[7] Y. Suzuki and A. Takeuchi, Jpn. J. Appl. Phys. **49** (2010) 106701.
[8] Y. Suzuki and A. Takeuchi, AIP Conf. Proc. **1365** (2011) 156.
[9] R. Sauneuf, J. M. Dalmasso, T. Jalinaud, J. P. Le Breton, D. Schirmann, J. P. Marioge, F. Bridou, G. Tissot, and J. Y. Clotaire: Rev. Sci. Instrum. **68** (1997) 3412.
[10] Y. Suzuki and A. Takeuchi, Jpn. J. Appl. Phys. accepted for publication.

Journal of Physics: Conference Series **463** (2013) 012020 doi:10.1088/1742-6596/463/1/012020

Mapping the subcellular localization of Fe_3O_4@TiO_2 nanoparticles by X-ray Fluorescence Microscopy

Y Yuan[1], S Chen[2], S C Gleber[2], B Lai[2], K Brister[3], C Flachenecker[4], B Wanzer[1], T Paunesku[1], S Vogt[2], and G E Woloschak[1]

[1]Department of Radiation Oncology, Northwestern University, Chicago, IL 60611, USA
[2]X-ray Sciences Division, Argonne National Laboratory, Argonne, IL 60439, USA
[3]Life Sciences Collaborative Access Team, Argonne National Laboratory, Argonne, IL 60439, USA
[4]Xradia , Pleasanton, CA 94588, USA

g-woloschak@northwestern.edu

Abstract. The targeted delivery of Fe_3O_4@TiO_2 nanoparticles to cancer cells is an important step in their development as nanomedicines. We have synthesized nanoparticles that can bind the Epidermal Growth Factor Receptor, a cell surface protein that is overexpressed in many epithelial type cancers. In order to study the subcellular distribution of these nanoparticles, we have utilized the sub-micron resolution of X-ray Fluorescence Microscopy to map the location of Fe_3O_4@TiO_2 NPs and other trace metal elements within HeLa cervical cancer cells. Here we demonstrate how the higher resolution of the newly installed Bionanoprobe at the Advanced Photon Source at Argonne National Laboratory can greatly improve our ability to distinguish intracellular nanoparticles and their spatial relationship with subcellular compartments.

1. Introduction

Metal oxide nanoparticles (NPs) initially developed by chemists and material scientists, are now being investigated as nanomedicines. The key advantage of metal oxide nanoparticles over small molecule therapies lies in their peculiar surface chemistries and physical properties. For example, titanium dioxide (TiO_2) nanoparticles smaller than 20 nm can form exceptionally stable bonds with orthosubstituted enediol molecules such as dopamine or 3,4-dihydroxyphenylacetic acid (DOPAC) [1,2], while molecules with free hydroxyl groups can form a more labile corona on the NP surface [3]. This allows for the subsequent conjugation of functional molecules such as antibodies or peptides that can direct the delivery of these NPs. We have synthesized 6-7 nm Fe_3O_4@TiO_2 NPs [3] that can bind the Epidermal Growth Factor Receptor (EGFR) by surface modifying NPs with short EGFR-binding peptides (Yuan et al submitted). These peptides are derived from the B-loop region of Epidermal Growth Factor (EGF), the native ligand of EGFR, and have been previously shown to bind and activate EGFR *in vitro* [4]. Since EGFR is often overexpressed in epithelial type cancers and ligand bound EGFR can translocate to the nucleus [5], we wish to use these EGFR-binding B-loop NPs to achieve both cellular and subcellular targeted delivery.

In order to directly map the subcellular distribution of metal oxide NPs we must be able to image cells treated with NPs at sub-micron resolutions. For that purpose we decided to use X-ray fluorescence microscopy in order to simultaneously map (i) intracellular and NP associated iron (Fe) and NP associated titanium (Ti) and (ii) different subcellular compartments by determining the distribution of intracellular trace metals and nanometal elements such as phosphorus (P), sulfur (S), and zinc (Zn) [6]. Using the X-ray fluorescence microprobe at the sector 2 beamline at the Advanced Photon Source (APS) at Argonne National Laboratory (ANL) we have mapped NPs inside cells in the past. We decided to do the same with cells treated with B-loop NPs and compare these elemental maps with the ones produced using a higher resolution instrument—the Bionanoprobe, which was installed at the Life Sciences Collaborative Access Team (LS-CAT) in the fall of 2011. This instrument can produce an X-ray beam spot size focused to 30 nm thus allowing detailed determination of the subcellular localization of metal oxide NPs within cells.

2. Materials and Methods

HeLa cervical carcinoma cells were grown at a density of 100,000 cells/mL overnight on 3.5 mm tissue culture dishes and then treated with 10 nM B-loop NPs for one hour at 4°C. The cells were then washed to remove loosely adhering NPs from the surface and chemically fixed in 4% formaldehyde in PEM buffer, detached from support using a cell scraper and resuspended in PBS. This suspension of cells was then mixed 1:1 with a molten low melting point 4% agarose in PBS, dehydrated in increasing concentrations of ethanol, paraffin embedded, and sectioned to a thickness of 7 μm.

Cell sections mounted on Si_3N_4 windows were raster scanned at the X-ray microprobe at sector 2-ID-E at the Advanced Photon Source at Argonne National Laboratory 10 keV hard X-rays produced by an undulator source were monochromatized through a single bounce Si<111> monochromator and focused by a Fresnel zone plate to a spot size of 0.7 μm x 0.5 μm. The sample was scanned with a step size of 200 nm and a dwell time of 2 seconds. The fluorescence spectrum at each scan position was acquired by an energy dispersive germanium detector and fitted using MAPS software [7] by comparing to sample spectra from NBS standards 1832 and 1833.

The same sample was then transferred to the BNP at 21-ID-D, which is also an undulator beamline, and scanned again as a comparison. A double crystal monochromator located five meters upstream of the BNP was used to generate 10 keV monochromatic radiation, which was then focused down on to the sample using a Fresnel zone plate with 70 nm outermost zone width, theoretically providing spatial resolution of 85.4 nm according to the Rayleigh criteria. The sample was scanned through the focused spot using 50 nm step size and 250 ms dwell time per step. At each scanning step, a full fluorescence spectrum was collected using a four-element silicon drift detector (Vortex-ME4®) at 90 degree of the incident beam and fitted using MAPS [7].

3. Results

HeLa cervical carcinoma cells were treated with 10 nM EGFR-binding B-loop NPs for 1 hour at 4°C. Under these conditions the NPs are able to bind to cell surface EGFR, but cannot be internalized due to the inhibition of energy-dependent receptor mediated endocytosis [8]. Since we have shown that even the uptake of bare NPs is energy-dependent [8] and the internalization of ligand bound EGFR is also a cell metabolism dependent process [9] we expect that with this experimental setup EGFR-binding NPs will be predominantly localized to the cell membrane.

XFM images of the intracellular distribution of trace metal elements show that the Fe and Ti from B-loop NPs co-localizes with S, an element that is distributed throughout the cell due to its incorporation in cellular proteins (Figure 1). Furthermore, the colocalization of S and Ti appears to be at the cell periphery implying that, at this treatment timepoint, the B-loop NPs are bound to cell surface receptors, but are not yet internalized. While the image from the X-ray fluorescence microprobe (Figure 1A) and Bionanoprobe (Figure 1B) show the same distribution pattern, the degree of detail discernable in the image obtained with the Bionanoprobe is significantly greater. The image in Figure 1A was done with a beam focused to a spot size of 0.7 μm x 0.5 μm while the image in Figure 1B was acquired with X-rays focused with the Fresnel zone plate setup of the Bionanoprobe instrument to a theoretical spot size of 85.4 nm according to the Rayleigh criteria. A tightly focused beam accompanied with the small step size (50 nm) in the image in Figure 1B allows identification of additional Ti rich spots at the cell periphery and allows us to resolve the morphology of the Ti (NP) deposits. Importantly, the new Bionanoprobe instrument allows acquisition of images with samples tilted at different angles for a total of 180 degrees which will allow us to view this 7 μm thick cell slice tomographically and determine if the Ti hotspots that appear to be inside the cell are actually on the surface of cell membrane. With the thickness of 7 μm and the processing we used (agarose and paraffin embedding) preservation of cell morphology in three dimensions was achieved even though the image presented in Fig. 1 represents only a single 2D projection of cellular elemental content.

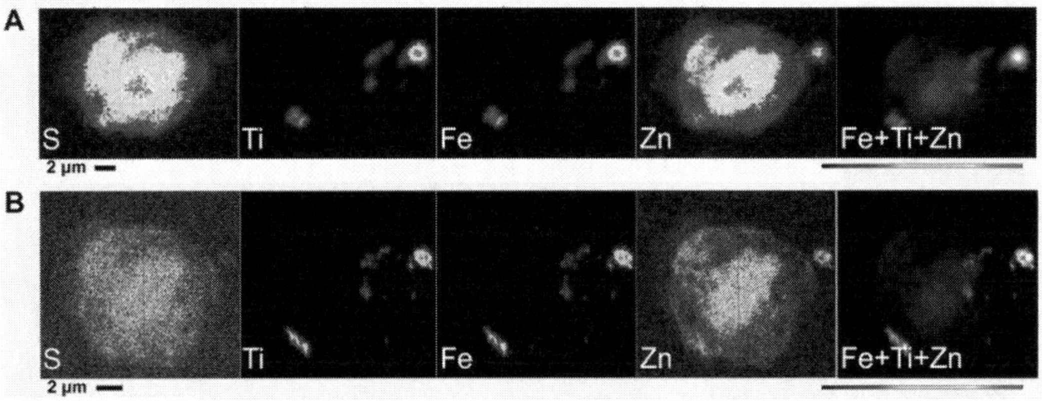

Figure 1. (A) XFM maps of S, Ti, Fe, and Zn distribution in B-loop NP treated HeLa cells scanned on the X-ray microprobe at sector 2-ID-E with a dwell time of 2 seconds and a step size of 200 nm. 3 element colocalization of Fe (red), Ti (green), Zn (blue) (B) The same HeLa cell scanned on the Bionanoprobe at sector 21-ID-D with a dwell time of 250 milliseconds and a step size of 50 nm. Images derived from the Bionanoprobe confirm that B-loop NPs are localized at or near the cell membrane at this treatment timepoint. Notably, in the images collected at higher resolution at the Bionanoprobe it is possible to distinguish additional and finer subcellular structures as well as individual clusters of NPs. The difference in the appearance of S and Zn is partly due to a difference in dwell times as well as to a difference in the angle (15 degrees) between the sample and the detector at the Bionanoprobe versus the microprobe leading to higher self-absorption of low Z elements.

4. Discussion

We have created EGFR targeted NPs by conjugating an EGFR-binding peptide to the surface TiO_2 molecules of Fe_3O_4@TiO_2 NPs. We have shown previously that these NPs colocalize with EGFR at the surface as well as within the nucleus of EGFR expressing HeLa cervical carcinoma cells at different treatment timepoints. Using the X-ray microprobe at sector 2-ID-E/D of the Advanced Photon Source at Argonne National Laboratory as well as the Bionanoprobe at sector 21-ID-D we wished to investigate the subcellular distribution of these NPs and showcase the differences between the two instruments. The difference in the degree of detail that can be visualized by the Bionanoprobe is impressive even though these images were not scanned at the highest resolution possible. In both cases, by mapping the distribution of intracellular elements such as S and Zn, we can distinguish the overall cell outline and cell nucleus, respectively. However, while the images derived from scans done at the microprobe show the distribution of NPs within cells, the higher resolution of the Bionanoprobe, especially when coupled with tomography, can be used to clarify the spatial relationship between internalized NPs and subcellular compartments. Reconstruction of 2-dimensional scans at different angles will allow for tomographic representation of the trace metal distribution within NP treated cells. This will greatly improve the ability to distinguish between NPs that are truly localized to the nucleus from cytoplasmic NPs that only appear to be in the nucleus in 2-dimensional projections.

Acknowledgments
This research was supported by the National Institutes of Health under the following Grant Numbers CA107467, EB002100, and U54CA119341. Y.Y. is supported in part by NIH/NCI training grant T32CA09560. Work at Argonne National Laboratory was supported by the U.S. Department of Energy, Office of Science, Office of Basic Energy Sciences, Contract No. DE-AC02-06CH11357. Implementation of the Bionanoprobe is supported by NIH ARRA grant SP0007167.

References

[1] Rajh T, Chen LX, Lukas K, Liu T, Thurnauer MC, Tiede DM. Surface restructuring of nanoparticles: An efficient route for ligand-metal oxide crosstalk. Journal of Physical Chemistry B. 2002;106(41):10543–52.
[2] Paunesku T, Rajh T, Wiederrecht G, Maser J, Vogt S, Stojicevic N, et al. Biology of TiO2-oligonucleotide nanocomposites. Nature Materials. 2003/04/15 ed. 2003;2(5):343–6.
[3] Arora HC, Jensen MP, Yuan Y, Wu A, Vogt S, Paunesku T, et al. Nanocarriers enhance Doxorubicin uptake in drug-resistant ovarian cancer cells. Cancer research. 2012 Feb 1;72(3):769–78.
[4] Komoriya A, Hortsch M, Meyers C, Smith M, Kanety H, Schlessinger J. Biologically active synthetic fragments of epidermal growth factor: localization of a major receptor-binding region. Proc Natl Acad Sci U S A. 1984/03/01 ed. 1984;81(5):1351–5.
[5] Lin SY, Makino K, Xia W, Matin a, Wen Y, Kwong KY, et al. Nuclear localization of EGF receptor and its potential new role as a transcription factor. Nature cell biology. 2001 Sep;3(9):802–8.
[6] Ortega R, Bohic S, Tucoulou R, Somogyi A, Devès G. Microchemical element imaging of yeast and human cells using synchrotron X-ray microprobe with Kirkpatrick-Baez optics. Analytical chemistry. 2004 Jan;76(2):309–14.
[7] Vogt S, Maser J, Jacobsen C. Data analysis for X-ray fluorescence imaging. Journal de Physique IV. 2003;104:617–22.
[8] Thurn KT, Arora H, Paunesku T, Wu A, Brown EMB, Doty C, et al. Endocytosis of titanium dioxide nanoparticles in prostate cancer PC-3M cells. Nanomedicine : nanotechnology, biology, and medicine. Elsevier Inc.; 2010 Sep;1:1–8.
[9] Sigismund S, Argenzio E, Tosoni D, Cavallaro E, Polo S, Di Fiore PP. Clathrin-mediated internalization is essential for sustained EGFR signaling but dispensable for degradation. Developmental cell. 2008 Aug;15(2):209–19.

Absolute zinc quantification at the sub-cellular level by combined use of hard X-ray fluorescence and phase contrast imaging techniques

Ewelina Kosior[1], Sylvain Bohic[1,2], Heikki Suhonen[1], Peter Cloetens[1]

[1]European Synchrotron Radiation Facility, 6 rue Jules Horowitz, 38000 Grenoble, France
[2]INSERM U-836, Chemin Fortuné Ferrini, 38706 La Tronche, France

E-mail: `ewelina.kosior@esrf.fr`

Abstract. Hard X-ray fluorescence microscopy and magnified phase contrast imaging are combined to obtain quantitative maps of the projected zinc mass fraction in whole cell of PC12 cell lines. The experiments were performed on freeze dried cells at the nano-imaging station ID22NI of the European Synchrotron Radiation Facility (ESRF). X-ray fluorescence analysis gives the areal mass of most major, minor and trace elements while quantitative phase contrast imaging provides maps of the projected mass. The combined method was validated on calibration samples by comparison with other alternative techniques such as Atomic Force Microscopy (AFM) and Scanning Transmission Ion Microscopy (STIM). Up to now, absolute quantification at the sub-cellular level was impossible using X-ray fluorescence microscopy but can be reached for the first time with the use of the proposed approach.

1. Introduction

Metals present as trace elements in biological systems play an important role in the cell metabolism [1], [2]. They are involved in important biosynthetic pathways including the conversion of electrochemical to chemical energy, the biosynthesis of DNA and an array of important metabolites [3]. The number of studies on trace elements is growing and nowadays their role in human health both in their natural occurance and via therapeutic drugs is recognized as crucial. Compounds containing metals are used for the diagnoses and treatment of diseases (gadolinium for imaging, platinum-based drugs for chemotherapy [4]) and to develop nanocomposites (for example functionalized titanium or gold nanoparticles) for gene therapy. Quantitative study of the distribution of trace elements at the sub-cellular level [5] provides important information for the study of the functions and pathways of metalloproteins and therapeutic approaches.

The relevant information about trace elements in biological specimens increases with the sampling resolution. To probe the sub-cellular complexity of metal ion homeostasis mechanisms, one requires sub-micrometer resolution together with a sub-femtogram absolute detection limits due to the decreasing quantity of sample probed. X-ray fluorescence (XRF) analysis with a nanoprobe is the most direct and sensitive method to quantify the distribution of metals and other elements at the sub-cellular level [6], [7]. It provides high sensitivity for transition metals and other relevant trace elements together with the capability of penetrating and mapping whole cells.

Most of the structural details of cells are undetectable in hard X-ray microscopy due to the weak absorption contrast between structures with similar transparency. However the various organelles show wide variation in X-ray refractive index. The technique exploiting the refractive index (its real part) is called phase contrast imaging. Methods based on phase contrast provide complementary information about cells by mapping the projection of the electron density hence to a good approximation the projected mass. Information about the projected mass combined with the average density of the sample allows to calculate its thickness [5], value which can be compared with the one obtained through the use of alternative techniques (STIM, AFM).

2. Methods

2.1. Sample preparation

PC12 cells, a clonal catecholaminergic cell line derived from rat pheochromocytoma [8] were cultured in RPMI 1640 medium supplemented with 10% (v/v) heat-inactivated horse serum, 5% (v/v) fetal bovine serum, penicillin (25 μg/mL) and streptomycin (25 μg/mL). Cultures were maintained according to standard protocols at 37°C in a 95% humidified incubator with 5% CO_2 as already described [9]. The cells were cultured directly on gelatin gel coated thin polycarbonate foil (2 μm thick). These targets were subsequently fastly cryofixed in liquid nitrogen, chilled isopentane (-160°C) and further lyophilized.

2.2. Experimental setup of ID22NI

X-ray fluorescence analysis was performed at the nano-imaging station ID22NI of the ESRF, using the intrinsic monochromaticity of the undulator of about $\delta\lambda/\lambda \simeq 0.015$. The experimental station is located at a distance of 63 m from the undulator source and at 37 m from the high power slits used as secondary source in the horizontal direction (25 μm slit opening).The synchrotron radiation is focused by an X-ray optical device consisting of two elliptically shaped mirrors acting in two orthogonal planes using the so-called Kirkpatrick-Baez geometry [10]. The mirrors are coated with a graded multi-layer. No other monochromator is used in the setup, resulting in a very high and unique X-ray flux (up to 10^{12} photons/s) at energies between 17 and 29 keV. In this work, the energy of the pink photon beam was set to 17.5 keV. For more details regarding the ID22 beamline refer to [11]. For the description of the techniques (X-ray fluorescence and phase contrast imaging) used at this beamline, refer to [5].

3. Analysis

PC12 cells were imaged by synchrotron X-ray fluorescence and X-ray phase contrast imaging. After scanning the sample and fitting the emitted fluorescence spectra, the intracellular distribution of the accessible chemical elements can be determined at a spatial resolution corresponding to the beam size (100 nm) (Fig. 1). By applying a phase retrieval algorithm [12], the relative phase map was generated. The quantitativeness of the phase imaging was validated through a comparison with AFM and STIM, see details in [5]. In order to combine both (fluorescence and phase contrast) images, corrections to the phase map were applied. The background of the phase map was fitted with a polynomial function to correct for low frequency fluctuations. Both images were aligned with a cross correlation technique and distortion corrections were applied to phase map. All the corrections were performed with the open source software ImageJ and under GNU octave code. The result of corrected phase map is presented on the Fig. 1. Once the phase map was corrected both images were divided pixel by pixel (fluorescence map by corrected phase map) in order to obtain the projected zinc mass fraction in the PC12 cell (Fig. 1). This provides truly quantitative information, not biased by changes in the sample thickness.

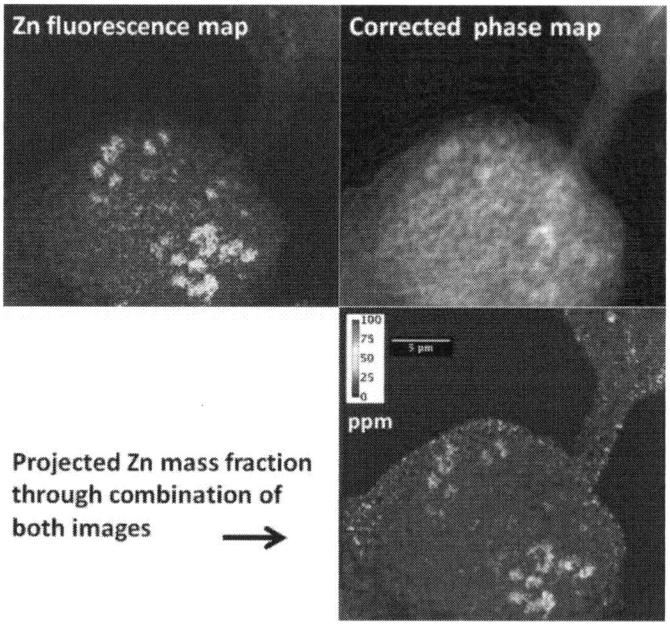

Figure 1. X-ray fluorescence of zinc (upper left) and phase contrast (upper right) image of PC12 cell. Dwell time of scanning was set to 150 ms and step size 150 nm. Both images were divided pixel by pixel in order to access absolute zinc mass fraction (bottom image); the scale is in ppm of dry weight.

4. Discussion

An important application of X-ray fluorescence microscopy is the determination of the distribution of the chemical elements at the sub-cellular level. However, the mapping of true element concentration at the sub-cellular level is a difficult task. Locally, the composition and the density or the thickness of different sub-cellular compartments can vary. To address this problem, X-ray fluorescence sub-cellular mapping and propagation-based X-ray phase contrast imaging for mass normalization were combined, for the first time, to provide high-resolution, truly quantitative, elemental distribution maps in cells [5]. Phase-contrast imaging with a partially coherent beam produces Fresnel interference fringes which makes possible to image samples with small variations in mass density. These variations would be undetectable using absorption contrast imaging.

The projection X-ray microscopy method used in the present work is optimized for relatively high energies (17 keV-29 keV) that match well with efficient excitation of the X-ray fluorescence. As we used a full-field CCD based approach; our method is more dose efficient, provides a larger field of view and is much faster than scanning methods to obtain the phase information (like the one in [13]). For comparison, the phase images are acquired with a few seconds of exposure time and contain 4 million pixels (1 μs/pixel), whereas the fluorescence scanning data requires about 6 hours of exposure time for 100000 pixels (250 ms /pixel).

Alternatively scanning differential phase contrast imaging has been performed by measuring the deviation of the beam by the sample with a position sensitive detector [14]. However the results obtained were done at low energy (2.5 keV) that limits the range of elements accessible through X-ray fluorescence. Also Compton to Rayleigh scattering analysis has been used to obtain more accurate X-ray fluorescence quantification in 2D [15] or 3D [16] measurements.

However a limitation in this approach comes from the estimation of the matrix mean atomic number which not necessarily match the sample matrix studied and variation at the sub-cellular scale. In addition the thickness needs to be determined from thickness calibration curves on the basis of the total Compton scatter intensity. On the overall these average values based on Compton signal seem not suited to provide absolute quantification at the sub-cellular level.
On the overall this work is encouraging the use of X-ray phase imaging to improve the elemental quantification in 2D and 3D X-ray fluorescence imaging.

5. Acknowledgements
We would like to thank the team from CENBG Bordeaux: Richard Ortega, Guillaume Devès and Asuncion Carmona for sample preparation and for a STIM measurement that was used for validation of our method as well as the team from Surface Science Laboratory of the ESRF: Florence Marchi and Jean Francois Guillet for AFM experiments.

References
[1] Meares, C., Wensel, T., 1984. Metal chelates as probes of biological systems. Acc. Chem. Res. 17 (6), 202-209.
[2] Bush, AI., Curtain, CC., 2008. Twenty years of metallo-neurobiology: where to now? Eur. Biophys. J 37 (3), 241-5.
[3] Sundberg, RJ., Martin RB, 1974. Interactions of histidine and other imidazole derivatives with transition metal ions in chemical and biological systems. Chem. Rev. 74 (4), 471-517
[4] Lippard, S. J., Berg, J. M., 1994. Principles of Bioinorganic Chemistry. University Science Books: Mill Valley, CA.
[5] Kosior, E., Bohic, S., Cloetens, P., Suhonen, H., Ortega, R., et. al., 2012. Combined use of hard X-ray phase contrast imaging and X-ray fluorescence microscopy for sub-cellular metal quantification. J. Struct. Biol. 177, 239-247
[6] Bohic, S., Simionovici, A., Snigirev, A., Ortega, R., Devès, G., et.al., 2001. Synchrotron hard x-ray microprobe: fluorescence imaging of single cells. Appl. Phys. Lett. 78 (22), 3544 - 3546.
[7] Ortega, R., Bresson, C., Fraysse, A., Sandre, C., Devès, G., et.al., 2009. Cobalt distribution in keratinocyte cells indicates nuclear and perinuclear accumulation and interaction with magnesium and zinc homeostasis. Toxicology Lett. 188, 26-32.
[8] Greene, L. A. and Tischler, 1976. A. S. Proc. natn. Acad. Sci. U.S.A. 73, 2424 2428.
[9] Carmona, A., Devès, G., Ortega, R., 2008. Quantitative micro-analysis of metal ions in subcellular compartments of cultured dopaminergic cells by combination of three ion beam techniques. Anal Bioanal Chem 390: 1585-94.
[10] Hignette, O., Cloetens, P., Rostaing, G., Bernard, P., Morawe, C., 2005. Efficient sub 100nm focusing of hard x rays. Rev. Sci. Instrum. 76.
[11] Martinez-Criado, G., Bohic, S., Cloetens, P., Kosior, E., Suhonen, H., et. al. 2012. Status of the hard X-ray microprobe beamline ID22 of the European Synchrotron Radiation Facility. J. Synchr. Rad. 19, 10-18
[12] Cloetens, P., Ludwig, W., Baruchel, J., V.Dyck, J., V.Landuyt, J., et.al., 1999. Holotomography: Quantitative phase tomography with micrometer resolution using hard synchrotron radiation x rays. Appl. Phys. Lett. 75, 2912-2914.
[13] Holzner, C., Feser, M., Vogt, S., Hornberger, B., Jacobsen, C., et.al., 2010. Zernike phase contrast in scanning microscopy with X-rays. Nature Physics 6,883-887.
[14] de Jonge, M., Hornberger, B., Holzner, C., Paterson, D., Jacobsen, C., et.al., 2008. Quantitative phase imaging with a scanning transmission x-ray microscope. Phys. Rev. Lett. 100, 163902.
[15] Vincze, L., Janssens, K., Adams, F., Rivers, M.L., Jones, K.W., 1995. A General Monte-Carlo simulation of ED-XRF spectrometers: Part II, Polarized Monochromatic Radiation, homogeneous samples. Spectrochimica Acta B, 50B, 127-147.
[16] Golosio, B., Simionovici, A., Somogyi, A., Lemelle, L., Chukalina, M., et.al., 2003. Internal elemental microanalysis combining X-ray fluorescence, Compton and transmission tomography. J. of Appl. Phys. 94, 145-156.

X-ray fluorescence microscopy of light elements in cells: self-absorption correction by integration of compositional and morphological measurements

E Malucelli[1], S Iotti[1, 2], M Fratini[3,8], C Marraccini[9], A Notargiacomo[3], A Gianoncelli[4], I Bukreeva[3], A Cedola[3], J Maier[5], G Farruggia[1, 2], C Cappadone[1], L Merolle[1], F Wolf[6], V Trapani[6] and S Lagomarsino[7]

[1] Dipartimento di Farmacia e Biotecnologie, Univ. di Bologna, It
[2] Istituto Nazionale Biostrutture e Biosistemi, Rome, It
[3] Institute of Photonics and Nanotechnology - CNR, Rome, It
[4] Sincrotrone Trieste, Area Science Park, Basovizza Trieste, It
[5] Dipartimento di Scienze Bionediche e Cliniche L. Sacco , Univ. di Milano, Milan, It
[6] Istituto Patologia Generale, Univ. Cattolica del Sacro Cuore, Rome, It
[7] Institute Chemical and Physical processess, c/o Phys. Dpt. Univ. Sapienza, Rome, It
[8] Fermi Center, Piazzale del Viminale 1 00187 Roma
[9] Dipartimento di Scienze della Vita, Univ. di Modena e Reggio Emilia, Modena, It

E-mail: stefano.lagomarsino@cnr.it

Abstract. We present here a new methodology for quantitative mapping of light elements in cells, based on combination of compositional and morphological information, derived respectively by X-ray Fluorescence Microscopy (XRFM), Atomic Force Microscopy and Scanning Transmission X-ray Microscopy (STXM). Since XRFM of light elements (carbon, nitrogen, oxygen, sodium and magnesium), are strongly influenced by self-absorption, we developed an algorithm to correct for this effect, using the morphological and structural information provided by AFM and STXM. Finally, the corrected distributions have been obtained, thus allowing quantitative mapping.

1. Introduction

X-ray Fluorescence Microscopy (XRFM) is a powerful method to measure elemental content and distribution in many kind of materials. In recent years considerable attention has been given to application of XRFM in cell biology, in particular for its ability to map trace elements in cells [1, 2]. Fluorescence micro-tomography has been also applied to diatoms for quantitative determination of element distribution in 3D [3]. In a recent paper [4] we have demonstrated that it is possible to merge compositional and morphological information to quantitatively derive the element concentration (magnesium in that case) combining XRFM with Atomic Force Microscopy (AFM). In this paper we aim to demonstrate that morphological data derived by AFM and by Scanning Transmission X-ray Microscopy (STXM) are crucial for mapping also main cell components, i.e. carbon (C), nitrogen (N) and oxygen (O), whose fluorescence radiation is strongly affected by the absorption in the cell itself (self-absorption effect). We therefore developed a specific algorithm to correct for self-absorption, and applied this procedure to obtain the true element distribution C, N, O, Na and Mg.

2. Experimental and results

2.1. *Sample preparation*
A colon carcinoma cell line (LoVo) was used in this study. The cells were cultured in RPMI medium (Sigma) supplemented with 10% PBS (phosphate buffered saline), 2 mM glutamine, 100 U/ml penicillin, and 100 µg/ml streptomycin sulphate. At 50-80% confluency, the cells were briefly rinsed in 150 mM KCl, then fixed in ice-cold methanol/acetone 1:1 and air dried.

2.2. *XRFM and AFM measurements*
The XRFM measurements on LoVo cells were carried out using synchrotron radiation at the TWINMIC beamline at ELETTRA. Fresnel zone plate lenses were used to focalize the incoming 1.5 keV beam, monochromatized by grating monochromators. The final spot size onto the sample was evaluated to be around 500 nm. The samples were transversally scanned in the zone plate focus, and at each scan step eight full fluorescence spectra were simultaneously acquired by 8 Si-drift detectors (SDD) placed radially with respect to the sample. The angle formed by the SDDs and the sample plane was 20° [5]. The SDDs have a 30mm^2 active area and certified >90% sensitivity for the photon energy range 200–15000 eV [6]. The flux transmitted (T) by the sample was simultaneously measured with a CCD camera. Zone plate, sample and detectors were in vacuum, thus avoiding any absorption by air.
AFM topography maps were collected off-line by a Digital Instruments D3100 AFM equipped with a Nanoscope IIIa controller. The AFM was operated in air ambient using Tapping Mode at a resonance frequency of about 260 kHz. Commercial monolithic silicon tips were employed, with an apex curvature radius in the 5-10 nm range and a nominal force constant of 40 N m-1. The scan size used was 55 µm and the image resolution in pixels was 512x512. Images were collected without any real-time filtering or flattening, and were post-processed to eliminate all the possible artifacts [4].

2.3. *Fluorescence spectra analysis*
The fluorescence spectra were analyzed for each scan step with the program PyMCA [7], and deconvoluted into their components, considering the incident photon energy of about 1.5 keV. Under these conditions C, N, O, Na, Mg were detected in LoVo cells. The resulting fluorescence intensity maps provided information on the total elemental content in the volume illuminated by the beam at each scanning step. However, especially for the lightest elements, the fluorescence radiation experiences strong self-absorption due to penetration in the sample itself. Figure 1 shows the raw oxygen fluorescence maps of four of the eight detectors, whose position with respect to the sample is schematically indicated in the figure. The intensity inhomogeneity due to self-absorption is clearly detected. Indeed the sides of raw maps opposite to the detectors show low intensity due to the self-absorption, while the sides close to the detectors show high intensity. In the figure the color scale goes from blue (low intensity) to red (high intensity).

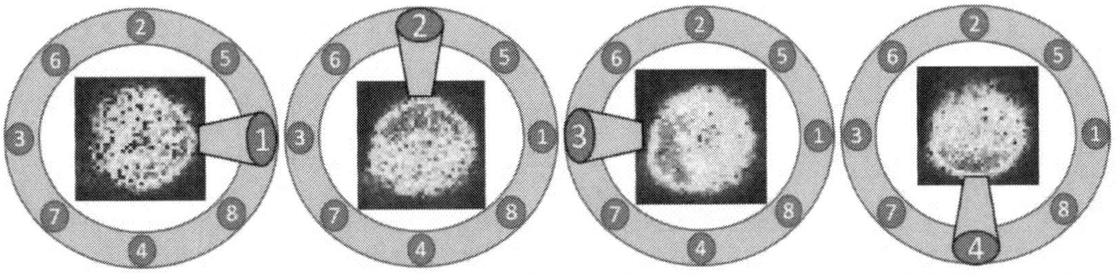

Figure 1. Raw fluorescence maps of oxygen derived from detectors 1, 2, 3 and 4. The red pixels correspond to high fluorescence intensity.

2.4. *AFM and fluorescence elemental maps alignment*

To register the raw elemental maps, obtained as described above, with the AFM maps Figure 2A, which has a different spatial resolution with respect to the first ones, we developed an automated operator independent alignment method. An accurate registration requires to align images with the same contrast and as much as possible with high signal to noise ratio. To this purpose we used as reference images the fluorescence raw maps (Figure2B) of the sum of the eight detectors. Fluorescence raw maps were obtained by the program PyMCA as the sum of all the channels of the spectrum acquired at each scanning step. AFM was registered onto fluorescence raw maps using the images registration software FLIRT (http://fsl.fmrib.ox.ac.uk/fsl/flirt/) using a transformation with 6 degrees of freedom (1 rotation, 2 translation, 2 scale, 1 skew i.e. oblique deformation). Registered images were then re-sampled interpolating with a tri-linear function. Figure 2C shows the AFM map after registration.

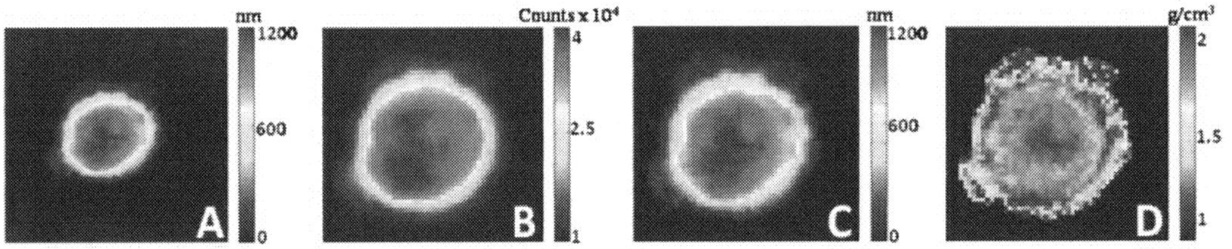

Figure 2. Panel A. AFM map. Panel B. Fluorescence raw map. Panel C. AFM registered onto fluorescence raw map. Panel D. Density map

2.5. XRFM self-absorption correction

Having a beam of energy E_0 and intensity I_0 incident at an angle α on the sample, at the depth z the intensity $I(z)$ is reduced to:

$$I(z) = I_0 e^{-\mu_m(E_0)\cdot\rho\cdot z\cdot cosec(\alpha)} \tag{1}$$

where $\mu_m(E_0)=\Sigma i\ w(i)\mu_m(i)(Eo)$ is the mass absorption coefficient at the energy E_0 of the sample composed of elements i, $\mu_m(i)$ and $w(i)$ are the mass absorption coefficient and the mass fraction of element i respectively, and ρ is the sample density. In the infinitesimal sample volume $dV=Sdz$ at this depth, a number of atoms will be ionized and will give rise to the emission of fluorescent radiation with energy E_i. For an accurate assessment of elemental map derived from XRFM, the self-absorption in the sample must be taken into account using this equation:

$$A(i) = \int_0^d exp\{-[\mu_m(E_0)cosec(\alpha) + \mu_m(E_i)cosec(\beta)]\rho z\}\, dz \tag{2}$$

Where $A(i)$ is the correction map for the element i, d the sample thickness, and β the exit angle of the fluorescence radiation collected by the detector. In this work we used tabulated values for $\mu_m(i)$ and approximate values, taken by literature, of $w(i)$ for major cell components (C, N and O) in dehydrated cells: ($w(O)$: 0.57, $w(N)$: 0.14, $w(O)$: 0.28) [8]. In equations (1) and (2) ρ is the density of the sample assessed using the relation $\rho=\mu_l/\mu_m(Eo)$ where μ_l is the linear absorption coefficient. We experimentally determined μ_l from the transmitted intensity T, which was measured at each scanning step, simultaneously to fluorescence detection. T is related to μ_l and to thickness t through the simple relation: $T = exp(-\mu_l t)$. We calculated the μ_l map simply inverting the above expression: $\mu_l = (1/t) * ln(1/T)$, and using the thickness map measured by the AFM previously registered onto transmission and fluorescence space. Figure 2D shows the density map obtained by measuring the transmission intensity. For an accurate evaluation of elemental distribution, the fluorescence maps must be corrected for self-absorption. To this purpose we elaborated an ad-hoc program, as much as possible automatic and operator independent, considering for each measuring point the path through the cell that the exiting fluorescence had to traverse in its way towards the detector. Each pixel of the AFM registered map was divided into small voxels of thickness dz. For each voxel the self-absorption from total path into the cell matrix has been calculated as shown in Figure 3. Then the contribution to fluorescence of each voxel has been summed up to obtain a correction map for each element according to equation 2. This procedure has been repeated for each detector. Figure 4 shows the final result, comparing the uncorrected fluorescence maps with the corrected ones. It must be underlined that a strong contribution to the nitrogen fluorescence comes from the substrate, composed of Si_3N_4. In the uncorrected map, the N fluorescence is indeed stronger outside the cell than in the cell, due to the absorption within the cell itself. We therefore subtracted the contribution from the substrate, corrected for absorption into the cell, to the total N fluorescence intensity, and then we applied the correction for self-absorption.

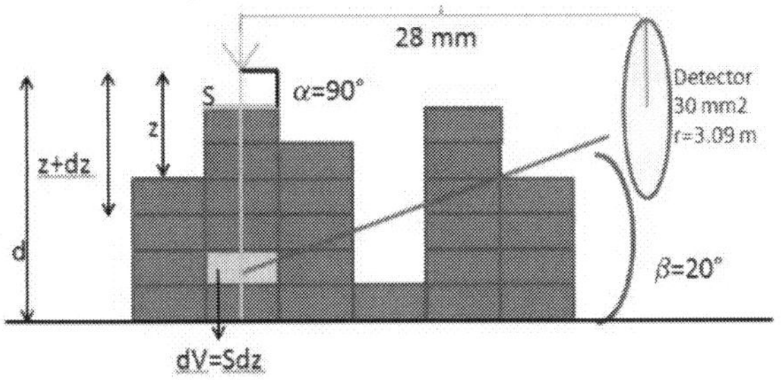

Figure 3. Scheme for the self-absorption correction in one voxel. The path which the fluorescence excited in the given voxel (in yellow) has to traverse through the cell, is calculated pixel by pixel taking into account the morphology measured by AFM.

3. Conclusions

In this work we have demonstrated that it is possible to assess the distribution of the major cell components, together with light metal elements, with XRFM, and provided that correction for self-absorption is a necessary requisite. We developed an ad-hoc analysis program for merging the morphology information obtained using AFM and STXM with XRFM. This procedure is a crucial preliminary step for the quantitative determination of element distribution and concentration.

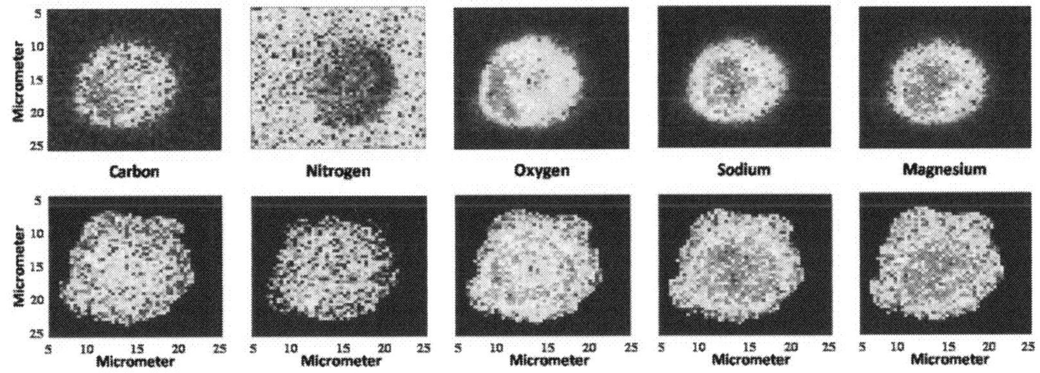

Figure 4. Uncorrected (top) and corrected (bottom) maps of C, N, O, Na and Mg.

References

[1] Bohic S, Simionovici A, Snigirev A, Ortega R, Devès G, Heymann D and Schroer C G 2001 *Appl. Phys. Lett.* **78** 3544–46

[2] Ortega R, Bresson C, Fraysse A, Sandre C, Devès G, Gombert C, Tabarant M, Bleuet P, Seznec H, Simionovici A, Moretto P and Moulin C 2009 *Toxicol. Lett.* **188** 26–32

[3] De Jonge M D, Holzner C, Baines S C, Twining B S, Ignatyev K, Diaz J, Howard D L, Legnini D, Miceli A, McNulty I, Jacobsen C J, and Vogt S., 2010 *PNAS*, **107** 15676-15680

[4] Lagomarsino S, Iotti S, Farruggia G, Cedola A, Trapani V, Fratini M, Bukreeva I, Notargiacomo A, Mastrototaro L, Marraccini C, Sorrentino A, McNulty I, Vogt S, Legnini D, Kim S, Gianoncelli A, Maier J A M and Wolf F I 2011 *Spectrochim Acta B* **66**, 834–40

[5] Gianoncelli A, Kaulich B, Alberti R, Klatka T, Longoni A, de Marco A, Marcello A and Kiskinova M 2009 Nuclear *Instr Meth Phys Res A* **608** 195-8

[6] R. Hartmann, K.H. Stephan, L. Struder, Nucl. Instr. and Meth. A 439 (2–3) (2000) 216.

[7] Solé V A, Papillon E, Cotte M, Walter P H and Susini J 2007 *Spectrochim Acta B* **62** 63-8

[8] Heldal M, Norland S and Tumyr O 1985 *Appl Environ Microbiol* **50** 1251-57

XMCD microscopy with synchronized soft X-ray and laser pulses at PETRA III for time-resolved studies

P Wessels[1], M Schlie[1], M Wieland[1], J Ewald[2], G Abbati[2], S Baumbach[2], J Overbuschmann[2], T Nisius[2], A Vogel[3], A Neumann[3], A Meents[4], J Viefhaus[4], H P Oepen[3], G Meier[3], T Wilhein[2] and M Drescher[1]

[1] Institut für Experimentalphysik, Universität Hamburg, Luruper Chaussee 149, 22761 Hamburg, Germany
[2] Institute for X-Optics, RheinAhrCampus Remagen, Südallee 2, 53424 Remagen, Germany
[3] Institut für Angewandte Physik, Universität Hamburg, Jungiusstraße 11, 20355 Hamburg, Germany
[4] Deutsches Elektronen-Synchrotron DESY, Notkestraße 85, 22607 Hamburg, Germany

E-mail: philipp.wessels@desy.de

Abstract. We present the setup of a new transmission X-ray microscope using zone plates at the soft X-ray beamline P04 at PETRA III, designed to utilize the X-ray magnetic circular dichroism (XMCD) contrast for imaging the magnetic orientation of nanostructured materials. We present the first results obtained with this setup at PETRA III using vacuum chambers with built-in multi-axis micropositioning stages that can be easily customized to mount a variety of components like condenser, samples, coils, zone plates etc. For further experiments the X-ray microscope will be equipped with a synchronized femtosecond laser system to implement a time-resolved detection of transient magnetic states in nanostructured media using a pump-probe scheme. The synchronization between laser and X-ray pulses has already been demonstrated at both the P04 and P11 beamlines at PETRA III with a jitter of $\sigma < 15$ ps.

1. Motivation and introduction

Studying nanoscale magnetic systems at their fundamental time- and length-scales is a fascinating and important research field with many applications in modern data storage technology. The performance of magnetic storage devices in terms of access times, data-density and stability may increase even more in the future with the development of new storage materials and ultra-fast physical mechanisms allowing the manipulation of magnetic nanostructures with femtosecond laser pulses [1–4]. Furthermore, the understanding of the fundamental physical properties and limitations in these processes may greatly contribute to the development of future computer concepts using magnetic nanostructures or even atoms instead of electronic elements as logical units (spintronic devices) [5, 6] or non-volatile memory technology [7–9].

Soft X-ray microscopy is a powerful and widely-used tool when it comes to imaging of magnetic nanostructures [10, 11] because it offers sufficient spatial resolution of a few 10 nm and it can be implemented in a pump-probe setup to add temporal sensitivity. Additionally,

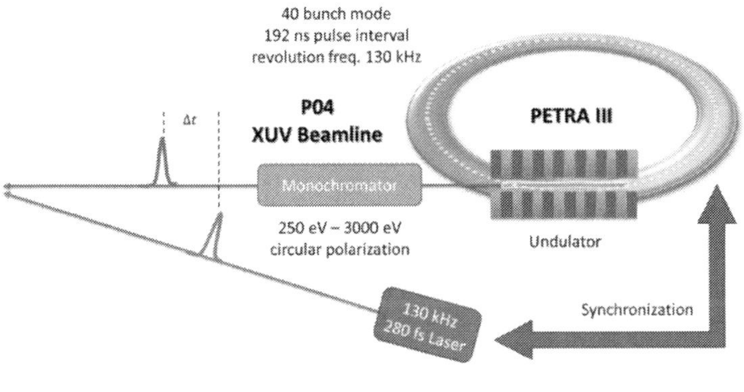

Figure 1. Schematic of the laser pump and soft X-ray probe experiment with a synchronized pulsed laser system at the P04 beamline of PETRA III at DESY.

soft X-ray microscopy enables a simultaneous detection of the magnetization via X-ray magnetic circular dichroism (XMCD) spectromicroscopy [12]. This contrast mechanism originates from the partly filled $3d$ shell of the transition metal elements because a difference in density of final states $\rho(E_f)$ in a magnetized medium acts as a detector for the magnetization component parallel to the propagation direction of the photons. For either spin-up or spin-down polarized electrons excited out of the initial $2p$ shell by the circular polarized light field [13, 14] the transition rate changes.

2. Experimental Setup

For time-resolved measurements, a magnetic system is excited by an ultrashort laser pulse and the magnetic response is successively probed by a soft X-ray pulse in the full-field transmission X-ray microscope (TXM). To perform a pump-probe experiment, the delay Δt between the pulses can be tuned electronically (figure 1).

The sample used here consists of cobalt-platinum (Co/Pt) multilayers with perpendicular magnetic anisotropy structured by electron-beam lithography and lift-off processing on a 200 nm thick silicon nitride (Si_3N_4) membrane.

The vacuum chambers in this experiment are part of an endstation system which can be adapted for different applications in, for example, imaging, spectroscopy or nonlinear physics at synchrotron radiation facilities, free-electron lasers (FEL) or in a laboratory. It features up to three vacuum chambers on a rail with viewports and large rectangular flanges for accessing the inner components (figure 2). The micropositioning stages inside can hold a variety of components like condenser, samples, coils, zone plates etc. and are designed for long travel ranges and high precision positioning.

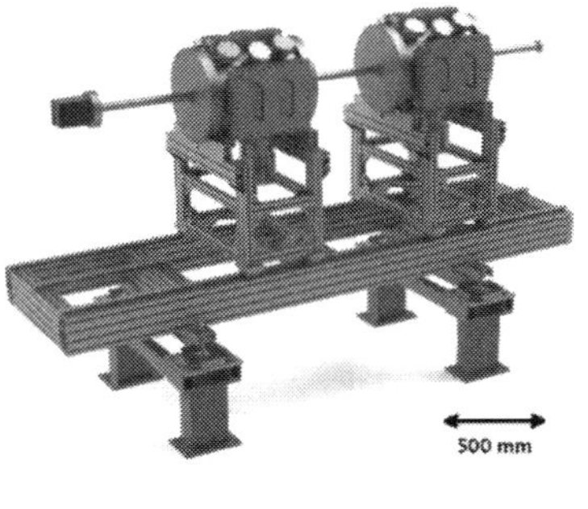

Figure 2. CAD image of the modular vacuum system.

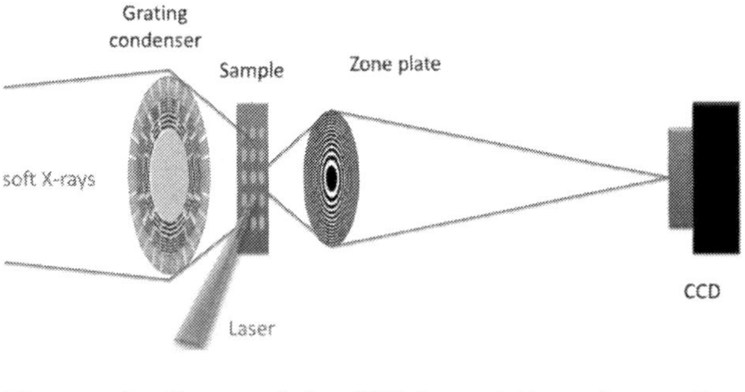

Figure 3. Setup of the TXM consisting of a grating condenser with gold central stop and a micro zone plate.

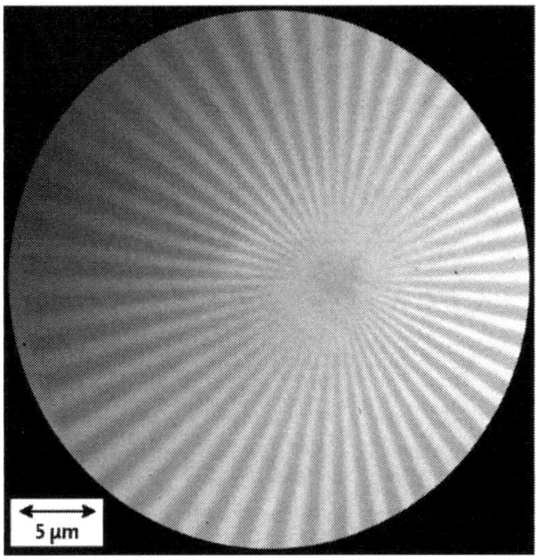

Figure 4. Siemens star with transmission of ≈ 75 %, diameter of D = 40 μm and n = 50 spokes imaged with the TXM (dark image subtracted).

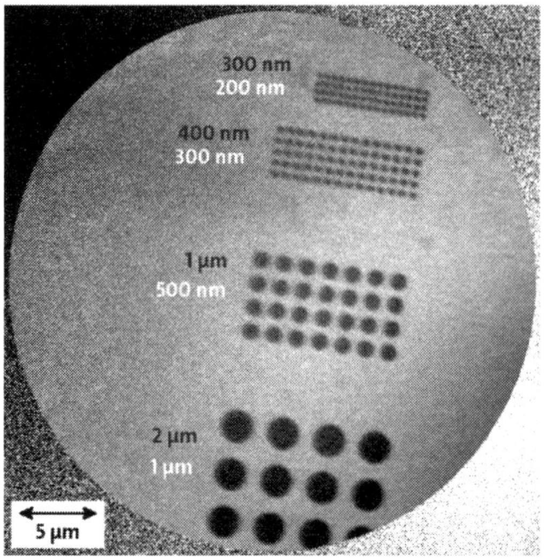

Figure 5. Flatfield corrected TXM image of a nanostructured magnetic cobalt-platinum (Co/Pt) multilayer system. The blue labels indicate the dot diameter while the white labels specify the minimum distance between two dots.

To investigate magnetic nanostructures, a full-field TXM based on a grating condenser [15] for Köhler-like illumination with an included gold central stop, a micro zone plate and a soft X-ray CCD (figure 3) has been set up and commissioned. The theoretical resolution of the microscope operating at the cobalt L_3 edge (778 eV) with 660X magnification is limited to $\Delta l = 1.22{\cdot}\Delta r = 61$ nm according to the outermost zone width of the micro zone plate measuring $\Delta r = 50$ nm. For best imaging results, the numerical apertures of condenser and zone plate are matched.

A phase-locked loop (PLL) synchronized chirped pulse amplification (CPA) laser system has been set up based on the commercial diode-pumped Yb:KGW laser PHAROS operating at a high repetition rate of 130 kHz (equal to the revolution frequency of the storage ring PETRA III) with a pulse duration of 280 fs and an output power of 6 W. The fundamental output wavelength of 1030 nm can be converted into the spectral range of 210 nm - 2700 nm via harmonic generation or optical parametric amplification (OPA). The laser system is mounted onto a mobile laser table with a footprint of only 1.2 m x 0.9 m.

3. Results

The TXM has been set up and tested successfully and in the first images (figure 4 and 5) features down to 130 nm can be resolved. The discrepancy to the expected theoretical resolution of 61 nm can be explained by drifts based on mechanical instabilities that have been overcome by now and an improved resolution could be demonstrated in a recent beam time (data not shown). The remaining gradient in the illumination in figure 4 was due to a misalignment of the condenser optic but can be fully eliminated via flat field correction.

As depicted in figure 6, the synchronization of the laser system has been demonstrated at beamline P11 with a residual jitter of $\sigma < 15$ ps (presumably limited by the photodiode) by measuring the delay of ≈ 7000 shots of the laser and the synchrotron pulses on the same photodiode. Given the fact that this jitter is short in comparison to the root mean square (RMS)

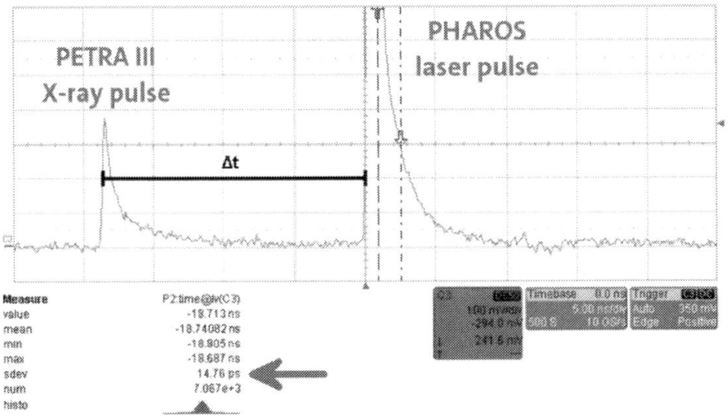

Figure 6. PHAROS (3rd harmonic) and PETRA III pulses synchronized on the same photodiode. The time scale is 5 ns/div and the voltage scale is 100 mV/div.

pulse length of the PETRA III pulses of about 44 ps [16], pump-probe experiments reaching a temporal resolution limited only by the synchrotron pulse duration become possible. Recent measurements suggest an even smaller jitter in the range of a few picoseconds (data not shown).

4. Summary and Outlook

We have set up and commissioned a new full-field TXM at the P04 beamline of PETRA III as well as a synchronized femtosecond laser system. The next steps are heading for the theoretical resolution and recording XMCD contrast images of magnetic nanostructures with the improved P04 beamline parameters offering a better beam profile and a higher resolving power $\frac{\lambda}{\Delta\lambda}$.

Acknowledgments

The authors would like to thank Dr. Pambos Charalambous (ZonePlates.com) for manufacturing of the grating condenser and zone plates as well as the scientific and technical staff of the beamlines P04 and P11 and the operators of PETRA III at DESY.

We gratefully thank the DFG Collaborative Research Centre 668 (SFB 668 - Magnetismus vom Einzelatom zur Nanostruktur) for financial support as well as the German Federal Ministry of Education and Research (BMBF - Bundesministerium für Bildung und Forschung) for financial support within the Nanofocus projects (05KS7GU4 and 05KS7UL1).

References

[1] Beaurepaire E, Merle J C, Daunois A and Bigot J Y 1996 *Phys. Rev. Lett.* **76** 4250
[2] Pfau B *et al.* 2012 *Nat. Commun.* **3** 1100
[3] Stanciu C D *et al.* 2007 *Phys. Rev. Lett.* **99** 047601
[4] Kirilyuk A, Kimel A V and Rasing T 2010 *Rev. Mod. Phys.* **82** 2731
[5] Wolf S A *et al.* 2001 *Science* **294** 1488
[6] Khajetoorians A A, Wiebe J, Chilian B and Wiesendanger R 2011 *Science* **332** 1062
[7] Parkin S S P, Hayashi M and Thomas L 2008 *Science* **320** 190
[8] Hayashi M, Thomas L, Moriya R, Rettner C and Parkin S S P 2008 *Science* **320** 209
[9] Loth S, Baumann S, Lutz C P, Eigler D M and Heinrich A J 2012 *Science* **335** 196
[10] Stöhr J and Siegmann H C 2006 *Magnetism: From Fundamentals to Nanoscale Dynamics* (Berlin: Springer)
[11] Fischer P 2011 *Mater. Sci. Eng. R Rep.* **72** 81
[12] Schütz G *et al.* 1987 *Phys. Rev. Lett.* **58** 737
[13] Fano U 1969 *Phys. Rev.* **178**(1) 131
[14] Fano U 1969 *Phys. Rev.* **184**(1) 250
[15] Vogt U, Lindblom M, Charalambous P, Kaulich B and Wilhein T 2006 *Opt. Lett.* **31** 1465
[16] Deutsches Elektronen-Synchrotron DESY 2012 Machine Parameters PETRA III at HASYLAB website URL `http://hasylab.desy.de/facilities/petra_iii/machine/parameters/index_eng.html`

Journal of Physics: Conference Series **463** (2013) 012024 doi:10.1088/1742-6596/463/1/012024

Efficient "water window" soft x-ray high-*Z* plasma source

T Higashiguchi[1], T. Otsuka[1], W Jiang[2], A Endo[3], B Li[4], P Dunne[4] and G O'Sullivan[4]
[1] Utsunomiya University, Yoto 7-1-2, Utsunomiya 3218585 Japan
[2] Nagaoka University of Technology, Kami-tomiokamachi 1603-1, Nagaoka, 9402188 Japan
[3] HiLASE Project, Institute of Physics AS, CR, Na Slovance 2, 18221 Prague 8, Czech Republic
[4] School of Physics, University College Dublin, Belfield, Dublin 4, Ireland

E-mail: higashi@cc.utsunomiya-u.ac.jp

Abstract. Unresolved transition array (UTA) is scalable to shorter wavelengths, and we demonstrate a table-top broadband emission "water window" soft x-ray source based on laser-produced plasmas. Resonance emission from multiply charged ions merges to produce intense UTAs in the 2 to 4 nm region, extending below the carbon K edge (4.37 nm). An outline of a microscope design for single-shot live cell imaging is proposed based on a bismuth (Bi) plasma UTA source, coupled to multilayer mirror optics.

1. Introduction

Development of shorter wavelength sources in the extreme ultraviolet (EUV) and soft x-ray spectral regions has been motivated by their application in a number of high profile areas of science and technology. One such topic is the challenge of three-dimensional imaging and single-shot flash photography of microscopic biological structures, such as cells and macromolecules, *in vivo* [1]. For x-ray microscopy, the x-ray source should emit a sufficient photon flux to expose the image of the biological sample on the detector. To date the most practical light source of high-power, high-brightness x-rays has been radiation from synchrotrons and more recently from free electron lasers (FEL) [2]. Table-top sources using ethanol sprays and liquid nitrogen droplets are being developed for use with zone plates for transmission microscopy. Recently $\lambda = 2.48$ nm narrowband emission from a liquid-nitrogen-jet laser-plasma was successfully combined with state-of-the-art normal-incidence multilayer condenser optics and 20-nm zone-plate imaging optics to demonstrate laboratory water-window x-ray microscopy with resolution less than 40 nm [3]. The total collected energy, however, is low, when one combines the narrow line emission with the low reflectivity of the collector mirror. As a result long exposures are needed to record an image and there is not yet published evidence of single-shot exposures using a laboratory-scale source. To overcome the low efficiency imposed by line sources, we propose using high power water-window emission from laser-produced high-*Z* plasmas, analogous to the scheme used for efficient, high-volume manufacturing EUV sources.

Before discussing the high-power water window source, it is important to summarize the characteristics of efficient UTA light sources used in the 5 to 15 nm region. All are based on $n = 4$–$n = 4$ ($4d$–$4f$ and $4p$–$4d$) transitions that overlap to generate an intense UTA. For efficient 13.5-nm operation, which corresponds to a photon energy $h\nu \approx 92$ eV, it is important to produce an optimum plasma electron temperature of 30–50 eV. The rare-earth elements of gadolinium (Gd, $Z = 64$) and terbium (Tb, $Z = 65$) produce strong emission near $\lambda = 6.7$ nm ($h\nu \approx 183$ eV) which is maximized at electron temperatures in the 100–120 eV range depending on initial focusing conditions [4-6]. The spectral behavior of Gd and Tb plasmas is expected to be largely similar to that of Sn plasmas, because in the temperature range of interest, both are dominated by $4d$ open-shell ions. Although the conversion efficiency (CE) from the input laser energy to the output in-band EUV emission energy depends on the bandwidth (BW) of the reflection coefficient of the multilayer mirror (MLM), the maximum CEs have been observed to be higher than 1%. Because it moves to shorter wavelength with increasing atomic number, Z, the $n = 4–n = 4$ UTA is expected to lie in the water window if higher Z elements from $Z = 79$ (Au) to $Z = 83$ (Bi) are used [7]. Higher Z elements such as uranium also emit in the water window but their radioactivity prohibits their use. Much of the previous work on high-Z plasmas has concentrated on the production of quasicontinuum spectra at moderate laser intensities, employing electron temperatures below 100 eV [8,9].

In this paper, we report the demonstration on the efficient "water window" soft x-ray source by strong UTA band emission in laser-produced high-Z plasmas. Our proposed procedure for producing the water window emission is expected to be efficient and scalable in output yield. Our calculations show that a bismuth plasma at an electron temperature in the range 570 to 600 eV radiates strongly near 3.9 nm. At electron temperatures above 800 eV, strong UTA emission around 3.2 nm would be expected.

2. Experimental results and discussion

Because it moves to shorter wavelength with increasing Z, the $n = 4–n = 4$ UTA can be used for other applications, such as transmission x-ray microscopy for biological imaging in the water window (Fig. 1). We have made preliminary studies of the potential of Bi as the "water window" soft x-ray source.

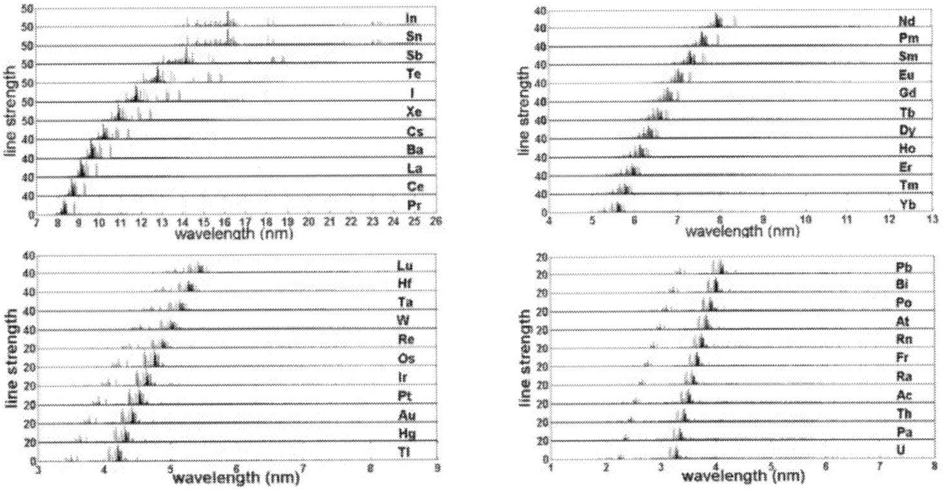

Figure 1. Calculated position of $n = 4–n = 4$ transitions in key ions in elements from indium ($Z = 49$) to uranium ($Z = 92$). The localization of emission near 3.2 and 3.9 nm in Bi is clearly evident.

Figures 2(a) and 2(b) show the time-integrated Bi spectra and the laser intensity dependence of the peak photon energies of $n = 4–n = 4$ and $n = 4–n = 5$ transitions. The position of the $n = 4–n = 4$ transition peak was observed to be unchanged. The position of the $n = 4–n = 5$ peak, on the other hand, shifts to higher photon energy with increasing laser intensity because of the rise in electron temperature and associated charge states. However, the strong emission expected around 3.2 nm, which originates from the $n = 4–n = 4$ UTA, was not observed in this laser intensity region [10].

The emission intensity of the $n = 4–n = 5$ transitions was compared with that of the $n = 4–n = 4$ UTAs. The strong emission at 3.15 nm due to the $n = 4–n = 4$ UTA in Bi plasmas may be coupled with a Sc/Cr MLM with a reflection coefficient of 15% [11]. The variation of this emission with electron temperature was calculated with Cowan's suite of atomic structure codes in order to predict its evolution with increasing laser flux [12]. The predicted spectral evolution as a function of electron temperature is summarized in Fig. 3(a). In this figure the spectra generated for each ion stage are based on excited state populations that allow for electron temperature assuming a Boltzmann distribution. These spectra are then weighted by an ion fraction appropriate to that temperature, calculated assuming collisional-radiative (CR) equilibrium to yield the distributions shown [13]. Our calculations show that high-*Z* plasmas, at an electron temperature lower than 700 eV, as shown in Fig. 3(b), radiate strongly near 3.9 nm. In the case of higher electron temperatures, from 800 to 1500 eV, the strongest emission is expected at around 3.2 nm [10], suitable for coupling to Sc/Cr MLMs [11]. Thus for an optimized source, we should produce a plasma at a higher electron temperature plasma of around 1 keV. The intensity of the Bi plasma emission in our experiment was compared with 2.48-nm nitrogen line emission from a Si_3N_4 planar target, in the same experimental setup, and was observed to be 1.2 times higher within a bandwidth of 0.008 nm (FWHM) [14] even though the plasma electron temperature was much lower than the optimum value.

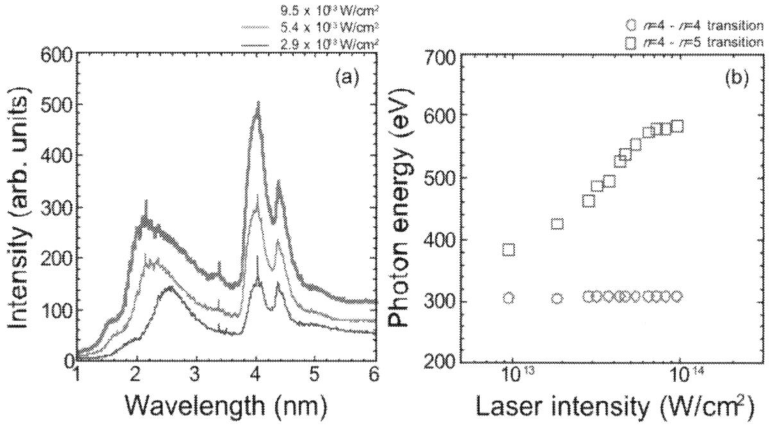

Figure 2. Spectral behaviors of the Bi plasmas in the laser intensity dependence on the observed emission spectra (a), the peak wavelength of the $n = 4$–$n = 4$ transition (circles) and the $n = 4$–$n = 5$ transition (rectangles) (b), respectively.

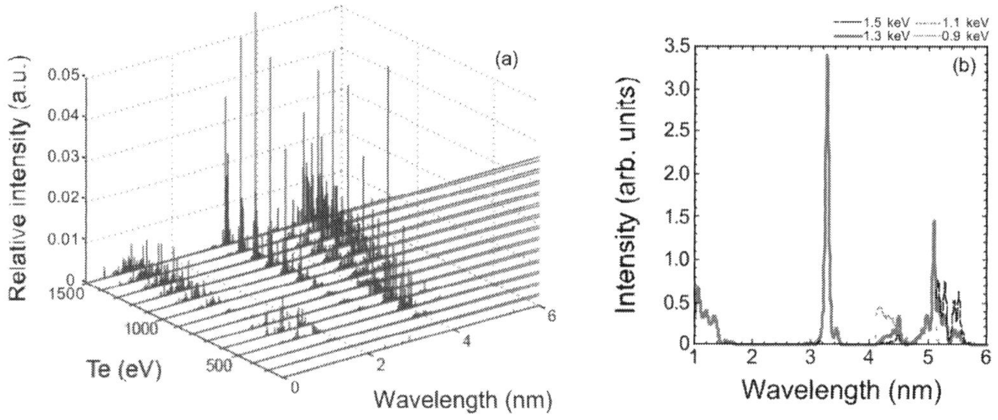

Figure 3. Calculated spectral variation as a function of electron temperature (a) and calculated spectra for electron temperatures higher than 900 eV (b), respectively.

3. Proposal of the UTA source microscope for single shot photograph

Taking the experimental and numerical results into account, we now propose a high brightness, high power water window source for single shot imaging at the laboratory scale [10]. Our proposed method has the advantage that the EUV energy efficiency and atomic number dependence can be scaled based on the fundamental properties of the plasma source. To produce not only a Bi plasma with a high electron temperature of the order of 1 keV but also one that has low density and is optically thin, we should switch to a CO_2 laser operating at a wavelength of 10.6 μm due to the low critical density of 1 × 10^{19} cm^{-3} attainable, with a pulse duration to permit ionization to the appropriate ion stages (Bi^{36+}–Bi^{45+}) i.e., 150 ps, while at the same time maintaining a laser intensity of the order of 10^{13} W/cm^2. In addition, for this proof of principle, we propose a dot Bi target with a diameter less than 10 μm with a thickness less than 1 μm to generate a microplasma (to fulfill the requirements for high brightness and a point source) as the expected focal spot size will be the order of 100 μm in the case of the CO_2 laser wavelength of 10.6 μm. Because of the broadband nature of the emission, zone plate components cannot be used, so one possible solution would be to use a transmission planar x-ray nano-waveguide to image the sample [15]. To achieve high resolution in the recorded image, we should also switch the recording device from x-ray CCD cameras to sensitive EUV resists to overcome the resolution limitation of the CCD pixel size, coupled with Schwarzschild optics, consisting of Sc/Cr MLMs with a reflection coefficient of the order of 15% around 3.2 nm [11]. Although our proposal is based on a simple microscope construction, the key component is the UTA emitted at 3.2 nm from a hot dense Bi plasma point source, combined with Sc/Cr MLMs and sensitive EUV resists based on photochemical reaction [16,17].

4. Summary

In summary, we have demonstrated high-efficiency emission in the water window spectral region based on laser-produced Bi plasmas, and have proposed methods to increase it still further. Resonance emission from multiply charged ions merges to produce intense UTA, extending to wavelengths below the carbon K edge. The overall spectral behavior is well described by simulations. The experimental results also provide an outline for the design concept for single-shot cell imaging with a novel microscope optical system. The method presented here opens the way for applications in next-generation biological science.

Acknowledgments

Part of this work was performed under the auspices of MEXT (Ministry of Education, Culture, Science and Technology, Japan). One of the authors (T.H) also acknowledges support from The Canon Foundation and Research Grant (Basic Research) on TEPCO Memorial Foundation. We also are grateful to the Komatsu Ltd. and Gigaphoton Inc. for providing the picosecond laser system.

References

[1] Solem J C and Baldwin G C 1982 *Science* **218** 229
[2] Gorniak T, Heine R, Mancuso A P, Staier F, Christophis C, Pettitt M E, Sakdinawat A, Treusch R, Guerassimova N, Feldhaus J, Gutt C, Grübel G, Eisebitt S, Beyer A, Gölzhäuser A, Weckert E, Grunze M, Vartanyants I A, and Rosenhahn A 2011 *Opt. Exp.* **19** 11059
[3] Legall H, Blobel G, Stiel H, Sandner W, Seim C, Takman P, Martz D H, Selin M, Vogt U, Hertz H M, Esser D, Sipma H, Luttmann J, Höfer M, Hoffmann H D, Yulin S, Feigl T, Rehbein S, Guttmann P, Schneider G, Wiesemann U, Wirtz M and Diete W 2012 Opt. Exp. **20** 18362
[4] Otsuka T, Kilbane D, White J, Higashiguchi T, Yugami N, Yatagai T, Jiang W, Endo A, Dunne P and O'Sullivan G 2010 *Appl. Phys. Lett.* **97** 111503
[5] Li B, Dunne P, Higashiguchi T, Otsuka T, Yugami N, Jiang W, Endo A and O'Sullivan G 2011 *Appl. Phys. Lett.* **99** 231502
[6] Cummins T, Otsuka T, Yugami N, Jiang W, Endo A, Li B, O'Gorman C, Dunne P, Sokell E, O'Sullivan G and Higashiguchi T 2012 *Appl. Phys. Lett.* **100** 061118
[7] Higashiguchi T, Otsuka T, Yugami N, Jiang W, Endo A, Dunne P, Li B and O'Sullivan G 2011 *SPIE Newsroom* "Shorter-wavelength extreme-UV sources below 10nm"
[8] Carroll P K and O'Sullivan G 1982 *Phys. Rev. A* **25** 275
[9] Zeng G-M, Daido H, Togawa T, Nakatsuka M, Nakai S and Aritome H 1991 J. *Appl. Phys.* **69** 7460
[10] Higashiguchi T, Otsuka T, Yugami N, Jiang W, Endo A, Li B, Dunne P and O'Sullivan G 2012 *Appl. Phys. Lett.* **100** 014103
[11] Sakano K and Yamamoto M 1999 *Proc. SPIE* **3767** 238
[12] Cowan R D (University of California Press, Berkeley, 1981) "The Theory of Atomic Structure and Spectra"
[13] Colombant D and Tonon G F 1973 *J. Appl. Phys.* **44** 3524
[14] Bertilson M, Hofsten O. von, Vogt U, Holmberg A and Hertz H M 2009 *Opt. Exp.* **17** 11057
[15] Kruger S P, Giewekemeyer K, Kalbfleisch S, Bartels M, Neubauer H and Salditt T 2010 *Opt. Exp.* **18** 13492
[16] Gowa T, Takahashi T, Oshima A, Tagawa S and Washio M 2011 *Radiat. Phys. Chem.* **80** 248
[17] Oyama T G, Oshima A, Washio M and Tagawa S 2011 *AIP Advances* **1** 042153

Development of real-time x-ray microtomography system

H Takano[1], M Morikawa[1], S Konishi[1], H Azuma[1], S Shimomura[1], Y Tsusaka[1], S Nakano[2], N Kosaka[2], K Yamamoto[2] and Y Kagoshima[1]

[1]Center for Novel Material Science under Multi-Extreme Conditions, Graduate School of Material Science, University of Hyogo, Kamigori, Hyogo 679-1297, Japan
[2]NITTO DENKO Corporation, Umeda, Osaka 530-0001, Japan

E-mail htakano@sci.u-hyogo.ac.jp

Abstract. We have developed a four-dimensional (4D) x-ray microcomputed tomography (CT) system that can obtain time-lapse CT volumes in real time. The system consists of a high-speed sample rotation system and a high-frame-rate x-ray imager, which are installed at a synchrotron radiation x-ray beamline. As a result of system optimization and introduction of a "zoom resolution" procedure, a real-time 4D CT movie with a frame rate of 30 was obtained with a voxel size of 2.5 μm using 10 keV x-rays.

1. Introduction

High penetration is one of the greatest characteristic of x-ray. The internal structure of an object in its natural state can be observed by x-ray radiography. Particularly, development of x-ray computed tomography (CT) techniques has dramatically broadened their contribution not only to medical applications but also to a wide variety of scientific fields. The spatial resolution of x-ray CT has improved along with advances in the development of x-ray sources and x-ray optical devices. It is now possible to realize a system with a spatial resolution of a few tens of nanometers using a synchrotron radiation x-ray source. However, CT measurement usually requires a long measurement time because a number of projection images are required for reconstruction of a CT volume. Therefore, most x-ray micro-CT systems are generally applied to static objects. If CT measurements can be sequentially performed within a short time period, the system may be applicable to objects in motion, and the movement can be obtained as time-lapse volume data. This would introduce a new dimension—time—to conventional three-dimensional CT data. This four-dimensional (4D) x-ray CT, which would directly observe in-situ processes or dynamic states, is a promising tool for measuring moving subjects.

Some x-ray CT approaches for observing dynamic phenomena have already been developed. On a macro scale, a multislice CT apparatus that can scan a human body in 0.35 s is commercially available for medical use [1]. Also, a simultaneous multiprojection system using an electron-induced x-ray source realized 4D CT observation with a very fast rate of 500 Hz, although the system is based on limited-angle-projection CT [2]. On a micro scale, a system that applies motion-synchronized image acquisition is available for repetitive phenomena such as breathing [3]. This concept is also applicable to very fast phenomena such fuel sprays [4]. On the other hand, a direct observation system for nonrepetitive phenomena has also been developed. The TOMCAT group has developed a fast CT system and applied it to 4D CT experiments [5, 6]. In addition, a 4D CT system using a phase contrast technique based on Talbot interferometry has been developed [7]. These systems have been installed at synchrotron beamlines providing highly intense x-ray irradiation. We have previously developed a

4D CT system using x-ray undulator emission of SPring-8 [8]. The system consists of a high-speed sample rotation stage and a high-frame-rate x-ray camera. In this study, we introduce the most recent version of the system, which has been upgraded and optimized to capture dynamic phenomena in real time. Also, a *zoom reconstruction* process [9] was applied to reduce spatial resolution degradation through a tomographic reconstruction procedure. As a typical application of real-time x-ray CT, bubbling in fluid is seen. Finally, the results of a real-time observation of melting paraffin are presented, along with a movie of the process.

2. 4D CT System and Measurement Conditions

The system is constructed at beamline 24XU (Hyogo-BL) of SPring-8 [10]. Hard x-ray emission from an undulator source is monochromatized to 10 keV by a silicon double-crystal monochromator and is incident on a sample with an intensity of 7×10^{12} photons/s in a beam of size 1.7 mm (horizontal) × 1.0 mm (vertical). The sample is mounted on a high-speed rotation stage with a maximum rate of 17.3 revolutions/s. Wobbling of the rotation axis of the stage (a modified sample spinner for powder diffraction experiments) is less than 4 μm. While a sample is in continuous rotation, a high-frame-rate x-ray imager continuously collects projection images of the sample without any synchronized signals. Then, the number of projection images for CT reconstruction is determined from the rotation rate and frame rate of the image collection. The sample can be heated by a halogen spot heater with a maximum power of 75W at a focal diameter of 2 mm, and the temperature can be controlled by a radiation thermometer. The sample can be monitored during measurement by a video microscope with a long focal length that is also used for sample positioning. The high-frame-rate x-ray imager is a visible-ray conversion-type system. It consists of a phosphor screen, an optical relay lens system, and a fast CMOS camera. The phosphor screen is composed of P46 powder and is 10 μm thick. The relay lens system consists of camera lenses of fixed focal lengths (Nikon F-mount) connecting to opposite direction with each other so that system magnification is easily variable. A fast CMOS camera is used (FASTCAM SA5, Photron Inc.). The maximum frame rate is 7000 images/s for a full frame 1024×1024 image matrix with a pixel size of 20 μm. Images with a bit depth of 12 bits are stored on a 32 Gbyte internal memory and are transferred to external storage through a Gigabit Ethernet interface after measurement.

The spatial resolution of the imager is 7 μm, which was estimated by an edge response of a CT-reconstructed image of a test object. The performance is restricted by the resolving power of the phosphor screen since the estimation was done with a much better effective pixel size of 0.95 μm using a high-magnification relay lens system and a high-resolution CCD camera. It is important for a 4D-CT system to optimize the effective pixel size from the viewpoint of both temporal and spatial resolution. For our system, we set it to 5 μm so that the system can achieve the best temporal resolution given the spatial resolution of the imager. In this configuration, x-ray images with good image quality can be captured with a frame rate of 4000 frames/s. When the number of projection images over a projection angle of 180° is set to 300 for one CT volume, the sample rotation rate is 6.67 revolutions/s, and then the measurement time becomes 75 ms.

Figure 1 shows a reconstructed cross-sectional image of a wooden tip that was obtained with this measurement configuration. Figure 1(a) is an original image reconstructed using a conventional convolution back projection procedure with a Shepp-Logan filter, and figure 1(b) is an enlarged image of the indicated area in figure 1(a). The image has a sufficient resolution since structures could be resolved to near the Nyquist spatial frequency. However, there is some spatial resolution degradation through conventional CT reconstruction, and the zoom reconstruction procedure is an effective way of reducing this degradation [9]. When we applied the procedure, sinogram data was oversampled in both dimensions (position and rotation) with linear interpolation. Figure 1(c) is the reconstructed image from a sinogram with a data matrix double that of the original. The processed image showing better spatial resolution demonstrates the capability of this procedure. With the doubling of the data matrix, the effective pixel size and hence the voxel size of the CT volume became 2.5 μm. Furthermore, in the construction of time-lapse CT volume data, projection datasets for each CT volume, consisting of projection images over 180°of the projection angle, were extracted from the obtained images every 90° of the projection angle. Consequently, the interval of the time-lapse CT volumes with the measurement conditions mentioned above became 37.5 ms (26.6 CT volumes/s).

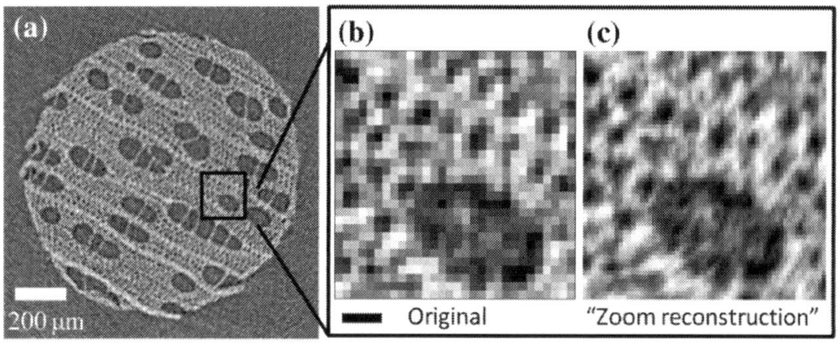

Figure 1. CT reconstructed cross-sectional image of wooden tip. Measurement time for CT volume and the number of the projections are 75 ms and 300, respectively. (a) Original image. (b) Enlarged image of area indicated in (a). (c) Enlarged image of area indicated in (a) obtained by applying the zoom reconstruction procedure. The scale bar in the enlarged images is 25 µm.

3. 4D Observation Results

The 4D-CT system we constructed was applied to observe dynamic phenomena. Figure 2 shows typical images of volume rendering for time-lapse CT volumes. The sample is a water solution with baking powder (commercially available hotcake mix) in a glass capillary with an outer diameter of 1 mm. The measurement configuration was identical to that mentioned in the previous section. 549 CT volumes were extracted from 82,500 projection images with an image matrix of 340 × 200. Since the volume rendering was done for CT volumes inside the glass capillary with inverted contrast, bubbles in the fluid are rendered as objects in the image. The sample heater was not used in this measurement; therefore, the bubble yielding and growth was only due to irradiation by x-rays.

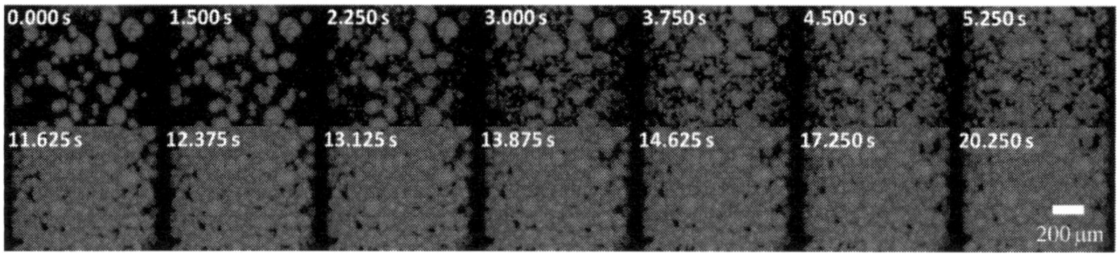

Figure 2. Volume rendered images extracted from time-lapse CT volumes obtained by a 4D-CT measurement of bubbling process in a fluid (water solution of baking powder). A real-time movie is available online as supplementary data (Movie 1).

The last measurement was performed using a test system with a different fast CMOS camera (FASTCAM SA-X, Photron Inc.). The camera has a high frame rate of 12,500 for an image matrix of 1024 × 1024 with a high sensitivity of ISO 25,000. Consequently, projection images utilizing a full dynamic range of 12 bit could be obtained with a frame rate of 5,000 for an effective pixel size of 5 µm. Figure 3 shows the result of the 4D-CT measurement during the melting of paraffin pieces in a glass capillary with an outer diameter of 1.5 mm. 478 CT volumes, each consisting of 333 projection images, were extracted from the total number of projection images (79,920), obtained with a sample rotation rate of 7.5 revolutions/s. Under these conditions, a frame rate of 30 frames/s was obtained for the obtained time-lapse CT volumes. The images were also extracted from volume rendering for CT volumes inside the glass capillary with inverted contrast. The sample was heated by the spot heater at 35W. Time-lapse crumbling and bubbling due to melting can be observed.

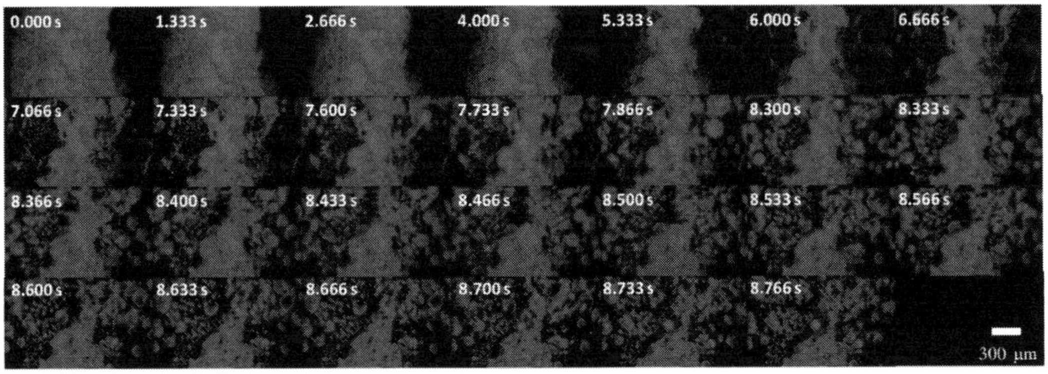

Figure 3. Volume rendered images extracted from time-lapse CT volumes obtained by 4D-CT measurement of melting paraffin pieces. A real-time movie is available online as supplementary data (Movie 2).

Acknowledgments

The synchrotron radiation experiments were performed at the BL24XU of SPring-8 with the approval of the Japan Synchrotron Radiation Research Institute (JASRI) (Proposal No. 2012A3203, 2012A3247, and 2011B3203, 2011B3247). The authors wish to acknowledge Kentaro Uesugi of JASRI for providing the back projection software for CT reconstruction.

References

[1] http://www.toshiba-medical.co.jp/tmd/english/index.html
[2] Bieberle M, Barthel F, Mentz H J, Mayer H G and Hampel U 2011 *Appl. Phys. Lett.* **98** 034101
[3] Sera T, Yokota H, Fujisaki K, Fukasaku K, Tachibana H, Uesugi K, Yagi N and Himeno R 2008 *Phys. Med. Biol.* **53** 4285
[4] Liu X, Im K S, Wang Y, Wang J, Tate M W, Ercan A, Schuette D R and Gruner M 2009 *Appl. Phys. Lett.* **94** 084101
[5] Marone F, Mokso R, Modregger P, Fife J, Pinzer B, Thüring T, Mader K, Mikuljan G, Isenegger A and Stampanon M 2011 *AIP Conf. Proc.* **1365** 116
[6] Mokso R, Marone F, Haberthur D, C Schittny J, Mikuljan G, Issenegger A and Stampanon M 2011 *AIP Conf. Proc.* **1365** 38
[7] Momose A, Yashiro W, Harasse S and Kuwabara H 2011 *Opt. Express* **19** 8423
[8] Takano H, Yoshida K, Tsuji T, Koyama T, Tsusaka Y and Kogoshima Y 2009 *J. Phys. Conf. Ser.* **186** 012049
[9] Mizutani R, Takeuchi A, Osamura R Y, Takekoshi S, Uesugi K and Suzuki Y 2010 *Micron* **41** 90
[10] Tsusaka Y, Yokoyama K, Takeda S, Takai K, Kagoshima Y and Matsui J 2001 *Nucl. Instrum. & Methods A* **467-468** 670

Design of Point Focusing Total-Reflection Zone Plate

Takuya Tsuji[1, 2], Hidekazu Takano[2], Yoshiyuki Tsusaka[2] and Yasushi Kagoshima[2]

[1]Japan Synchrotron Radiation Research Institute, 1-1-1 Kouto, Sayo, Sayo, Hyogo 679-5198, Japan

[2]Center for Novel Material Science under Multi-Extreme Conditions, Graduate School of Material Science, University of Hyogo, 3-2-1 Kouto, Kamigori, Ako, Hyogo 678-1297, Japan

E-mail: t.tsuji@spring8.or.jp

Abstract. We designed point focusing total-reflection zone plates (TRZPs): a double linear type TRZP in Kirkpatrick-Baez arrangement, a conical TRZP, and an open conical TRZP. In this paper, we show detail characteristics of the open conical TRZP, which is the most promising candidate for practical use, and evaluate the effect of fabrication error and misalignment on focused beam profiles by calculation.

1. Introduction

Many optical devices for focusing hard X-rays have been developed, among which the most extensively used ones are Fresnel zone plates (FZPs) [1] and Kirkpatrick-Baez (KB) mirrors [2]. Improvements in nanofabrication technology have been reducing the focused beam size closing to diffraction limits. A focused beam size less than 10 nm has been achieved using new nanofocusing devices, such as a multilayer Laue lens [3, 4] and multilayer KB mirrors [5]. However, all of these optical devices require highly advanced techniques for precisely fabricating the structures at the nanometer level.

We developed total-reflection zone plate (TRZP) as a new optical device for hard X-ray nanofocusing. The TRZP consists of a varied-space grating on a substrate. The positions of the nth zone boundary r_{n+} are given by

$$r_{n+} = \frac{-n\lambda\cos\theta \pm \sqrt{n^2\lambda^2 + 4nf\lambda\sin^2\theta}}{2\sin^2\theta}, \quad (1)$$

where λ is the incident X-ray wavelength, θ is the grazing incident angle, and f is the focal length. The effective size of the zone patterns on the substrate is greatly reduced by using grazing incidence geometry, which considerably facilitates the fabrication. Furthermore, unlike with conventional FZP, the focusing size of the TRZP is not limited by dynamical diffraction effect. We reported on the development of a linear type TRZP that achieved a one-dimensional focusing beam size of 14.4 nm at 10 keV [6]. We also proposed the possibilities of two-dimensional focusing TRZPs [7]. In this paper, we describe detail characteristics of the point focusing TRZPs by calculation study.

2. Point focusing TRZP

Table 1 shows a schematic of our proposing point focusing TRZP candidates and their calculated focused beam size. One candidate is a double linear type TRZP with operating KB geometry (Table 1(a)). This system can be easily constructed using two linear type TRZPs that we already developed. However, the diffraction efficiency is about 10 %, so the beam flux diffracted by the double TRZP system is reduced to about 1 %. Also, the focal length of the upstream TRZP is longer than that of the downstream TRZP, which restricts the reduction ratio of the upstream TRZP and thus the focused beam size. Furthermore, high stability is required for nanofocusing with the KB arrangement.

Another candidate is a conical TRZP, in which ring zone patterns are formed on a conical substrate (Table 1(b)). The conical TRZP focuses the beam in the sagittal direction in relation to the conical shape surface and in the meridional direction in relation to the zone pattern. The radius of curvature R of the inner surface is given by

$$R = f\sin(2\theta). \tag{2}$$

The conical TRZP has great potential for nanofocusing because of its large numerical aperture (NA), which provides high flux and a small symmetrical beam. Fabricating the substrate surface and zone patterning is technically very difficult since the R we designed is about 50 μm. Therefore, we examined to develop an open conical TRZP (Table 1(c)). The open conical shape greatly simplifies the surface fabrication and zone patterning process. In addition, the alignment of the open conical TRZP should be easy. In the following section, we focus on the open conical TRZPs and show simulation results for the focused beam profile and effects of figure error on the substrate inner surface and of misalignment on the focused beam size by calculating the Fresnel-Kirchhoff diffraction integral.

Table 1. Point focusing TRZPs with incident angle of 6 mrad and X-ray energy of 10 keV.

	(a) Double linear type TRZP	(b) Conical TRZP	(c) Open conical TRZP
	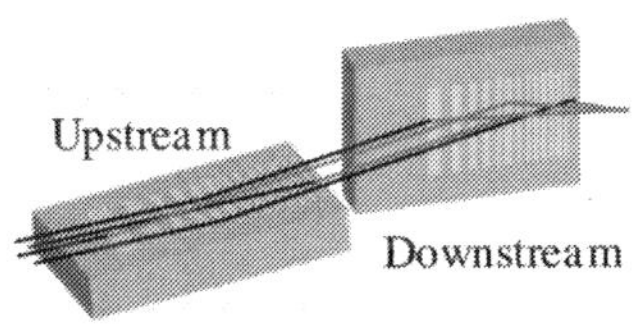	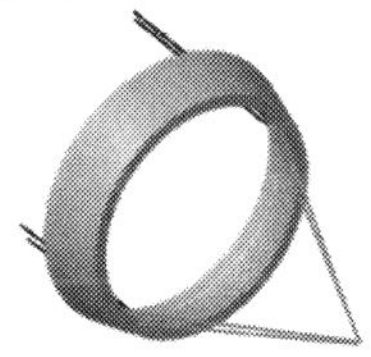	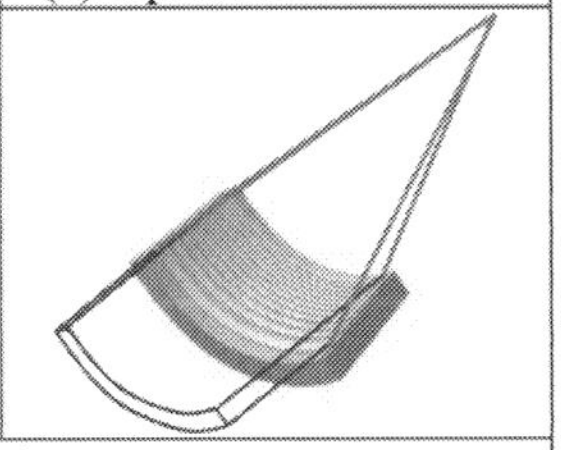
Focal length	15(up), 4.16 mm(down)	4.16 mm	
Focusing size	34.9 × 14.7 nm	2.7 nm	12.2 × 4.5 nm

3. Estimated characteristics of open conical TRZP

Figure 1 shows the calculated intensity distribution at the focal plane for a quarter conical TRZP with a radius of curvature on the upstream side of 50 μm, meridional length of substrate of 2.4 mm, focal length of 4.16 mm, number of zones of 1000, incident X-ray energy of 10 keV, and grazing incident angle of 6 mrad. The calculated focused beam spot size (FWHM) was 12.2 (meridional) × 4.5 nm (sagittal), which is larger than that of the conical TRZP. This result comes from the difference in the NAs between meridional and sagittal directions.

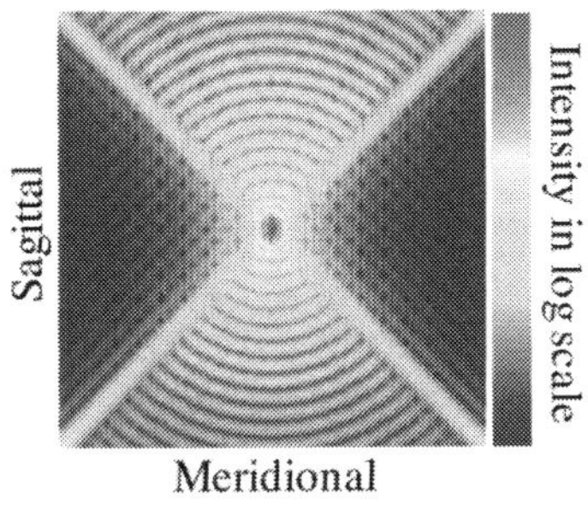

Figure 1. Intensity distribution of beam at focal plane focused by quarter conical TRZP. Calculation area: 400 × 400 nm.

Next, we describe the effect of error in the radius of curvature of the substrate inner surface on the focused beam spot size. When the radius of curvature differs from its designed value, the focal position in the sagittal direction is shifted from that in the meridional direction under the designed alignment conditions. It is, however, possible to achieve point nanofocusing by optimizing both the incident angle and X-ray energy. First, the optimal incident angle can be calculated by substituting the actual radius of curvature into Eq. 2. The optimal incident X-ray energy is calculated by substituting the optimal incident angle into Eq. 1 and solving for λ. Figure 2 shows the relationship between the optimal incident angle and incident X-ray energy calculated for the open conical TRZP. When the TRZP had a 10% larger radius of curvature than that of the design, the optimal incident angle and X-ray energy were calculated to be 6.6 mrad and 8.26 keV. Figure 3 shows the intensity distribution at the focal plane at the optimal value of incident angle and energy. A calculated beam spot size of 13.5 (M) × 5.0 nm (S) can be expected, which is only 10% larger than the designed one.

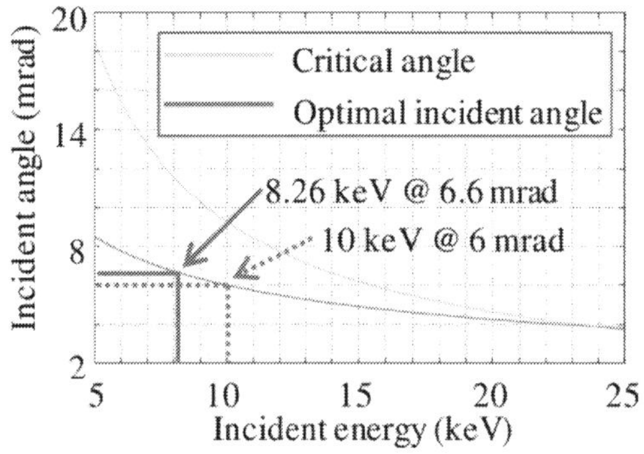

Figure 2. Relation between optimal incident angle and incident X-ray energy.

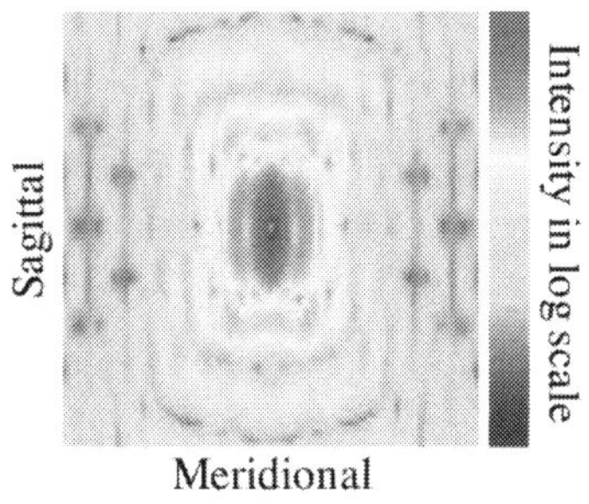

Figure 3. Intensity distribution at focal plane with optimal incident angle and energy. Calculation area: 50 × 50 nm.

We describe the effect of the misalignment of in-plane rotation angle θ_z of the device shown in Fig. 4(a). Figures 4 (b), (c), and (d) show the intensity distribution at the focal plane with misalignment angles of 0, 0.1 and 0.2 mrad, respectively. The misalignment of the in-plane rotation had almost no effect on the meridional beam size. Figure 5 shows the sagittal beam size and beam flux dependence on the in-plane rotation misalignment. At 0.2 mrad, the sagittal beam size increased only 25%, but the integrated flux was reduced by 42%, which is a substantial effect. However, it is easy to align the in-plane rotation angle with an accuracy of ± 0.1 mrad by measuring the sagittal beam size and to find the optimal position.

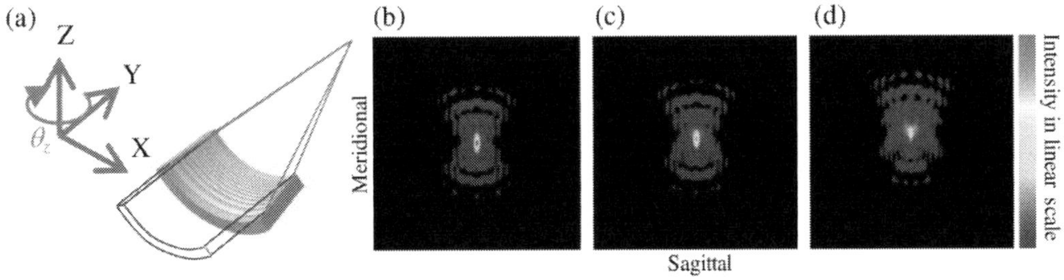

Figure 4. (a) Misalignment of in-plane rotation. Intensity distribution at focal plane at misalignment angle of (b) 0, (c) 0.1 and (d) 0.2 mrad. Calculation areas: 200 × 200 nm.

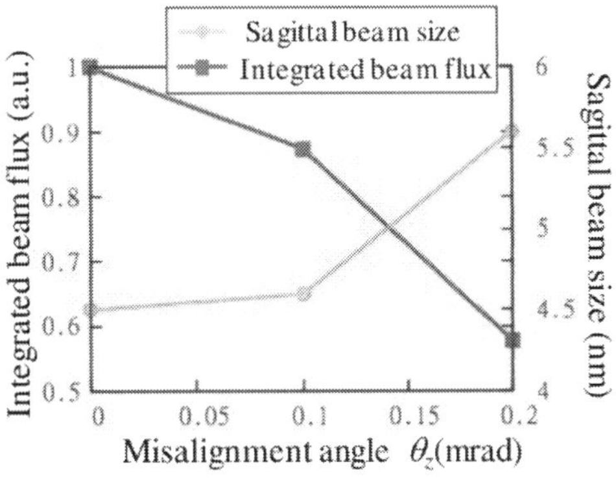

Figure 5. Relation between sagittal beam size and integrated beam flux and misalignment angle.

4. Conclusion

In this paper, we examined point focusing TRZPs from the view point of practical use. We also evaluated the effect of fabrication errors and misalignment for open conical TRZPs by calculating diffraction integrals and describing countermeasures against these errors. We plan to fabricate a prototype point focusing TRZP as further work.

5. Acknowledgment

This work was supported by JSPS KAKENHI Grant Number 23510121.

References

[1] Suzuki Y, Takeuchi A, Takano H and Takenaka H 2005 *Jpn. J. Appl. Phys.* **44** 1994-1998.
[2] Mimura H, Yumoto H, Matsuyama S, Sano Y, Yamamura K, Mori Y, Yabashi M, Nishino Y, Tamasaku K, Ishikawa T and Yamauchi K 2007 *Appl. Phys. Lett.* **90** 051903.
[3] Kang H C, Maser J, Stephenson G B, Liu C, Conley R, Macrander A T and Vogt S 2006 *Phys. Rev. Lett.* **96** 127401.
[4] Koyama T, Takenaka H, Ichimaru S, Ohchi T, Tsuji T, Takano H and Kagoshima Y 2011 *AIP Conf. Proc.* 1365 24-27.
[5] Mimura H, Handa S, Kimura T, Yumoto Y, Yamakawa D, Yokoyama H, Matsuyama S, Inagaki K, Yamamura K, Sano Y, Tamasaku K, Nishino Y, Yabashi M, Ishikawa T and Yamauchi K 2010 *Nat. Phys.* **6** 122-125.
[6] Takano H, Tsuji T, Hashimoto T, Koyama T, Tsusaka Y and Kagoshima Y 2010 *Appl. Phys. Express* **3** 076702.
[7] Takano H, Tsuji T and Kagoshima Y 2011 *Proc. of SPIE* **8139** 81390D-2.

Status of the Nanoscopium Scanning Hard X-ray Nanoprobe Beamline of Synchrotron Soleil

A Somogyi, CM Kewish, M Ribbens, T Moreno, F Polack, G Baranton, K Desjardins, and JP Samama

Synchrotron Soleil, L'Orme des Merisiers, Saint-Aubin - BP 48, 91192 GIF-sur-YVETTE CEDEX

E-mail: somogyi@synchrotron-soleil.fr

Abstract. The Nanoscopium 155 m-long scanning hard X-ray nanoprobe beamline of Synchrotron Soleil (St Aubin, France) is dedicated to quantitative multi-modal 2D/3D imaging. The beamline aims to reach down to 30 nm spatial resolution in the 5-20 keV energy range. Two experimental stations working in consecutive operation mode will be dedicated to coherent diffractive imaging and scanning X-ray nanoprobe techniques. The beamline is in the construction phase, the first user experiments are expected in 2014. The main characteristics of the beamline and an overview of its status are given in this paper.

1. Introduction

The Nanoscopium scanning hard X-ray nanoprobe beamline is dedicated to quantitative multi-modal imaging on the submicron and nano-length scales. Two experimental stations, CSI and Nanoprobe, are dedicated to coherent scatter imaging and X-ray nanoprobe techniques, respectively, in the 5-20 keV energy range. Typical scientific applications cover fields such as biology and life sciences, and earth sciences, environmental sciences and geobiology. All of these scientific communities are aiming to study heterogeneous samples at multiple length scales; to obtain information on the specimen structure, its major, minor, and trace element content, and the chemical speciation. We will enable this multiscale and multitechnique imagine by scanning in fast continuous and high precision step scan modes. Cryogenic cooling of radiation-sensitive, fragile samples will be implemented at the beamline.

2. Characteristics and status of the Nanoscopium beamline

2.1. Source, frontend and first beam-shaping elements

The Nanoscopium beamline [1] shares a 12 m-long straight section of the Soleil storage ring with the future Tomography beamline. The dedicated source optics providing double low vertical beta functions [2] was installed in 2011. This was followed by the implementation of a prototype cryo-cooled U18 $Pr_2Fe_{14}B$ permanent magnet undulator [3] for the commissioning period of Nanoscopium, in canted geometry, with the undulator source of Tomography.

The optical scheme of the beamline can be seen in figure 1. The 800 x 800 μm^2 fixed-opening front end (FE) aperture, the primary slits, and the beamline filters are situated in the FE Hutch. The size of the FE aperture is matched to the maximum beam acceptance of the nano-focusing optics of the experimental stations.

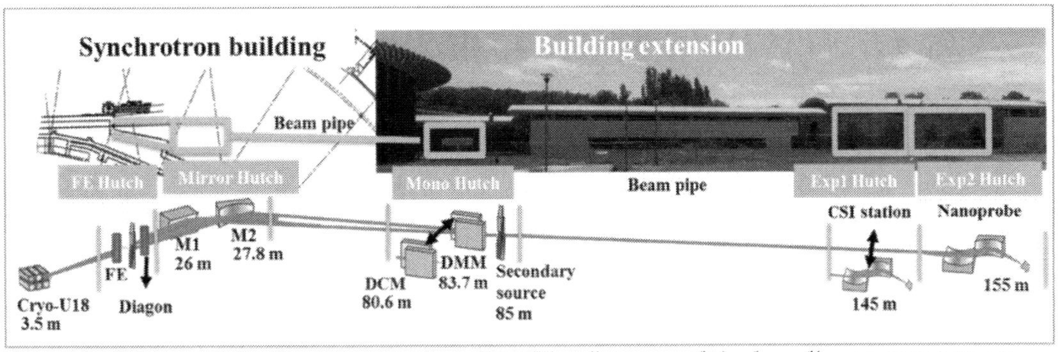

Figure 1. Scheme of the Nanoscopium beamline. The distances of the beamline components are given from the center of the straight section. The source is a cryocooled U18 undulator, on the left. Proceeding downstream, the other components are the front-end (FE), filters and slits, a Diagon, white-beam focusing mirrors M1 and M2, double-crystal and double-multilayer monochromators (DCM and DMM, respectively) and then the experiment hutches CSI and Nanoprobe. See text for details.

The FE aperture was centered on the emission axis of the undulator using a hard X-ray version of the Diagon device [4], which images Bragg diffraction from a water-cooled Si220 crystal at an angle of 45±1 degrees relative to the incoming white beam. The emission cone of the undulator is imaged at 4.57±0.08 keV or at 13.7±0.2 keV defined by the 1st- or 3rd-order Si(220) reflections, respectively (figure 2). This also allows a comparison between the theoretical and measured undulator emission, for diagnostic purposes.

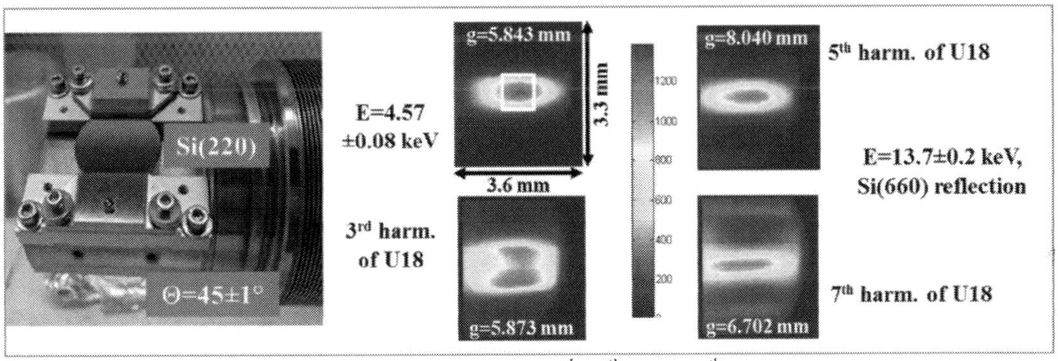

Figure 2. Te "Diagon" device and the measured 3rd, 5th, and 7th undulator harmonics. The corresponding gap values, g, are indicated in the figure. The ~ 4.6 x 4.6 μm^2 effective pixel size is determined by the Basler camera of the Diagon. The white rectangle shows the size and position of the 800 x 800 μm^2 FE aperture aligned at the beam axis. The beam shape was obtained by scanning the FE diaphragm in the ±2 mm range.

2.2. Main optical elements of the beamline

The two 400 mm long, fixed curvature mirrors, M1 and M2 (Winlight), work at 2.5 mrad grazing incidence angle. They are situated in the Mirror Hutch inside the synchrotron experiment hall (figure1). The mirrors will create an image of the source at the location of the secondary source, 85 m from the center of the straight section. M1 is a sagittally-focusing mirror with 81.7 mm sagittal radius with a Rh coated optical surface. The Si and Rh coatings of the M2 tangentially-focusing mirror (with 21.4 km measured tangential radius) ensures the rejection of higher order harmonics from the undulator spectrum. The cumulative effect of the measured tangential residual shape-errors of M1 and

M2 [5] (figure 3a) on the horizontal beam-profile (figure 3b) was calculated by ray-tracing calculations [6] at the secondary source (85 m) and at the positions of the focusing optics of the CSI (145 m) and Nanoprobe (155 m) stations. The homogenous beam profile shown in figure 3c overfilling the apertures of the focusing optics (typically 200-500 μm) is one of the requirements for stable beamline operation.

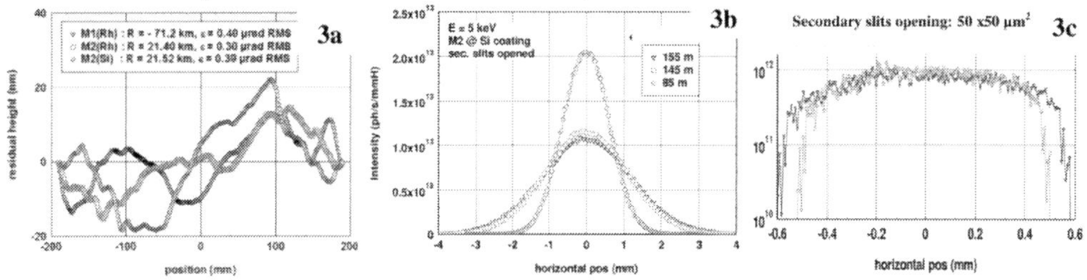

Figure 3a. Measured (by long trace profiler, LTP) tangential residual height of the coated M1 and M2 prefocusing mirrors [5]. **3b.** The horizontal beam profile at 85 m (secondary source), at 145m (CSI station) and at 155 m (Nanoprobe) calculated by ray tracing taking into the measured mirror quality. **3c.** Horizontal beam profile at the position of the focusing optics of the CSI (red curve) and Nanoprobe (black curve) stations with 50 x50 μm^2 secondary source size. See text for details

The two interchangeable monochromators, a Si(111) double-crystal (DCM, CINEL) and a double-multilayer (DMM) one, together with the secondary source will be situated in the Monochromator Hutch in the building extension (figure 1). In keeping with the all-horizontal geometry, both monochromators will be used in π-polarization mode. The secondary source will be determined by high precision slits and pinholes, in the μm-mm opening range. They will be situated after the monochromators, 85 m from the center of the straight section. All the equipments, except the DMM, are under construction; their installation is expected for the beginning of 2013. The DMM, allowing for high flux measurements is in the design phase.

2.3. Experimental stations and methods

Two consecutive experimental stations will be situated in the 10 m-long concrete hutches in the external building at ~60 and ~70 m distances from the secondary source (figure 1). The experimental setup optimized for scanning coherent scatter imaging methods, such as ptychography [7] can be found in the first, CSI, Hutch. Quantitative 2D/3D images based on electron density contrast, and nanostructure orientation, will be offered by these coherent imaging methods. Scanning X-ray spectro-microscopy methods will be situated in the second, Nanoprobe, Hutch offering elemental mapping at trace levels (by X-ray fluorescence, XRF), speciation mapping (X-ray absorption near edge structures, XANES), phase gradient mapping (scanning differential phase contrast imaging) with down to ~30 nm spatial resolution.

The use of achromatic total-reflection Kirkpatrick-Baez (K-B) mirrors at both end-stations will ensure high flux in the probing nanobeams, and facilitating easy energy tuning in the full energy range without realignment of the beamline. The ultra-high optical quality of these elliptically figured K-B mirrors is crucial both for wave-front preservation and for diffraction limited focusing. Wave-optical calculations [8] have been carried out to specify the required surface quality and alignment tolerances of the KB mirrors in the 5-20 keV energy range using four figures-of-merit to characterize the focused wavefield; namely, focal spot intensity full-width at half-maximum (FWHM) in the lateral (beam size) and longitudinal (depth-of-field) directions, integrated intensity in the FWHM, and finally the presence of significant features, or "side-lobes" in the vicinity of the focus. The measured optical and mechanical characteristics of the Rh-coated K-B mirror-pair (mirrors manufactured by JTEC Corp., Osaka, Japan [9]), and the ultra-high precision UHV positioning system (designed and manufactured by Bruker ASC) indicate tolerances ensuring diffraction-limited focusing at the Cryo-Nanoprobe station. This system is to be delivered at the end of 2012.

Journal of Physics: Conference Series **463** (2013) 012027 doi:10.1088/1742-6596/463/1/012027

In order to offer multiscale and multimodal imaging, both continuous scanning with millisecond dwell times and step-scan mode operation with nm positioning precision will be implemented at the beamline with simultaneous multiple detections. This will enable imaging large fields of view with moderate spatial resolution and sensitivity, followed by high spatial resolution and high sensitivity mapping of small volumes of interest. Fast multi-technique experiments, namely scanning X-ray fluorescence spectrometry combined with absorption, differential phase contrast, and dark-field imaging will provide statistically significant quantitative imaging possibilities. The corresponding distributed fast-acquisition system for synchronized multi-technique experiments (FLYSCAN) is being prototyped at SOLEIL (see K. Medjoubi *et al* [10]).

3. Conclusion and Outlook

The Nanoscopium beamline is in the final phase of construction. The external building will be finished in Nov. 2012, which will be followed by the installation and commissioning of the optical elements, such as monochromators and secondary slits, and the experimental stations in 2013. The beamline will be open for users in 2014.

Acknowledgments

The authors would like to acknowledge the Colleagues and Support Groups of Soleil for their significant contribution, continuous help, and outstanding technical support to the Nanoscopium project.

References

[1] Somogyi A *et al* 2011 in XRM2010, The 10th International Conference on X-ray Microscopy, Eds.: I. McNulty, C. Eyberger and B. Lai, *AIP Conf. Proc.* **1365** 57
Somogyi A *et al* 2010 in 10th International Conference on Radiation Instrumentation, Eds.: R. Garrett, I . Gentle, K. Nugent and S. Wilkins, *AIP Conf. Proc.* **1234** 395
[2] Loulergue A *et al* 2010 in Proceedings of IPAC'10, Kyoto, Japan, *WEPEA011* 2496
Brunelle P *et al* 2011 in Proceedings of IPAC2011,San Sebastian Spain, *WEPC050* 2124
[3] Benabderrahmane C *et al.* 2012 *Phy. Res. Sec.* A **669** 21
[4] Desjardins K *et al* 2007 CP879 in *Synch. Rad. Inst.*: Ninth International Conference, Eds: J. Y. Choi and S. Rah, American Institute of Physics
[5] M. Thomasset et al., Rapport de métrologie du Miroir Nanoscopium M1 v.4, Soleil internal report 19.07.2012; Rapport de métrologie du Miroir Nanoscopium M2 v.1, Soleil internal report 19.07.2012.
[6] Moreno T *et al* 2001 *J. de Physique IV* (Proceedings) **11** Pr2
[7] Thibaut P *et al* 2008 *Science* **321** 379;
Dierolf M *et al* 2010 *Nature* **467**, 436-440
[8] Kewish CM *et al* 2007 *Applied Optics* **46** 2010
Kewish CM *et al* 2007 *NIM* A **582** 138
[9] Yamauchi K *et al* 2002 *Rev. Sci. Instrum.* **73** 4028
[10] Medjoubi K *et al* submitted to this prceedings;
Medjoubi K *et al* submitted to for publication

Journal of Physics: Conference Series **463** (2013) 012028 doi:10.1088/1742-6596/463/1/012028

Hard X-ray Imaging Microscopy using X-ray Guide Tube as Beam Condenser for Field Illumination

Yoshio Suzuki, Akihisa Takeuchi, Kentaro Uesugi, Yasuko Terada

JASRI/SPring-8, Koto 1-1-1 Sayo, Hyogo 679-5198, Japan

Hiromoto Nakazawa,

NIMS, Tsukuba, Ibaragi 305-0047, Japan

Sumito Ohzawa, Tomoki Aoyama, Hajime Nii, Katsumi Handa

Horiba, Ltd. Minami-ku, Kyoto 601-8510, Japan,

yoshio@spring8.or.jp

Abstract. An optical system for illumination of object in x-ray imaging microscopy is developed. The optical system is a beam condenser consisting of a single-bounce conical-shape mono-capillary (x-ray guide tube: XGT) made of Pyrex glass. The XGT condenser was tested at the beam line 47XU of SPring-8 using a Fresnel zone plate as an objective lens. Comparing with the microscope without beam condenser, the flux density is improved by a factor of 12-20 in the x-ray energy range of 6-8 keV. Test patterns with a 50 nm-structure are clearly resolved at 8 keV with an exposure time less than 1 s.

1. Introduction

The illumination optics for imaging microscope is as important as an objective lens. The imaging properties, resolution and contrast, is determined not only by the performance of objective but also by the illumination condition. Off-axis parallel beam illumination is sometimes used in x-ray imaging microscopy with synchrotron radiation x-ray sources. However, the parallel beam illumination is generally improper way for absorption-contrast imaging because of its asymmetric feature and non-uniformity of imaging properties. Although beam diffuser (random phase plate) is often used to improve the imaging properties in off-axis parallel beam illumination, the beam diffuser is also an origin of stray light, which causes reduction of image contrast. A hollow cone illumination is preferable for imaging microscope with Fresnel zone plate (FZP) objective. One of the best illumination optics for the FZP microscope with low-emittance synchrotron radiation light sources is hollow-cone Köhler illumination which was realized with a rotating double-mirror system developed by Niemann et al [1]. We have already developed Köhler illumination system for FZP imaging microscope using concentric periodic-grating optics or combination of linear grating (sector condenser zone plate) [2-4] instead of the rotating double-mirror optics, because the alignment of condenser zone plate is much easier than that of rotating double-mirror system.

The sector condenser zone plate is now usually used for imaging microscopy and microtomography experiments at the beam line 47XU of SPring-8. However, the diffraction efficiency of condenser zone plate is 20-10% in the energy range of 6-12 keV. Improvement of efficiency is one of the most important issues for the microscopy and microtomography experiments. The chromatic aberration of the grating condenser optics is also a disadvantage of diffraction optics. The x-ray guide tube (XGT), a total-reflection-mirror optics consisting of inner surface of tapered glass capillary [5. 6], is a candidate of high-efficiency beam condenser because of high reflectivity of external-total-reflection in the hard x-ray region. We have tried to apply the XGT to an illumination system for FZP microscope. In this report, the design and optical properties of the XGT condenser is described, and the results of performance test at the beam line 47XU of SPring-8 are shown.

2. Experimental Setup

The schematic drawing of x-ray imaging microscope optics with the XGT condenser is shown in Fig. 1. The XGT used in the experiment is a single-bounce conical-shape mono-capillary. Some x-ray imaging microscopes operated in the synchrotron radiation facilities are equipped with an axisymmetric total-reflection mirror whose reflecting surface is ellipse of rotation or paraboloid for the condenser system [7, 8]. We chose conical-shape capillary in order to achieve the hollow-cone Köhler illumination. The details of the XGT structure are shown in Fig. 2. The conical-shape tapered-XGT is made of Pyrex glass tube drawn in a furnace under the numerically controlled condition. The XGT usually tend to bend, and the error in straightness was about 0.1 mm or more. In order to correct the bend of XGT, we used a precise V-groove (straightness better than 10 μm), and the XGT is mildly pressed on the V-groove. Hence, the straightness of XGT is estimated to become better than 10 μm. The length of XGT is about 210 mm. The entrance diameter and the exit diameter are 790 μm and 590 mm, respectively. When an illuminating x-ray beam is parallel, the convergent angle of output beam is calculated to be 1.9 mrad from the measured shape of the capillary. The angular divergence of illuminating beam is estimated to be about 10 μrad at the beam line 47XU of SPring-8. Therefore, the parallel beam illumination to the XGT condenser is approximately satisfied.

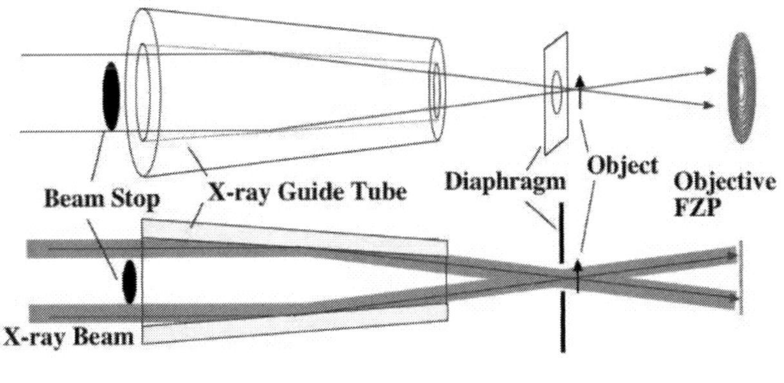

Fig. 1. Schematic diagram of optical system.

A center beam stop made of 0.5 mm-diameter lead ball is placed in front of the XGT condenser in order to stop the direct beam on the object and the objective FZP, and a diaphragm with a diameter of 50 μm is placed in front of the object to define the illuminating field. The hollow-cone Köhler illumination is achieved by using the conical mono-capillary XGT combining with the beam stop and the diaphragm. In the present experiment, the objective zone plate is conventional FZP fabricated at NTT Advanced Technology by electron-beam lithography and reactive ion etching. The zone material is tantalum with a thickness of 0.5 μm. The outermost zone width of FZP is 50 nm, and the diameter is 155 μm. Therefore, the focal length is 50 mm at 8 keV, and 37.5 mm at 6 keV. The numerical aperture (*NA*) is 1.55×10^{-3} at 8 keV, and 2.07×10^{-3} for 6 keV x-rays, respectively. The spatial resolution depends on the convergent angle of illuminating beam. The detection limit for periodic object (grating-like like object) is λ/NA in case of normal-incidence parallel-beam illumination, and the best resolution (detection-limit of $0.5\lambda/NA$) is attained for the convergent angle of $2NA$, In contrast, when a convergent angle of illuminating beam is equal to the NA of objective FZP, the maximum field of view is obtained, and the detection limit for periodic object becomes $2\lambda/(3NA)$ in the full-pitch periodic length. The convergent angle of the XGT condenser, 1.9 mrad, is matched with the numerical aperture of the used FZP to optimize the field of view at the x-ray energy region of 6-8 keV. The object and field-defining diaphragm are placed about 320 mm from the middle point of XGT.

The x-ray microscope setup in this experiment is essentially the same as that of our previous experiment except for the XGT condenser system [2-4]. The imaging detector is placed 7 m from the objective FZP. Therefore, the magnification of x-ray optics is about 140 at 8 keV, and 187 at 6 keV. The total length of x-ray microscope is limited by the length of the experimental hutch at the beam line 47XU. The imaging detector is an indirect-sensing-type x-ray camera consisting of a fine-powder phosphor screen (P43: Gd_2O_2S: Tb+, 10 μm thick), optical lens system (magnification of about 4), and a cooled charge-coupled-device (CCD) camera. The format of the CCD is 4000 x 2624 pixels, and the pixel size is 5.9 μm. The CCD is operated in a 2 x 2 binning mode. Therefore, the effective pixel of CCD is 2000 x 1312 with a pixel size of 11.8 μm. Then, the convergent pixel size of imaging detector defined by the optical lens system is 3.1 μm, and the effective pixel size in the x-ray microscope is 22.1 nm for 8 keV x-rays, and 16.6 nm for 6 keV x-rays, respectively. The XGT and the FZP were set in a He-flow vessel in order to suppress the contamination and the radiation damage.

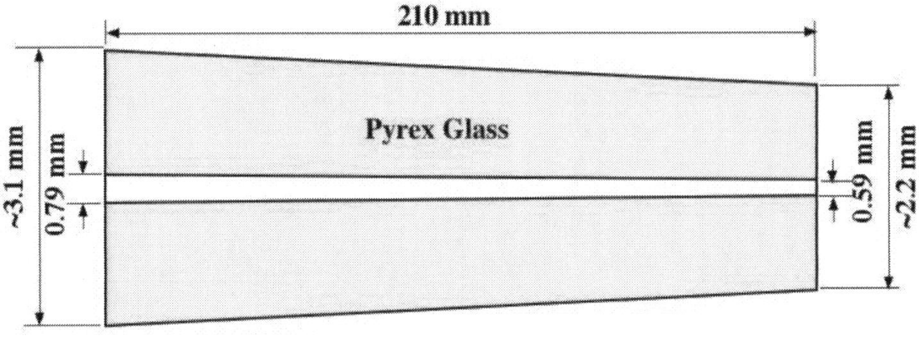

Fig. 2. Structure of conical-shaped XGT. Cross-sectional drawing.

3. Results

Fig. 3 shows an example of x-ray microscopy image taken at an x-ray energy of 8 keV with a magnification of about 140. A weak beam diffuser made of a sheet of letter paper is used for suppression of speckle noises. The specimen is resolution test patterns made of tantalum with a thickness of 0.5 μm (x-ray chart: XRESO-50HC, NTT Advanced Technology). The innermost structure of the Siemens star pattern shown in the magnified image of Fig. 3 is 50 nm-line and 50 nm-space. The finest patterns are observable in the measured image as shown in the figure. The exposure time of the image shown in Fig. 3 is 0.7 sec that is about 1/12 of the exposure time without the XGT condenser. Fig. 4 shows an x-ray microscopic image of the same x-ray chart taken at 6 keV. The magnification is 187, and the diffuser is not used in this case. The exposure time is 5 s, which is much longer than that at 8 keV, because of absorption loss by x-ray windows and air. However, the improvement of flux density is about 20, comparing with the exposure time without the XGT condenser. The finest patterns of 50 nm line and space are clearly observable in the measured image. No radiation damages are detected after a continuous 10 hours exposure to the x-ray beam at the beam line 47XU of SPring-8.

The XGT condenser is considered to be very useful for illumination system of full-field imaging microscopy in the hard x-ray region for its high efficiency and achromatic feature.

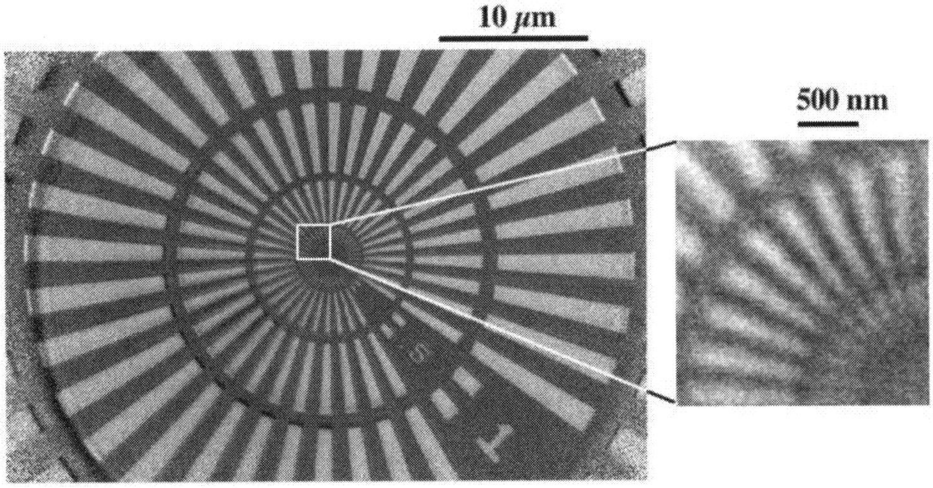

Fig. 3. X-ray microscopy image of resolution test pattern. X-ray energy: 8 keV. Exposure time: 0.7 s.

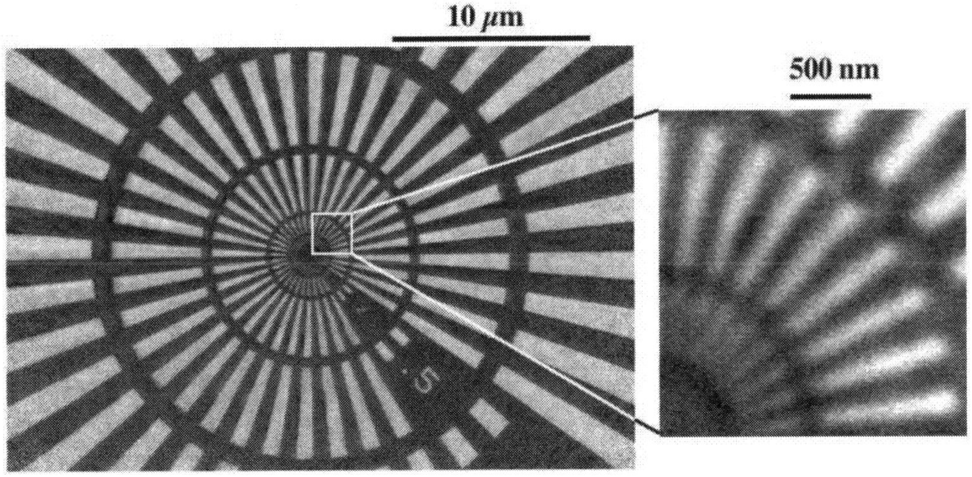

Fig. 4. X-ray microscopy image of test pattern taken at 6 keV. Exposure time is 5 s.

4. Acknowledgements

The experiment has been done with an approval of SPring-8 Proposal Review Committee (Proposal No. 2011A1322).

References

[1] B. Niemann, et al., AIP Conf. Proc. **507** (2000) 440.
[2] A. Takeuchi, K. Uesugi, and Y. Suzuki, J. Phys. Conf. Ser. **186** (2009) 012020.
[3] A. Takeuchi, Y. Suzuki, and K. Uesugi, AIP Conf. Proc. **1365** (2011) 301.
[4] Y. Suzuki, A. Takeuchi, K. Uesugi, and M. Hoshino, AIP Conf. Proc. **1365** (2011) 160.
[5] H. Nakazawa, J. Appl. Cryst. **16** (1983) 239.
[6] H. Nozaki and H. Nakazawa, J. Appl. Cryst. **19** (1986) 453.
[7] S. Rehbein, et al., Phys. Rev. Lett. **103** (2009) 110801.
[8] G. Yin, et al., Appl. Phys. Lett. **89** (2006) 221122.

Optomechanical Design of a Multilayer Laue Lens Test Bed for 10-nm Focusing of Hard X-rays

Deming Shu[1], Evgeny Nazaretski[2], Jungdae Kim[2a], Hanfei Yan[2], Kenneth Lauer[2], Brian Mullany[2], Dennis Kuhne[2], Jörg Maser[1], and Yong S. Chu[2]

[1]Advanced Photon Source, Argonne National Laboratory, Argonne, IL 60439, USA
[2]NSLS-II, Brookhaven National Laboratory, Upton, NY 11973, USA

Email: shu@aps.anl.gov

Abstract. A multilayer Laue lens (MLL) test bed was designed at the Argonne National Laboratory for the National Synchrotron Light Source-II (NSLS-II) project at Brookhaven National Laboratory. Argonne and Brookhaven scientists and engineers collaborated to carry out this work in order to achieve NSLS-II's R&D goals of 10-nm focusing of hard x-rays. The test-bed design was based on the experience gained from the Argonne nanopositioning system designed for the CNM/APS hard x-ray nanoprobe project and an MLL test bed at Advanced Photon Source (APS) beamline 26-ID [1-3]. Optomechanical design of the MLL test bed, as well as the preliminary test results for the alignment apparatus and sample base stages' mechanical vibration performances are presented in this paper.

1. Introduction

In 2006, H. C. Kang et al. reported on the development of hard x-ray nanofocusing with linear multilayer Laue lenses (MLLs). Using tilted partial MLL structures, a one-dimensional focus as small as 16 nm with efficiencies up to 44% was demonstrated using synchrotron x-rays with a photon energy of 19.5 keV [1]. These MLLs are intrinsically suitable for focusing hard x-rays to a few nanometers.

To perform ideal 2D focusing, the first prototype of a precision multidimensional alignment apparatus was designed and constructed for an MLL test bed at the APS and tested at beamline 26-ID in 2008 [2,3]. With this test bed, MLL optics demonstrated the focusing capability of hard x-rays to a 2-D focus of 25 nm horizontal x 27 nm vertical FWHM at a photon energy of 12 keV, and of 25 nm horizontal x 40 nm vertical FWHM at a photon energy of 19.52 keV [4].

In 2010, a second MLL test bed was designed at Argonne National Laboratory (ANL) for the National Synchrotron Light Source-II (NSLS-II) nanopositioning development project at Brookhaven National Laboratory (BNL). This work was carried out as collaboration between Argonne and Brookhaven scientists and engineers, in an effort to achieve NSLS-II's R&D goals of achieving 10-nm focusing of hard x-rays. The second MLL test bed has been installed, and commissioning began at a long imaging beamline experimental station (I13L) at the Diamond Light Source (DLS), UK, in April 2012. It will be used as an initial platform for further nanofocusing developments in collaboration with the DLS staff. Figure 1 shows a 3-D model for the MLL test bed for 10-nm focusing of hard x-rays. It includes a precision MLL multidimensional alignment apparatus (1) and a sample stages group (2). Optomechanical design of the second MLL test bed for NSLS-II is presented in this paper.

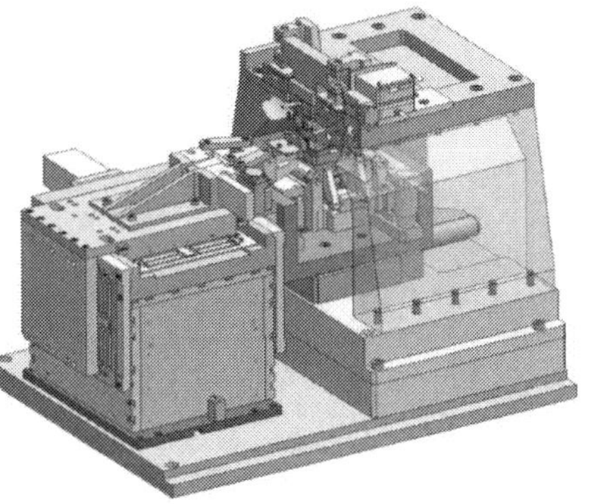

Figure 1. A 3-D model for the MLL test bed for 10-nm focusing of hard x-rays. It includes a precision MLL multidimensional alignment apparatus and a sample stages group.

2. Design of the multidimensional alignment apparatus for MLL 2-D focusing

To perform 2-D focusing, the MLL test bed needs to provide capability to align precisely two MLLs with respect to each other in a total of eight degrees of freedom. Figure 2 (left) shows a schematic diagram of precision motions required for alignment of two MLL sections for a 2-D x-ray focusing. As shown in Figure 2 (right), the multidimensional alignment apparatus for the second MLL test bed has the same mechanical design as the first one [3]. To minimize the thermal energy dissipation on the alignment apparatus, the PI^{TM} M-663 stages applied for the second MLL test bed are customized to relocate their encoder interpolators from the stages to the controllers.

In the alignment apparatus, the upstream MLL lens is mounted on the lower group of stages, which provides precision positioning in five degrees of freedom that include three translational and two angular adjustments as shown in Figure 3 (left). The downstream MLL lens is mounted on the upper group of stages that provide two translational and one rotational degree of motion as shown in Figure 3 (right). Table 1 summarizes the design specification for each adjustment [2].

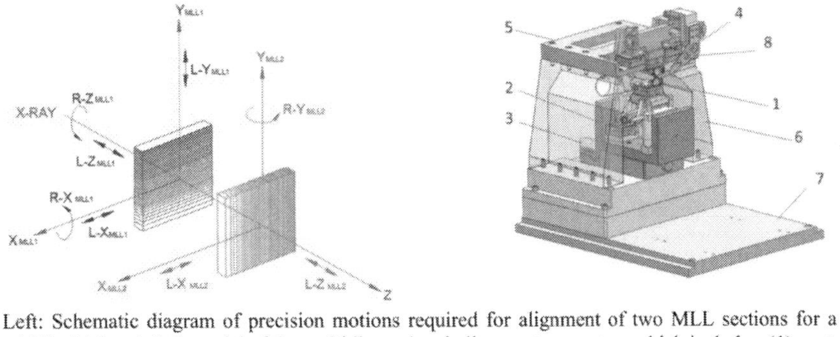

Figure 2. Left: Schematic diagram of precision motions required for alignment of two MLL sections for a 2-D x-ray focusing test-bed [2]. Right:: A 3-D model of the multidimensional alignment apparatus, which includes: (1) upstream MLL, (2) nested flexural bearing structure, (3) PI^{TM} M-126 motorized horizontal linear stage, (4) downstream MLL, (5) upper stages group, (6) support structure, (7) base plate, and (8) OSA and stages.

TABLE 1. Design specifications for alignment motions for a 2-D x-ray focusing prototype with two MLLs in a crossed configuration.

Motion	Driver	Resolution	Travel Range
$L\text{-}X_{MLL1}$	Ultrasonic piezomotor with linear optical encoder	0.1 micron	15 mm
$L\text{-}Y_{MLL1}$	DC motor with gearhead and shaft optical encoder	0.1 micron	5 mm
$L\text{-}Z_{MLL1}$	Stepping motor with gearhead and shaft optical encoder	1 micron	25 mm
$R\text{-}X_{MLL1}$	Differential driven DC motor with gearhead and shaft optical encoder	0.4 arcsec	3 degrees
$R\text{-}Z_{MLL1}$	Ultrasonic piezomotor with linear optical encoder	0.4 arcsec	3 degrees
$L\text{-}X_{MLL2}$	Ultrasonic piezomotor with linear optical encoder	0.1 micron	15 mm
$L\text{-}Z_{MLL2}$	Dual ultrasonic piezomotor with linear optical encoder	0.1 micron	15 mm
$R\text{-}X_{MLL2}$	Differential driven ultrasonic piezomotor with linear optical encoder	0.3 arcsec	3 degrees

Figure 3. Left: A 3-D model of the nested flexural bearing structure for angular adjustments along the axes of XMLL1 and ZMLL1. The nested flexural bearing structure is operated by two PI™ M-110 motorized linear stages (1) and (2), and two PI™ M-663 PZT-driven linear stages (3) and (4). The linear stages (1) and (2) are mounted on a U-shaped base frame (5). Cooperated with the outer flexural bearing structures (6) and (7), linear stages (1) and (2) provide linear positioning in the Y direction and tilting adjustment around the X-axis for another U-shaped frame (8), which is the base for a 4–bar flexural bearing structure (9) nested in the outer flexural structure. The 4–bar flexural bearing structure is operated by the PZT-driven linear stage (3) to provide a precise angular positioning around the Z-axis. The linear stage (4), mounted on the top of the 4–bar flexural bearing structure, aligns the upstream MLL (10) in the X direction. Right: A 3-D model of the upper stages group, which includes three PI™ M-663 PZT-driven linear stages. A pair of linear stages (1) and (2) drive a carriage (3) through a set of flexural bearings (4) to perform linear positioning in the Z-axis direction with tilting adjustment around the Y-axis. The linear stage (5) provides the alignment in the X direction for the downstream MLL (6).

3. The flexural sample base stages for the MLL test bed

As shown in Figure 1, the sample stages group (2) for the MLL test bed provides one angular and three linear scanning capabilities for analyzing the x-ray focal spot at the location of a specimen. As in the flexural sample base stages T8-31/32 developed in 2008 at APS [5,6], the design of the vertical and horizontal flexural stages T8-33/34 for the NSLS-II sample stages group are based on experiences from developing high-stiffness laminar overconstrained weak-link mechanisms for ultrahigh-resolution monochromators and nanopositioning stages at the APS [6,7]. The T8-33/34 linear stages are guided with the same fishbone-shaped multiple parallelogram weak-link modules used for T8-31/32 [6]. Each of the 2-mm-thick 118 mm x 116 mm weak-link modules is a bound laminar structure with eight thin-metal weak-link sheets configured and manufactured by chemical etching and lithography techniques. The finite element simulation for a 1.61-mm displacement of the carriage shows that, while a 17-N force is applied on the carriage, the maximum Von Mises stress is 974 MPa, which is about 81% of the tensile yield strength of the stainless steel 17-7 PH. The simulation shows that, if the carriage is driven by the same level of push-pull force, the weak-link linear stage will have a travel range of 3.22 mm.

The T8-31 and T8-32 flexural stages were driven by PI Nexline™ N-214 and N-110 PZT actuators. To extend the stage driving force and load capacity, T8-33 and T8-34 stages were designed to be driven by PI Nexline™ N-216 and N-111 PZT actuators with optional high vacuum compatibility as shown in Figure 4 (left). Two sets of T8-33/34 stages were constructed in 2010, one for the APS Sector 2 hard x-ray microscope station with high vacuum compatibility, and one for the NSLS-II MLL test bed. To improve closed-loop positioning repeatability, external capacitive or optical sensors were added on the T8-33/34 stages for the NSLS-II MLL test bed as shown in Figure 4 (right).

Figure 4. Left: A photograph of the sample stages group. Right: A photograph of the sample stages group with MicroSenceTM capacitive position sensors. In both photographs, the original vertical axis tip-tilting stage and associated SmarAct™ SLC-1720-S linear stage as shown in Figure 1 are replaced with a Pizesystem Jena™ Tritor 38-SG scanner and a SLC-1720-S linear stage for high-speed sample scanning applications.

4. Preliminary test and summary

To ensure that the mechanical structural stability of the MLL test bed is suitable for 10-nm focusing of hard x-rays, vibration tests were performed for both the alignment apparatus and the sample base stages. Vibration spectra were measured with PCB™ 356 and 393 accelerometers for the alignment apparatus. Preliminary measurement on the top of the upstream MLL stage showed that the apparatus does not amplify the vibration from the mounting base. Although the sensor used on the top of the upstream MLL stage is a compact sensor (PCB-356) with a noise level higher than 10 nm in the frequency range below 8 Hz, we did not observe any peak above its noise floor [2].

Preliminary vibration test using Vibra-Metrics™ 1030 accelerometers demonstrated that the T8-33 vertical linear weak-link stage with 8-kg load does not amplify the vibration from the mounting base, similar to the T8-31 stage test [5]. Also like the T8-31 stage, the T8-33 vertical linear weak-link linear stage has demonstrated high tilting stiffness and high straightness of trajectory with nanometer-level positioning sensitivity. Its tilting stiffness is better than many commercial motor-driven linear bearing-guiding vertical stages with similar structure sizes, and the amount of the tilt is repeatable within 1-μrad range over the stage's full 3 mm travel range.

The second MLL test bed has been installed, and commissioning began at a long imaging beamline experimental station (I13L) at the DLS, UK, in April 2012, as shown in Figure 5.

Figure 5. A photograph of the MLL test bed for 10-nm focusing of hard x-rays to be commissioned at a long imaging beamline experimental station (I13L) at the Diamond Light Source (DLS), UK. It will be used as an initial platform for further nanofocusing developments in collaboration with the DLS staff.

5. References

[1] H. C. Kang et al., Phys. Rev. Lett. 96, 127401 (2006).
[2] D. Shu, H. Yan, and J. Maser, to be published in Proceedings of SRI-2008, Nucl. Instrum. Methods A, Saskatoon, Canada, 2008.
[3] U.S. Patent granted No. 7,597,475, D. Shu, H. Yan, and J. Maser, Oct. 2009.
[4] H. Yan et al., Optics Express 19 15069-15076 (2011).
[5] D. Shu and J. Maser, Proc. SPIE 7424 74240D-1-9 (2009).
[6] U.S. Patent granted No. 7,994,688, D. Shu and J. Maser, Aug. 9, 2011.
[7] D. Shu, J. Maser, Y. Chu, H. Yan, E. Nazaretski, S. O'Hara, S. Kearneya, J. Anton, J. Quintana, and Q. Shen, American Institute of Physics CP1365,144-147 (2011).

[a] Jungdae Kim's current address: Department of Physics, University of Ulsan,Ulsan 680-749, South Korea.

Acknowledgments

The authors would like to thank Dr. Christoph Rau and his team from the Diamond Light Source, and Steven Kearney and Jayson Anton from ANL and the University of Illinois at Chicago for their help with the MLL test-bed development. This work was supported by the U.S. Department of Energy, Office of Science, under Contract No. DE-AC02-06CH11357. Work at Brookhaven is supported by the Department of Energy, Office of Science, under contract DE-AC-02-98CH10886.

Quantitative analysis methods for three-dimensional microstructure of the solid-oxide fuel cell anode

X Song,[1] Y Guan,[1] G Liu,[1*] L Chen,[1] Y Xiong,[1] X Zhang,[1] and Y Tian[1*]
[1] National Synchrotron Radiation Laboratory, University of Science and Technology of China, Hefei, Anhui 230029, People's Republic of China
[*] E-mail: liugang@ustc.edu.cn; ychtian@ustc.edu.cn

Abstract. The electrochemical performance is closely related to three-dimensional microstructure of the Ni-YSZ anode. X-ray nano-tomography combined with quantitative analysis methods has been applied to non-destructively study the internal microstructure of the porous Ni-YSZ anode. In this paper, the methods for calculating some critical structural parameters, such as phase volume fraction, connectivity and active triple phase boundary (TPB) density were demonstrated. These structural parameters help us to optimize electrodes and improve the performance.

1. Introduction

Solid-oxide fuel cells (SOFCs) have gained great interest among scientists for its clean, economic and environmentally friendly features. The porous composite material of nickel (Ni) and yttria-stabilized zirconia (YSZ) is widely used as the SOFC anode because of its strong electronic conductivity and high catalytic activity. The oxygen (air) flows through the cathode pore channel and reacts with electrons to form oxygen anions. Electrons participating in the reactions must be supplied by the cathode electronic conductor. Oxygen anions are transported to anode by the electrolyte and react with fuel (H2) to produce electrons [1-3]. The electrochemical reactions are active at the TPB where all the three phases coexist [2, 4]. So the TPB length affects the SOFC performance, which indicates that the electrochemical properties of SOFC depend on the electrode microstructure. The reactants are unable to be transported to the reactive sites by isolated phase networks. Therefore, only the TPBs consisting of three connected phases (active TPBs) can keep the electrochemical reaction going on.

Non-destructive imaging method based on X-ray computed tomography with synchrotron radiation (SR-XCT) is a powerful tool to study the porous electrodes [5-8]. Due to the tunable photon energy, a specific element can be identified by spectroscopic imaging which can indicate the presence of a particular atomic species by taking two imaging below and above the element absorption edge [9, 10]. Moreover, SR-XCT is capable of imaging large volume materials due to its large depth of focus and high penetrating power [11-13]. The quantitative analysis of the 3D microstructure is critical for illustrating how and where the triple phases interconnect based on the detailed 3D microstructure information. So quantitative analysis methods were developed for computing the microstructure parameters, such as volume fraction, connectivity of special phase and active TPB density.

2. Experiment

Imaging experiments were performed using Xradia nanoXCT-S100 system on the beamline U7A at National Synchrotron Radiation Laboratory (NSRL) in Hefei, China [14]. The sample was fetched from the used SOFC anode and polished to several tens of microns, then mounted on the tungsten tip holder that was rotated from -90° to 90° at 1° intervals. A position adjacent to the top of specimen (less than twenty microns) was chosen to image. A series of 181 X-ray transmission radiographs was collected and reconstructed as a 3D volume using reconstruction software of Xradia. Ni, YSZ and pore phase can be identified from the images by comparing Ni absorbing contrast in the two imagings taken at photon energy 8.3 keV (below the Ni edge) and 8.38 keV (above the Ni edge) [15]. Since the focal length of a Fresnel zone plate is proportional to the incident X-ray energy [16]. The focal length is 24.101 mm for 8.3 keV and 24.333 mm for 8.38 keV in the imaging system, respectively. We only change the object distance by moving the sample stage for the different photon energy imaging to meet the thin lens formula [17]. The imaging magnification ratio of 8.3 keV to 8.38 keV is about 1.0098 due to object distance changed, which can be negligible in the experiments. According to their different grey levels in two volume data, the individual virtual slices were allowed to be segmented into Ni, YSZ and pore phase.

3. Computational method

A method applied to calculate the phase connectivity and active TPB density is described briefly in this section. Each voxel of the saved 3D volume data was labelled as 1 (Ni), 2 (YSZ) or 3 (Pore). The connectivity of each phase can be estimated in the same implementation procedure. Here take the Ni phase for example. One side of the image is defined as input and the opposite is defined as output. Ni voxels in the input side are chosen as start points and assigned a mark 'CI'.

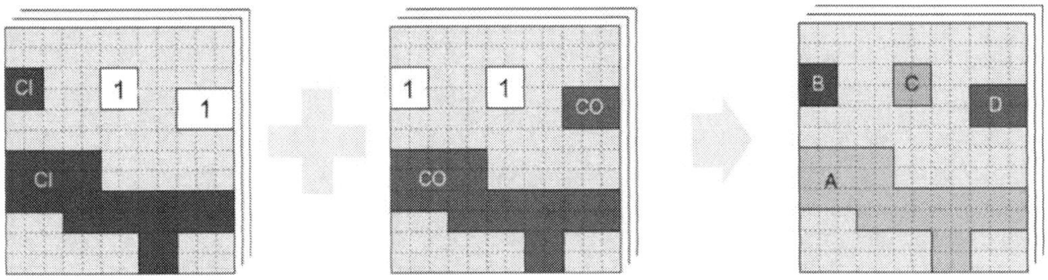

Figure 1. Schematic diagram of the 'burning' method for Ni phase.

Ni voxels adjacent to these marked voxels will be assigned the same mark (see Fig. 1), which is so-called the 'burning' method [18]. All voxels with this mark 'CI' are considered to form a spanning cluster from the input to the output side. Similarly, another spanning cluster assigned another mark 'CO' is found from the output to the input side. By comparing these two spanning clusters, the overlapped regions could be identified as the connected phase (area 'A' in Fig. 1). The remained regions in two spanning clusters are identified as dead-end phase (area 'B' and 'D' in Fig. 1). Ni voxels not in spanning cluster are identified as isolated phase (area 'C' in Fig. 1).

If the neighbouring four voxels in 3D structure contain three labels, 1 (Ni), 2 (YSZ) and 3 (Pore), the common edge of these three voxels is labelled as TPB. The above labels ('connected', 'dead-end' and 'isolated') is applied to identify whether the TPB components are touching the connected networks of each phase. If the TPB lies on connected networks of each phase (Ni, YSZ, pore), it is labelled 'active'. Total active TPBs are obtained via inspecting the whole volume in three directions x, y and z.

4. Results and discussion

A cuboid volume was cropped from the total reconstructed 3D volume and every virtual slice was segmented into three phases (Ni, YSZ and pore) by their gray levels. Figure 2 shows a 3D visual representation of the cropped volume, with Ni phase in red, YSZ phase in yellow and pore phase in blue. The cuboid volume is $8.1\times7.4\times8.1\ \mu m^3$ and the voxel size was $58.3\times58.3\times58.3\ nm^3$.

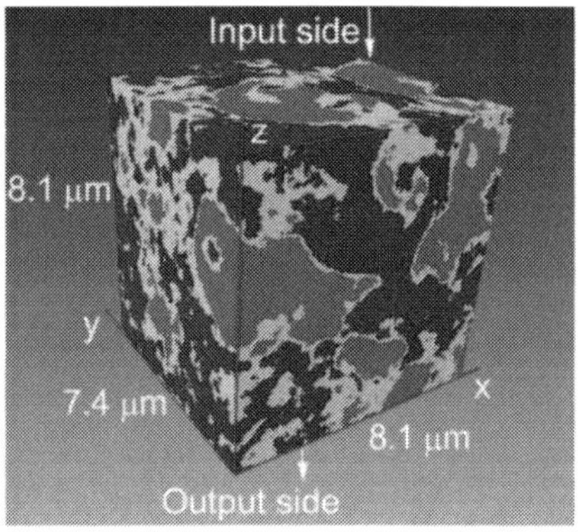

Figure 2. The 3D image view of the anode volume with Ni shown in red, YSZ in yellow, pore in blue. Both input and output sides are perpendicular to the z axis.

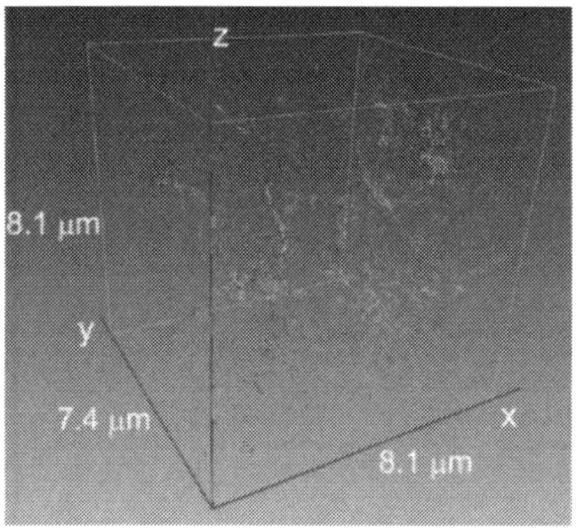

Figure 3. The 3D view of electrochemically active TPBs (green) and other TPBs (red).

The connectivity of each phase plays an important role in the electrochemical reaction of H_2 with O^{2-} to produce electrons. It is expected that only the TPBs that lie on connected networks of each phase (Ni, YSZ and pore) are electrochemically active. Thus, analysing the connectivity of each phase is an indispensable step to obtain the quantity of electrochemically active TPBs. As shown in Table 1, the volume fraction and connectivity of special phase are calculated.

Table 1. Percentages of the different compositions.

	Ni	YSZ	Pore
Volume fraction (%)	23.13	34.87	42.00
Connectivity (%)	92.53	98.49	98.98
Dead-End rate (%)	1.96	0.11	0.21
Isolated rate (%)	5.51	1.40	0.81

The volume fraction is calculated via dividing the volume of a given phase by the total volume. The connectivity, dead-end rate and isolated rate are calculated via dividing the number of voxels with corresponding labels (as mentioned above) by the total given phase volume. Analysing the dead-end and isolated rate of anodes fabricated under different preparation conditions will disclose the variation of microstructures and provide a guide for experiments.

In Figure 3, the active TPBs are marked green to distinguish from the other TPBs (red) in the entire volume. The total TPB density is 6.807×10^5 m cm^{-3}and the active TPB density is 5.598×10^5 m cm^{-3}. These values show slightly less than other reported results [3-5], it may due to that the microstructure of Ni-YSZ anode has been degraded after long-term used. On the basis of the above computation, 82.239 % of the total TPB density is active.

5. Conclusion

In this study, quantitative analysis methods were demonstrated to calculate SOFC microstructural parameters such as volume fraction, connectivity and active TPB density. The results show that the proposed methods help to detect the microstructures of anodes and optimize structure.

Acknowledgments

This work was supported by grants from the 985 project of the State Ministry of Education, the National Natural Science Foundation of China (11275204), and the Knowledge Innovation Program of the Chinese Academy of Sciences (KJCX2-YW-N43).

References

[1] Jiang S P and Badwal S P S 1999 *Solid State Ionics* **123** 209-24
[2] Adler S B 2004 *Chem. Rev.* **104** 4791-843
[3] Wilson J R et al. 2009 *Microsc. Microanal.* **15** 71-7
[4] Wilson J R et al. 2006 *Nat. Mater.* **5** 541-4
[5] Grew K N et al. 2010 *J. Electrochem. Soc.* **157** B783-B92
[6] Guan Y et al. 2011 *J. Power Sources* **196** 1915-9
[7] Shearing P R et al. 2010 *Electrochem. Commun.* **12** 1021-4
[8] Guan Y et al. 2010 *J. Synchrotron Radiat.* **17** 782-5
[9] Scheckel K G, Hamon R, Jassogne L, Rivers M and Lombi E 2007 *Plant Soil* **290** 51-60
[10] Guan Y et al. 2011 *J. Power Sources* **196** 10601-5
[11] Yang Y et al. 2010 *J. Microsc-Oxford* **240** 14-20
[12] Li W J et al. 2009 *Appl. Phys. Lett.* **95** -
[13] Chen J, Wu C Y, Tian J P, Li W J, Yu S H and Tian Y C 2008 *Appl. Phys. Lett.* **92** -
[14] Tian Y C et al. 2008 *Rev. Sci. Instrum.* **79** -
[15] Izzo J R et al. 2008 *J. Electrochem. Soc.* **155** B504-B8
[16] Wang J et al. 2012 *Appl. Phys. Lett.* **100**
[17] Liu Y J et al. 2012 *Anal. Bioanal. Chem.* **404** 1297-301
[18] Bentz D P and Garboczi E J 1991 *Cement. Concrete. Res.* **21** 325-44

Development of fast parallel multi-technique scanning X-ray imaging at Synchrotron Soleil

K. Medjoubi[1], **N. Leclercq, F. Langlois, A. Buteau, S. Lé, S. Poirier, P. Mercère, C. M. Kewish and A. Somogyi**
Synchrotron Soleil, BP48 Saint-Aubin, Gif sur Yvette, 91192, France
E-mail: kadda.medjoubi@synchrotron-soleil.fr

Abstract A fast multimodal scanning X-ray imaging scheme is prototyped at Soleil Synchrotron. It permits the simultaneous acquisition of complementary information on the sample structure, composition and chemistry by measuring transmission, differential phase contrast, small-angle scattering, and X-ray fluorescence by dedicated detectors with ms dwell time per pixel. The results of the proof of principle experiments are presented in this paper.

1. Introduction

Scanning hard X-ray nanoprobe techniques provide the possibility of introducing several detectors around the sample (figure 1). As such, complementary information can be registered simultaneously while the sample is raster scanned in the focused beam; elemental maps can be obtained by using an energy-dispersive X-ray fluorescence (XRF) detector while the defocused beam is registered in transmission by a fast and sensitive two-dimensional (2D) detector providing differential phase contrast (DPC), dark field and transmission images.

The Nanoscopium hard X-ray nanoprobe beamline of Soleil will be dedicated to such multimodal scanning imaging in the 5 to 20 keV energy range [1]. It aims to offer 2D/3D quantitative information on the elemental state composition and electronic density of the sample with high spatial resolution and detection sensitivity. The targets are 30 nm in 3D spatial resolution, and parts-per-million (ppm) to parts-per-billion (ppb) trace element detection. Fast scanning imaging will allow the probing of large fields of view while high precision step-by-step scanning will provide precise local information. The main user communities of the beamline are biology, life sciences, geo-biology, and environmental sciences.

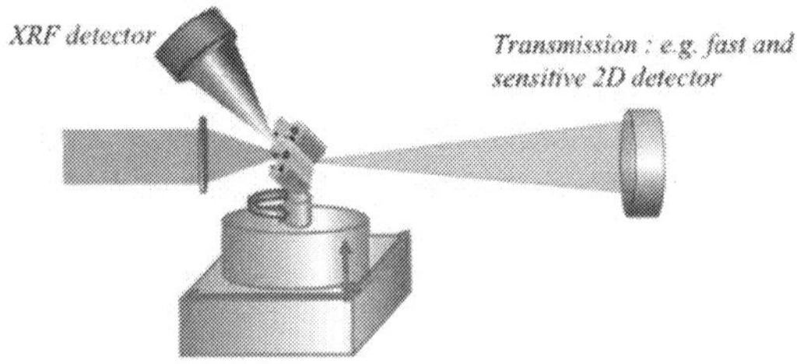

Figure 1. Principle of multi-technique scanning hard X-ray nanoprobe measurements. The X-rays are incident from the left, and the red arrows indicate the degrees of freedom in sample motion required for 3D imaging.

[1] Corresponding author.

Fast simultaneous data acquisition by high data through-put detectors with kHz readout frequencies results in large data volumes of the order of terabytes (1 TB = 2^{40} bytes) during a user experiment. In order to tackle the related technical challenges, a transverse project, named as FLYSCAN, has been prototyped at Synchrotron Soleil. The project aims to provide a flexible, general hardware and software architecture, which is adaptable to all Soleil beamlines for hardware-synchronized fast multi-detector experiments.

2. The FLYSCAN Architecture

The architecture of the FLYSCAN scheme is presented in figure 2. A hardware trigger signal synchronizes the acquisition of the detectors and the readout of the encoded motor positions. All these data are read out simultaneously at each pixel of the continuous scan with an exposure/dwell time determined by the period of the trigger signal.

The data collection of each detector is handled by its own independent TANGO device [2], which takes care of the data conversion and data output into temporary files. An asynchronous process, piloted by the Merger TANGO device, merges these temporary files and the metadata of the experiment conditions into a unique Nexus file [3]. The possibility of reading this file during the acquisition process permits on-line data processing. At the end of the acquisition, the final Nexus file is copied to the central storage space of Soleil, where it is available for off-line data processing.

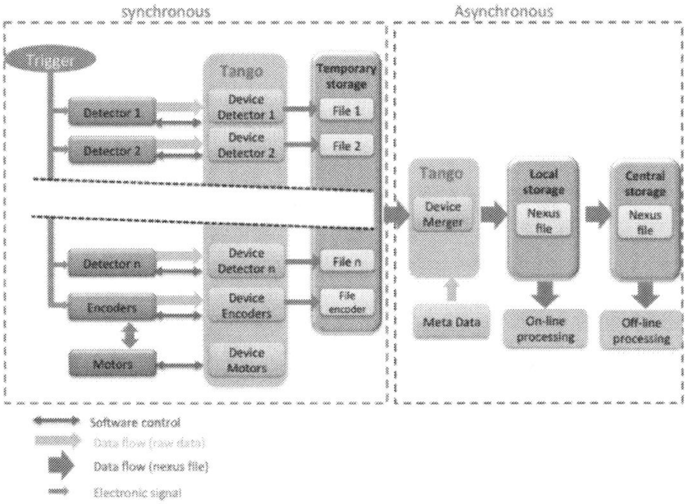

Figure 2. Scheme of the FLYSCAN architecture developed for hardware synchronized multi-technique measurements, divided into synchronous and asynchronous processes.

3. Experimental set-up

Test experiments were carried out to demonstrate the feasibility of the FLYSCAN architecture, at the Metrologie bending magnet beamline [4] using a temporary scanning microprobe set-up. The 14 keV X-ray beam was focused by a Fresnel zone plate (FZP) into a 1.6 µm full-width at half-maximum (FWHM) measured beam size with 6×10^6 photons per second intensity.

During the experiments the XY sample positions transversal to the beam were recorded for each trigger signal. The corresponding defocused transmitted beam pattern was measured at each scan position by a photon counting pixel detector placed 4.41 m behind the sample. The single module XPAD [5] detector has a 500 µm-thick silicon sensor composed of 120 × 560 array of 130 × 130 µm^2 pixels, which can be read out as a full frame in 1.6 ms.

The X-ray fluorescence (XRF) signal emitted by the sample was detected by a silicon drift detector (SDD) having 80 mm^2 active area (VITUS H80, KETEK GmbH). The detector was placed perpendicularly to the incident beam in the horizontal plane to minimize the detection of elastic X-ray scattering from the sample. A high-speed digital multichannel analyser (MCA) (xMAP, XIA LLC) was used in double-buffering mode to process the signal of the SDD.

A 5µm thick silicon photodiode (AXUV36, IRD Inc.) was used to monitor the flux of the incident beam, for data normalization. The counter card, the XPAD, the xMap and the ADC were controlled by their own TANGO devices. The data acquisition of each device was started by an external trigger signal.

The sample was scanned using a continuous scan in the horizontal direction while step scan mode was used in the vertical direction. In order to reduce further the total scanning time we performed bi-directional horizontal scans.

In each pixel of the reconstructed images, the DPC, the transmission and the dark field information were obtained from the recorded XPAD frame [6]. The DPC was obtained by calculating the centroid of the illumination pattern. The transmission was given by the total intensity recorded in the image. The dark field was evaluated by integrating the scattered intensity, outside the first order beam. Elemental maps were obtained from the intensities of selected fluorescence peaks, defined by integrating regions of interest (ROIs) of the XRF spectra. Each reconstructed image was rescaled by the recorded encoder positions to correct for motor acceleration/deceleration, and backlash.

4. Results

Large field of view fast scanning imaging has been demonstrated on a food moth (figure 3a) by measuring an image of 1000 × 1000 pixels. The pixel size is 1.6 × 2 μm^2. We used 12 ms dwell time/pixel, comprising 10 ms exposure time, required to obtain sufficient intensity in the XPAD image, and 2 ms readout time. The obtained data set consists of one million XPAD frames and intensity monitor values, and 2 million motor positions. All the data are merged into a single Nexus file of 140 Gbytes.

The vertical DPC images, both uncorrected and with encoder position correction are shown in figures 3b and 3c, respectively. The phase map was reconstructed from the vertical and

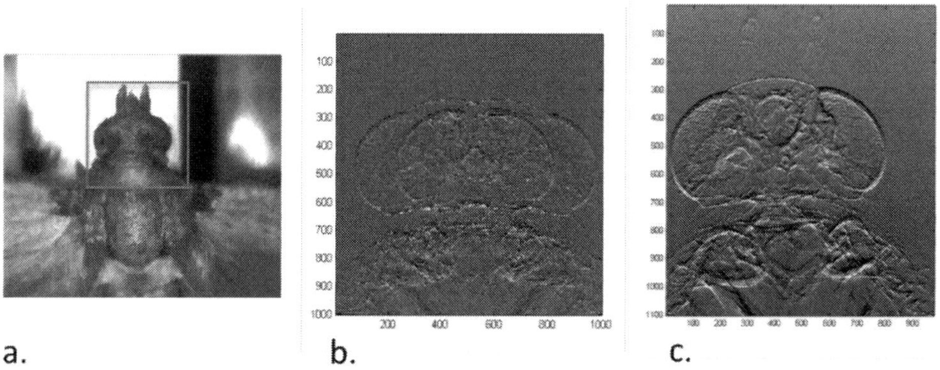

Figure 3. Scanning X-ray images of the head of a moth using the FLYSCAN scheme. The scanned area is indicated by the red rectangle in Fig 3a. Vertical differential phase contrast image,before (Fig. 3b) and after (Fig. 3c) correcting with the encoded motor positions. For scale reference, the width of images (b-c) is 1.6 mm.

horizontal DPC images using a Fourier integration technique [7]. Further details on the phase, transmission, DPC, and dark field images can be found in Medjoubi et. al. [8].

Simultaneous multi-technique fast scanning imaging was performed by the XPAD and the SDD detectors. The test sample was a copper wire of 230 μm diameter containing an insulating sheath. The simultaneously recorded absorption, DPC, dark field and XRF maps can be seen in Figure 4. The image contains 500 × 500 pixels with 2.6 × 2 μm^2 pixel size. The dwell time/pixel was 22 ms (20 ms exposure time and 2 ms readout). The total acquisition time including communication and data storage overhead was ~1 hour.

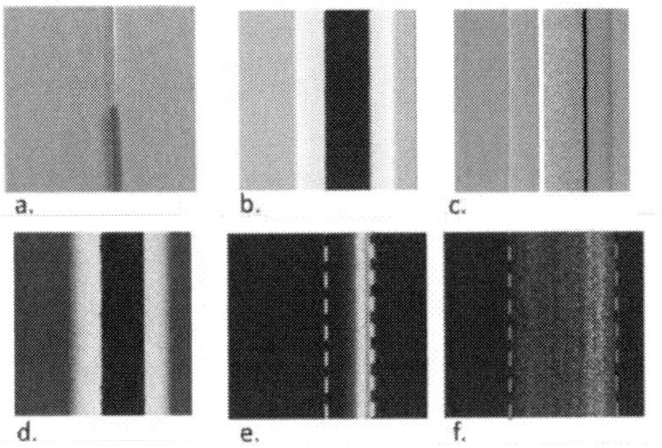

Figure 4. (a) Photograph of insulated copper wire, (b) transmission, (c) horizontal DPC, (d) dark field image, (e) copper, and (f) selenium distribution. The dashed lines represent the edges of the wire and the insulating sheath. For scale reference, the width of images (b-f) is 1 mm.

Figures 4b, 4c and 4d show the measured transmission, DPC, and dark field images. The images have been corrected with the recorded encoder positions. In the Cu map shown in figure 4e the self-absorption of the Cu-Kα fluorescence line is evident. Selenium is distributed in-homogenously in the insulating sheath, the line structures of the sheath and the self-absorption effect is clearly visible (Figure 4f).

5. Conclusion

A fast, parallel multi-technique continuous scanning scheme has been developed and tested at Synchrotron Soleil. The FLYSCAN architecture ensures the flexible integration of individual, externally triggered detectors, each of which is controlled by its own TANGO device. The minimum dwell time available in our current scheme was 1.6 ms, limited by the XPAD detector readout time.

We demonstrated the collection of large field of view images by the FLYSCAN scheme. The measured data and metadata are recorded in a single Nexus file. On-line data processing during the acquisition is possible for data visualisation. The simultaneous recording and combining of X-ray fluorescence, absorption, phase contrast and dark field images opens the possibility for quantification [9].

Acknowledgement

The authors are grateful to G. Baranton and P. Da Silva from Soleil for their assistance with the synchrotron experiments, and to the XPIX collaboration (SOLEIL/CPPM-CNRS/Neel Institute) providing the XPAD detector. We acknowledge V. Kaiser (ANKA / Softwareschneider GmbH) for providing the xMAP TANGO device and our colleagues from the Controls groups ICA/ECA from Soleil who helped with the test experiments.

References

[1] Somogyi, A., Kewish, C. M., Polack, F., & Moreno, T. (2011), *AIP Conf. Proc.*, **1365**, 57-60

[2] http://www.tango-controls.org

[3] Poirier, S., Maréchal, C., Ounsy, M., Buteau, A., Martinez, P., Gagey, B., Pierrot, P., Mederbel, M., & Rochat, J. M., (2009). *Proceedings of ICALEPCS09*, Kobe, Japan

[4] Idir, M., Mercere, P., Moreno, T., Delmotte, A., Dasilva, P. and Modi, M.H. (2010). *AIP Conf. Proc.*, **1234**, 485-488.

[5] Medjoubi, K., Bucaille, T., Hustache, S., Bérar, J. F., Boudet, N., Clemens, J. C., Delpierre, P., & Dinkespiler, B. (2010). *J. Synchrotron Rad.* 17(4), 486-495.

[6] Menzel, A., Kewish, C. M., Kraft, P., Henrich, B., Jefimovs, K., Vila-Comamala, J., David, C., Dierolf, M., Thibault, P., Pfeiffer, F., and Bunk, O. (2010). *Ultramicroscopy*, **110**, 1143-1147.

[7] Kottler, C., David, C., Pfeiffer, F. & Bunk, O. (2007). *Opt. Express*, **15**, 1175-1181.

[8] Medjoubi, K., Leclercq, N., Langlois, F., Buteau, A., Lé, S., Poirier, S., Mercère, P., Sforna, M., Kewish, C. M. & Somogyi, A. (2012) submitted

[9] Kosior, E., Bohic, S., Suhonen, H., Ortega, R., Devès, G., Carmona, A., Marchi, F., Guillet, J.F., Cloetens, P. (2012), *Journal of Structural Biology*, **177**, 239–247

Nanoscale spectroscopy and tomography with the HZB X-ray microscope: Applications in materials and life sciences

P Guttmann, S Rehbein, S Werner, K Henzler, B Tarek, G Schneider

Helmholtz-Zentrum Berlin für Materialien und Energie GmbH, Institute for Soft Matter and Functional Materials, Albert-Einstein-Str. 15, 12489 Berlin, Germany

E-mail: peter.guttmann@helmholtz-berlin.de

Abstract. The full-field transmission X-ray microscope at the electron storage ring BESSY II allows nanoscale spectroscopy as well as correlative fluorescence and nanoscale tomographic imaging of frozen-hydrated adherent cells. Reconstructions of tilt series permit the visualization of the sub-cellular structures in mammalian cells and plant cells. Quantitative information about package densities of cell organelles in the cytoplasm is possible. We present an overview of recent results in materials and life sciences obtained with the HZB X-ray microscope.

1. Introduction

The full-field transmission X-ray microscope (TXM) at the electron storage ring BESSY II operated by the Helmholtz-Zentrum Berlin (HZB) was installed at the undulator beamline U41-FSGM [1] in 2007. It is in full user operation since spring 2008. It was the first TXM in the soft X-ray energy range operating at an undulator source with a focusing spherical grating monochromator (FSGM) and an elliptically shaped glass capillary mirror as condenser [2]. This set-up delivers a spectral resolution which is an order of magnitude higher than for conventional TXMs using a zone plate monochromator. Therefore, nanoscale spectroscopy applications like NEXAFS investigations are feasible [3]. On the other hand with this set-up no pinhole close to the specimen is necessary. The free space for specimen rotation permits to use flat sample supports. Additionally, the partially coherent object illumination delivers an improved contrast transfer function of the microscope compared to incoherent systems [4]. The schematic optical layout and the workflow for nanoscale spectroscopy and nanoscale tomography are depicted in Fig. 1.

The spatial resolution of the HZB TXM was determined to be 11 nm (half-pitch) for 2D imaging [5]. This resolution cannot be transferred directly to the 3D resolution in nanoscale tomography which is affected by the small depth of focus of the zone plate objective. In practice, the achieved 3D resolution is about 36 nm (half-pitch) [4,6].

The sample stage is a modified CompuStage from FEI Company. Therefore, TEM holders can be used. The specimens are investigated under vacuum conditions. Depending on the application, users can choose between two different holders: A room temperature holder allows mounting either round standard EM grids with 3 mm in diameter or alternatively silicon nitrate membranes with 3.5 mm x 3.5 mm in wafer size. For biological samples a cryo high tilt holder (Gatan model 630, see Fig. 2) is employed. For this holder special rectangular grids suited for high tilt were designed: IFR-1 and HZB-2 types are shown in Fig. 2. Holy carbon films like the R2/2 coating from Quantifoil [7] turned out to be the best sample support film.

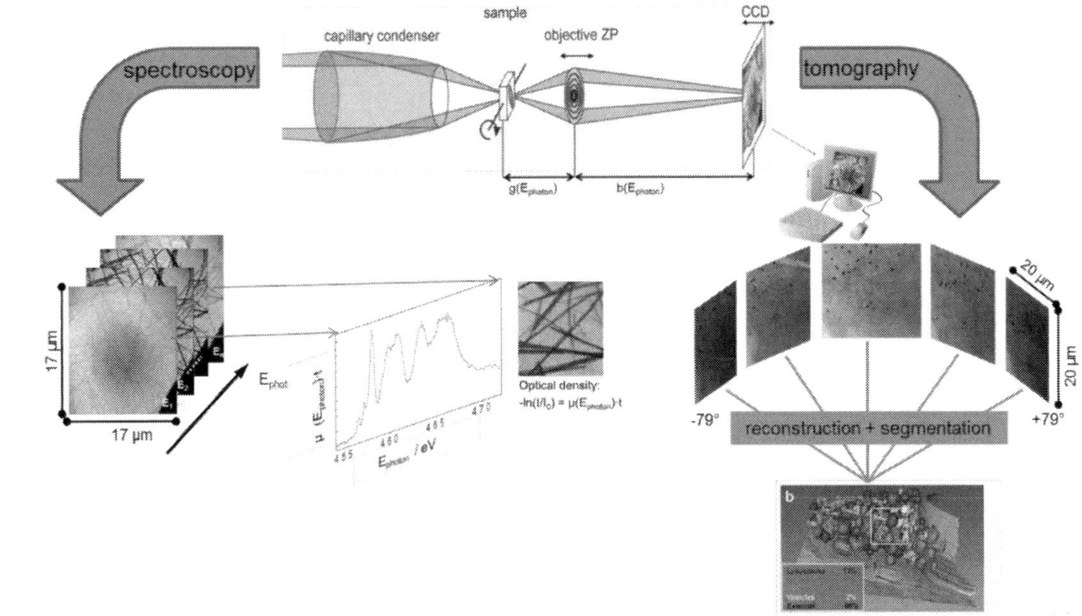

Fig. 1: Optical set-up of the HZB-TXM (upper) and workflow for nanoscale spectroscopy (lower left) and tomography (lower right) investigations.

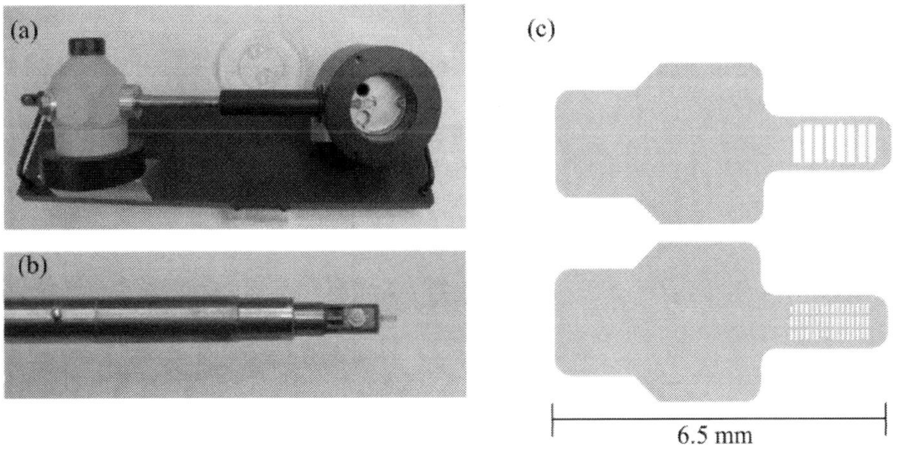

Fig. 2: The Gatan model 630 cryo high tilt holder: (a) in the transfer station, (b) showing the tip with a grid in working position. (c) Two types of high tilt grids employed for cryo tomography of cells.

Recently, a light microscope inside the X-ray microscope became accessible for users. The schematic set-up is displayed in Fig. 3. It is possible to switch between the light and the X-ray microscope working positions within 2 minutes (Fig. 4). For correlative microscopy, the sample is under the same viewing angle for light and x-ray microscopy. The light microscope delivers an overview of the samples on the grid and helps to preselect the objects without applying an X-ray dose. Additionally, the fluorescence mode can be used to identify labeled regions in cells as well as fiducial markers for correlative alignment.

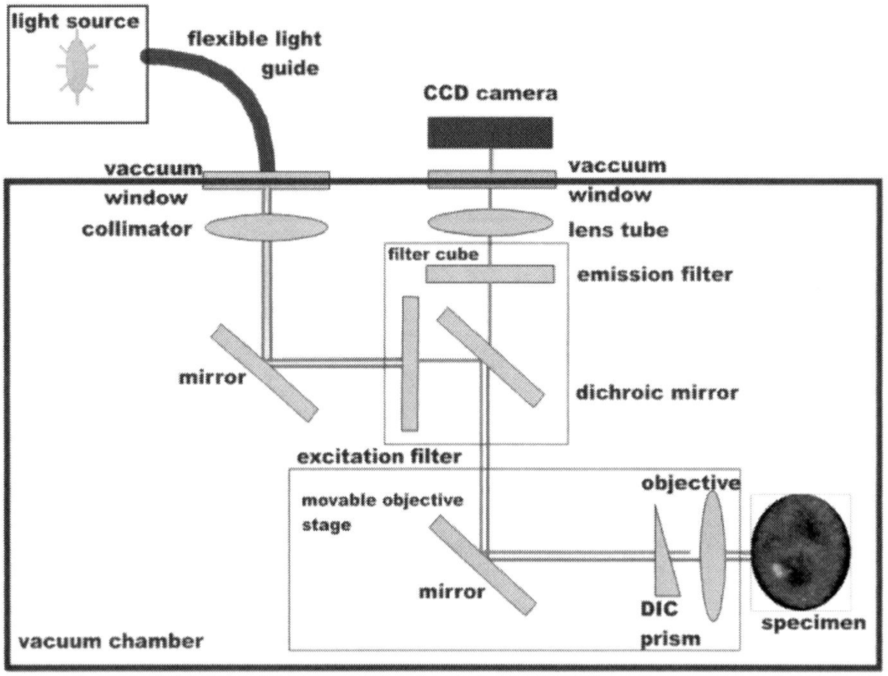

Fig. 3: Schematic layout of the incorporated light microscope. The filter cube allows switching between four different modes: bright field, differential interference contrast (DIC) and two fluorescent modes: GFP (green) and rhodamine (red). Details of the filters used in the fluorescent mode are given in Table 1.

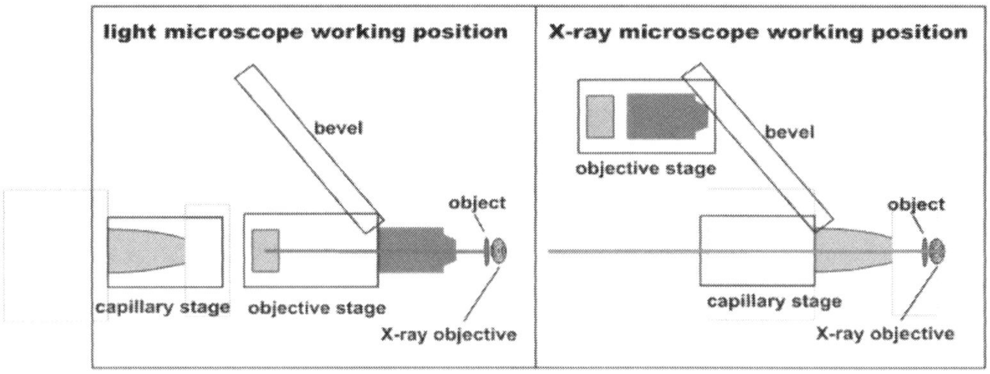

Fig. 4: Working position of the light microscope (left). Switching to the X-ray microscopy working position (right) requires to retract the light microscope objective along the optical axis with subsequent travel on the bevel to the home-position. Finally, the condenser (capillary stage) of the X-ray microscope has to be moved along the optical axis to the X-ray microscopy working position.

U41-TXM, HZB	
Microscope type	TXM
Source	Undulator beamline, linear polarization
Condenser/Objective aperture	Partially coherent illumination
Energy range	250 – 800 eV (- 1.5 keV)
Spectral resolution	up to $E/\Delta E = 10^4$
Zone plates for user operation	dr_n = 25 nm, 40 nm
Spatial resolution	2D: ≈10 nm half-period (depends on objective) 3D: ≈36 nm Rayleigh criterion
Sample format	1) Special designed rectangular grids (HZB-2 or IFR-1) having 2 mm x 1 mm sample area and open grid slits width 50 µm x 200 µm or 150 µm x 700 µm 2) 3 mm diameter EM grids or 3.5 · 3.5 mm^2 wafer, but for these no cryo and no tomography possible
Tomography capability	yes
Tilt range	-80°.... +80°
Cryo capability	yes
Detector type	Thinned, backside illuminated CCD, 1340 pixel x 1300 pixel
X-ray magnification	adjustable, value depends on the used objective
Light microscope	BF, DIC, fluorescence
Filter sets for light fluorescence microscope	Filter set 38 from Zeiss, GFP, ex 470, emm 525; Filter set 43 from Zeiss, Cy 3 (rhodamin), ex 545, emm 605
Raw data format	*.spe (WinView); 3.407 MB per image
Data analysis software capabilities available	ImageJ, Imod (eTomo)
Preparation Lab	BioLab (restricted access): safety level 2 (BioStoffV), genetically modified organisms (GMOs) belonging to safety level S1 (GenTG); incubator; plasma cleaner; plunge freezer (liquid ethane); fluorescence light microscope LEICA DMI6000B; cryo light microscopy stage (ordered)

Table 1: Main parameters of the U41-TXM and the infrastructure available for X-ray microscopy experiments at HZB.

2. Overview of the modes of operation and their applications

Two modes of operation are already outlined in Fig. 1: Nanoscale spectroscopy and tomography. The photon energy available for X-ray microscopy experiments ranges from 270 eV - 800 eV (see Table 1). Accepting a reduced photon flux, this energy range can be extended up to about 1.5 keV.

2.1. Nanoscale spectroscopy

For spectroscopy applications the CCD camera will be moved together with the objective zone plate to ensure the same image magnification at each photon energy. After collecting a data set of images, the images can be aligned using a cross-correlation method as implemented for example in the eTomo package [8] or the images can be directly imported into the aXis2000 software [9]. After the alignment, a NEXAFS spectrum for each pixel in the image can be obtained. The image field is 13 – 20 µm in square with a pixel size of 10 – 15 nm depending on the zone plate objective applied for imaging. Therefore, one image data set of nanoparticles already contains many different particles which makes statistical analysis possible. Typically, one data set is taken within 60 – 90 minutes with a few seconds exposure time per image. Compared to scanning X-ray microscopes (SXM's) the TXM allows a much faster data acquisition for the same amount of particles.

This mode of operation was used for the investigation of different TiO_2 nanoparticles like nanoribbons [10,11,3], carbon nanotubes, graphite layers [12] and especially synthesized microgels with TiO_2 particles [13]. Another example for using the tunability of the photon energy is the investigation of organic solar cells [14]. In most of these studies complementary methods like electron microscopy or atomic force microscopy are used to combine the findings for a more detailed interpretation.

2.2. Nanoscale tomography

Tomography studies can be performed using the high tilt grids in combination with the Gatan model 630 holder. This allows tilt angles in the range of up to ±79°. After taking a data set in typically 1° steps within about an hour with a few seconds exposure time per image, the images will be aligned. For this purpose, freely available software packages like eTomo [8] or BSoft [15] – both originally developed for electron microscopy – are used. These software packages deliver a reconstruction of the sample volume. Subsequent segmentation processes can be performed to display e.g. different organelles in cells. Besides materials sciences applications like tomography studies of stress migration in semiconductors [16] or the morphology of cement particles [17], the main application field for nanoscale tomography is cell biology. Due to the higher penetration depth of X-rays compared to electrons, tomography of whole cells in their native state is possible. Only cryo fixation without staining or sectioning is necessary to accomplish X-ray tomography studies of cells. The method of cryo fixation is well established and described for X-ray microscopy work in several publications [4,6,18-20]. Studies of mouse adenocarcinoma cells delivered already an atlas of mammalian cell ultrastructures in the cellular volume like mitochondria, nuclear membrane channels, granular structures in nucleoli and many other organelles [20].

The application of cryo X-ray tomography together with complementary methods solved the question whether the nucleation of hemazoin in the life cycle of *Plasmodium falciparum* occurs in the lipid droplet or at the lipid-water interface of the digestive vacuole. As a result it could be clearly determined that an oriented crystallization at the inner membrane in the aqueous rather than the lipid phase takes place [21].

The study of virus cell interactions is another field of interest: In a first attempt it was demonstrated that membranes and inner compartments of Vaccinia viruses could be resolved by cryo X-ray tomography [18]. In a second step these viruses could be identified in cells. Combining the use of the incorporated fluorescent light microscope together with subsequent investigations by electron microscopy, viral factories within cells could be defined and volume representations of different states of these viruses (immature and mature) were achieved [19].

Furthermore, tomography studies of cells treated with a fluorophore-tagged protein showing the same behavior as Herpes viruses show different aspects of changes within the cell nucleus [6].

Additionally, quantitative findings on the volume densities of different cell organelles within the cytoplasm of mammalian cells [3] or within e.g. algae [22] were procurable.

3. Outlook

Current nanoscale tomography studies pave the way to a better understanding of the interaction of nanoparticles with cells and help to understand their toxicological effects on cells. These investigations are important for humankind as nowadays many substances used e.g. for sun-milks or for paints contain nanoparticles. Developing new fiducial markers visible as well in the fluorescent light microscope as in the X-ray microscope will allow for correlative microscopy. So far, by default cryo samples were prepared by plunge freezing in liquid ethane or in a mixture of ethane and propane [6], but recently it was verified that high pressure freezing might be advantageous [24].

Acknowledgements
This work was funded in part by the Human Frontier Science Program Research Grant Ref. RGP0053/2005-C, the German Federal Ministry of Education and Research under contract number 05KS4BY1/7, the 7th framework program under grant agreement 226716 (WP23 NANOFOX) and BioStruct-X under grant agreement number 283570.

References

[1] G. Schneider, P. Guttmann, S. Rehbein, S. Werner, R. Follath, J. Struct. Biol. **177** (2012), 212-223
[2] P. Guttmann, X. Zeng, M. Feser, S. Heim, W. Yun, G. Schneider, Journal of Physics: Conference Series **186** (2009) 012064
[3] P. Guttmann, C. Bittencourt, S. Rehbein, P. Umek, X. Ke, G. Van Tendeloo, C.P. Ewels, G. Schneider, Nature Photonics **6** (2012), 25-29
[4] G. Schneider, P. Guttmann, S. Heim, S. Rehbein, F. Mueller, K. Nagashima, J.B. Heymann, W.G. Müller, J.G. McNally, Nature Methods **7** (2010), 985-987
[5] S. Rehbein, P. Guttmann, S. Werner, G. Schneider, Optics Express **20** (2012), 5830-5839
[6] C. Hagen, P. Guttmann, B. Klupp, S. Werner, S. Rehbein, T.C. Mettenleiter, G. Schneider, K. Grünewald, J. Struct. Biol. **177** (2012), 193-201
[7] http://www.quantifoil.com/
[8] eTomo is part of the IMOD package: http://bio3d.colorado.edu/imod/
[9] http://unicorn.mcmaster.ca/aXis2000.html
[10] C. Bittencourt, P. Krüger, M.J. Lagos, X. Ke, G. Van Tendeloo, C. Ewels, P. Umek, P. Guttmann, Beilstein J. Nanotechnol., accepted for publication
[11] P. Guttmann, C. Bittencourt, X. Ke, G. Van Tendeloo, P. Umek, D. Arcon, C.P. Ewels, S. Rehbein, S. Heim, G. Schneider, in: I. McNulty, C. Eyberger, B. Lai (Eds.), The 10th International Conference on X-ray Microscopy, AIP Conference Proceedings **1365** (2011), 437-440
[12] C. Bittencourt, A.P. Hitchock, X. Ke, G. Van Tendeloo, C.P. Ewels, P. Guttmann, Beilstein J. Nanotechnol. **3** (2012), 345-350
[13] K. Henzler, manuscript in preparation
[14] T. Mönch, P. Guttmann, M. Riede, L. Müller-Meskamp, K. Leo, manuscript in preparation
[15] http://lsbr.niams.nih.gov/bsoft/#
[16] E. Zschech, R. Huebner, D. Chumakov, O. Aubel, D. Friedrich, P. Guttmann, S. Heim, G. Schneider, J. Appl. Phys. **106**, 093711 (2009)
[17] S. Brisard, R.S. Chae, I. Bihannic, L. Michot, P. Guttmann, J. Thieme, G. Schneider, P.J.M. Monteiro, P. Levitz, American Mineralogist 97 (2012), 480-483
[18] J. Carrascosa, F.J. Chichón, E. Pereiro, M.J. Rodríguez, J.J. Fernández, M. Esteban, S. Heim, P. Guttmann, G. Schneider, J. Struct. Biol. **168** (2009), 234-239
[19] F.J. Chichón, M.J. Rodríguez, E. Pereiro, M. Chiappi, B. Perdiguero, P. Guttmann, S. Werner, S. Rehbein, G. Schneider, M. Esteban, J.L. Carrascosa, J. Struct. Biol. 177 (2012), 202-211
[20] W.G. Müller, J.B. Heymann, K. Nagashima, P. Guttmann, S. Werner, S. Rehbein, G. Schneider, J.G. McNally, J. Struct. Biol. 177 (2012), 179-192
[21] S. Kapishnikov, A. Weiner, E. Shimoni, P. Guttmann, G. Schneider, N. Dahan-Pasternak, R. Dzikowski, L. Leiserowitz, M. Elbaum, PNAS **109** (28) (2012), 11188-11193
[22] E. Hummel, P. Guttmann, S. Werner, B. Tarek, G. Schneider, M. Kunz, A.S. Frangakis, B. Westermann, submitted to PLOS One
[23] D. Drescher, P. Guttmann, T. Büchner, S. Werner, G. Laube, A. Hornemann, B. Tarek, G. Schneider, J. Kneipp, mauscript in preparation
[24] A. Weiner, S. Kapishnikov, E. Shimoni, S. Cordes, P. Guttmann, G. Schneider, M. Elbaum, J. Struct. Biol. (2012), available online 15 October 2012

Partial Coherence: a Route to Performing Faster Coherent Diffraction Imaging

Bo Chen[1,2*], Brian Abbey[1,3,4], Ruben Dilanian[1,2], Eugeniu Balaur[1,3], Grant van Riessen[1,3], Mark Junker[1,3], Chanh Q. Tran[1,3], Michael W.M. Jones[1,3], Ian McNulty[5], David J. Vine[5], Corey T. Putkunz[1,2], Harry M. Quiney[1,2], Keith A. Nugent[1,2,6]

[1] Australian Research Council Centre of Excellence for Coherent X-ray Science (CXS), Victoria, Australia
[2] School of Physics, The University of Melbourne, Victoria 3010, Australia
[3] Department of Physics, La Trobe University, Victoria 3086, Australia
[4] Melbourne Centre for Nanofabrication, 151 Wellington Road Clayton, Victoria 3168, Australia
[5] Advanced Photon Source, Argonne National Laboratory, Argonne, Illinois 60439, USA
[6] Australian Synchrotron, 800 Blackburn Rd, Clayton, Victoria 3168, Australia

E-mail: boc@unimelb.edu.au

Abstract. Coherent diffraction imaging (CDI) typically requires that the light source should be highly coherent both laterally and longitudinally. Beamlines at synchrotrons usually install a monochromator and slits to achieve a highly coherent source, leading to a large reduction of beam flux. We demonstrate that lateral and longitudinal partial coherence can be successfully included in a CDI reconstruction algorithm simultaneously, reducing the associated exposure time by two orders of magnitude. For the experimental case we present this allows the acquisition of CDI data in just 5 seconds compared to 20 minutes for full coherence. This significantly reduces the requirements on the stability of the imaging system as well as providing a route to imaging samples in real-time.

1. Introduction

With the development of modern X-ray sources such as third-generation synchrotrons and X-ray Free Electron Lasers (XFELs), many new forms of X-ray microscopy have been developed making it an area of active research. One of the most promising methods to be recently demonstrated is coherent diffraction imaging (CDI). Since its first demonstration using synchrotron radiation [1], CDI has been widely investigated and has found applications in material and biological sciences [2–4]. In principle, the resolution of CDI is not limited by the fabrication of X-ray optics, but by the experimental geometry. In practice, however, the resolution decreases as the 4th power with respect to incident flux, so much longer exposure times are required to achieve the highest resolution images [5].

Conventional CDI requires the source to be highly coherent both laterally and longitudinally. To achieve this, slits and a monochromator are installed at the beamline to give high coherence for the CDI experimental setup. The result of conditioning the beam in this way is a very significant loss of flux. Due to this decrease of flux, it usually takes tens of minutes to obtain high-resolution CDI experimental data, so that the requirement for stability of the experimental system is high, preventing the investigation of samples in real time. Recently, it has been found that the use of a modified reconstruction algorithm including *a priori* knowledge of the spectrum means that the monochromator can be removed from the system for CDI [6, 7]. In this method, the source is modeled as the combination of many monochromatic coherent modes of different frequency with the diffraction of each frequency component weighted and combined in the detector plane. This method still requires that the lateral coherence of the source is high. For a real source, however, the lateral coherence length is not always necessarily larger than the dimensions of sample. So, in many cases, the sample is illuminated with a partially coherent source, which will blur the diffraction pattern and cause the image reconstruction to fail. [8, 9].

In this paper, we demonstrate simultaneous use of partial lateral and partial longitudinal coherence in CDI and reduce the exposure time significantly to achieve real-time CDI.

2. Reconstruction Algorithm and Methods

The fundamental theory upon which partially coherent CDI is based can be found in [10, 11]. Here we provide a summary of the partially coherent CDI reconstruction algorithms. The notations are defined as follows. $\mathbf{r}$ and $\mathbf{r}'$ are any vectors in the sample and detector plane respectively, ν is the

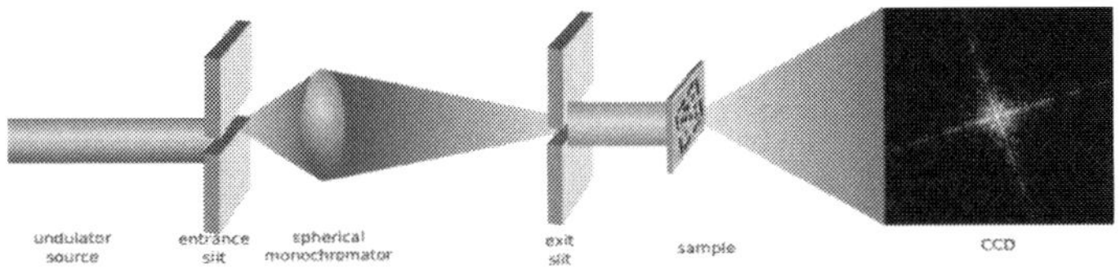

Figure 1. Schematic of the experimental setup. The diffraction pattern of the sample is shown on the right(in logarithmic scale).

frequency of the incident light with ν_0 the frequency at the peak of the spectrum, integer n is the mode number.

(i) Extract the coherent modes $\phi_n(\mathbf{r})$ from the source [12];

(ii) Fix at the peak spectrum $S(\nu_0)$, propagate the transmission function of the sample $T(\mathbf{r})$ with all the coherent modes $\phi_n(\mathbf{r})$, and calculate $I_n(\mathbf{r}', \nu_0) = |\phi_n(\mathbf{r})|^2$. The dominant coherent mode $\phi_0(\mathbf{r})$ results in the far field $\psi_0(\mathbf{r}')$;

(iii) Calculate the intensity $I(\mathbf{r}', \nu_0) = \sum_{n=0}^{N} \eta_n I_n(\mathbf{r}', \nu)$ with η_n the occupancy of coherent mode $\phi_n(\mathbf{r})$;

(iv) Scale the intensity with other frequency to calculate $I(\mathbf{r}', \nu)$;

(v) Sum the intensity with$I(\mathbf{r}') = \int S(\nu) I(\mathbf{r}', \nu) d\nu$ to get the total calculated intensity $I_c(\mathbf{r}')$;

(vi) Use the measured intensity $I_m(\mathbf{r}')$ and calculated intensity $I_c(\mathbf{r}')$ to constrain the intensity, and use the phase $\psi_0(\mathbf{r}')$; $\psi(\mathbf{r}') = \sqrt{I_m(\mathbf{r}')}\psi_0(\mathbf{r}')/\sqrt{I_c(\mathbf{r}')}|\psi_0(\mathbf{r}')|$

(vii) Propagate $\psi(\mathbf{r}')$ back to sample plane by inverse Fourier Transforming and divide $\phi_0(\mathbf{r})$ to get the transmission function of the sample $T(\mathbf{r}) = \mathrm{FT}^{-1}\{\psi(\mathbf{r}')\}/\phi_0(\mathbf{r})$

(viii) Use sample plane constraint to get a new guess $T(\mathbf{r})$;

(ix) Repeat ii-viii until $\Omega = \sqrt{\sum_{\mathbf{r}'}[I_m(\mathbf{r}') - I_c(\mathbf{r}')]^2}/\sqrt{\sum_{\mathbf{r}'} I_m^2(\mathbf{r}')}$ is less than a given value.

Sample plane constraint such as error-reduction(ER), hybrid-input-out (HIO) [13] can be used as what has been done in traditional CDI.

3. Experimental results

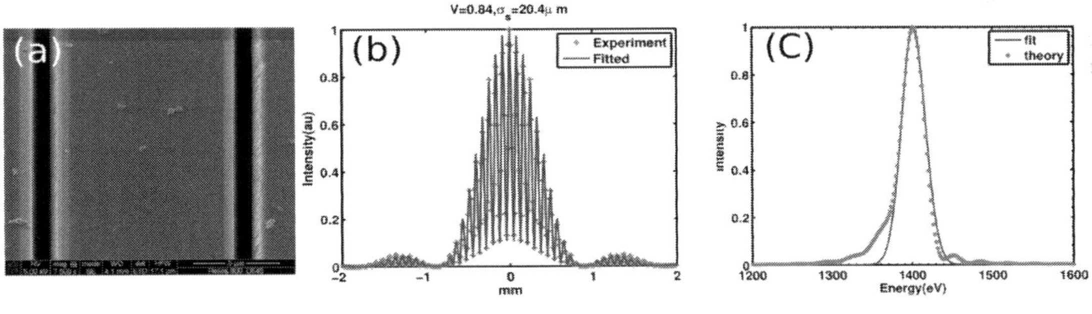

Figure 2. (a)The SEM image of the double slit fabricated with Focused Ion Beam (FIB).(b)The measured interference pattern of the double slit and fitted data when the incident beam is monochromatic and highly coherent.(c)The spectrum of the first order from the undulator at 2ID-B beamline of APS and a Gaussian fitting to this spectrum. The peak is at 1400eV. Theoretical data was calculated with SPECTRA developed by Takashi Tanaka and Hideo Kitamura, SPring-8/RIKEN.

Fig.1 shows a the schematic of the experimental setup. The experiment was carried out at the 2-ID-B beamline at the Advanced Photon Source (APS) [14] using conventional CDI. An x-ray beam with peak energy of 1.4 keV was used. The longitudinal coherence was controlled by the exit slit, while

slit setup(μm)	σ_s(μm)	σ(eV)	exposure time(s)
20/5	20.4	0.88	2
20/100	15.5	1.27	0.12
20/250	15.1	1.30	0.06
20/330	14.2	2.00	0.04
20/450	13.7	2.60	-
50/5	16.6	1.60	0.26
50/100	15.5	2.10	0.03
50/250	14.5	3.00	0.018
50/330	13.9	4.00	0.01
200/5	18.5	2.00	0.09
200/100	14.8	5.00	0.01
200/250	12.5	1.00	0.005
400/5	19.0	6.50	0.14
400/100	16.2	9.00	0.012
pink/5	12.3	15.08	10*
pink/100	10.8	15.08	1*
pink/250	9.9	15.08	0.5*
pink/330	9.5	15.08	0.4*
pink/450	9.0	15.05	0.5*

Table 1. The entrance and exit slit combinations used and the corresponding coherence length and bandwidth. The exposure time is for the diffraction data. "-" means the diffraction data is not available due to saturation, "*" means kapton film was used when taking the diffraction imaging. The thickness of the film is 51μm, the transmission at 1400eV is 3.94×10^{-4}, so the effective exposure time is the real exposure time by 3.94×10^{-4} . The effective exposure time for pink/450 is larger than pink/330, which may be due to the drop of the flux of the incident beam.

the lateral coherence was controlled by the entrance slit. A pair of Young's double slit and sample were installed in the same sample stage after a beam defining aperture (BDA) in the vacuum chamber. The CCD was sitting at 1057 ± 1mm downstream from the sample plane, with 2048×2048 pixels, each pixel $13.5 \times 13.5\mu m^2$. In our experiment, the separation of the double slits was $d = 11.75 \pm 0.25\mu m$, with the width of each slit $w = 1 \pm 0.03\mu m$.

In the experiment, we first closed the entrance slit to 20μm and used a monochromator to filter the beam and fitted the parameters of the slit. From the fitting, the real separation of the double slit is $d = 11.57\mu m$, the width of each slit $w = 1.02\mu m$, and the visibility is 0.84, so the coherence length $\sigma_s = 20.4\mu m$. In the case that the exit slit is fully open or the monochromator is not used, the longitudinal partial coherence is not negligible and must be included in the fitting. As a good approximation, we can fit the spectrum with a Gaussian function, even when there is no monochromator, as shown in Fig.2(b). In the experiment, we used several combinations of entrance and exit slit. For every combination, we first measured the interference pattern by translating the double slit into the beam, and then exchanging it with a fabricated sample, with the slits removed, to take diffraction data. To avoid saturation of the CCD, we keep the maximum counts to about 45000 for every frame of data; we summed 600 frames for every dataset presented here. The coherence properties of the source for every entrance and exit slit combination is fitted using the fitting methods [15] and is tabulated in Table 1. The fastest speed of the shutter is 0.005s, in some combinations, we had to attenuate the beam by installing a kapton film with thickness of 51μm in front of the CCD due to the shutter speed being too slow in comparison to the time for the CCD to saturate. This is equivalent to decreasing the flux by a factor of 3.94×10^{-4}. From the table we can see that the effective exposure time of partial coherence data is significantly smaller than that of the full coherence data. The exposure time of the 'pink'/5 setup is about 1/250 of that of 20/5 setup, which means the CDI experiment time was reduced from 20 minutes to just a few seconds. The effect of this major reduction in exposure time is to also decrease the stability requirements of the experimental setup.

The sample is fabricated on a gold film with thickness of 6μm with a focused ion beam (FIB). The size along the diagonal direction is 11.57μm. Because the size of holes in the sample is around 1μm, which is much smaller than the thickness, the walls of the triangular structure are not perpendicular to the surface and it barely extends through the film. The transmission through the gold sample at 1400eV is negligible, such that to a good approximation, it may be treated as a binary object with transmission function comprising only an amplitude component.

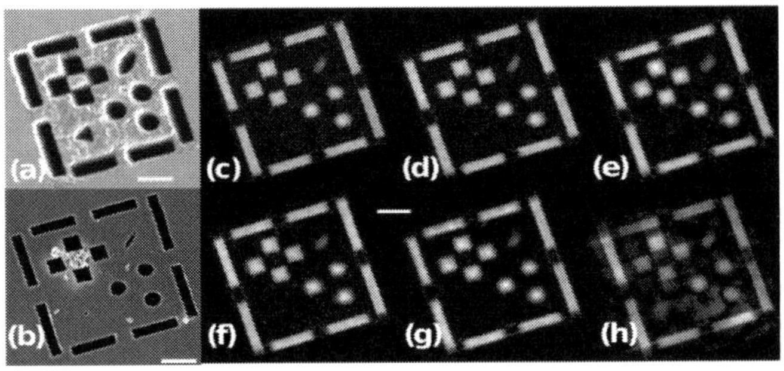

Figure 3. (a-b)The scanning electron microscopy image of the sample at (a) the front side and (b) the back side.(c-h) Reconstructed amplitudes of the sample for different entrance and exit slit combinations with our algorithm. (c) 20/5μm; (d) 20/250μm; (e) 50/5μm; (f) 200/100μm; (g) pink/5μm; (h) pink/100μm. Because the sample is thick relative to the size of the structures, the triangle is not fully penetrated. The scale bar is 2μm.

We reconstructed the sample with different slit combinations, as shown in Fig.3. We can see that by including the partial coherence in the reconstruction, the reconstruction works until at some instances, the data is blurred so much that the algorithms fails. The limit of the algorithm has been investigated elsewhere [11].

4. Conclusion

In this paper, we have demonstrated a partially coherent diffractive imaging method that includes both lateral and longitudinal coherence simultaneously in the reconstruction. This method leads to a two-orders of magnitude reduction in the exposure time for the sample examined compared to traditional CDI and makes much faster CDI experiments possible without loss of quality. It also reduces the stability requirement for the CDI system. With the use of recently developed fast-read detector technology, CDI it should be possible to image samples in real time in the future.

Acknowledgments

The authors acknowledge the support of the Australian Research Council Centre of Excellence for Coherent X-ray Science. Use of the Advanced Photon Source was supported by the US Department of Energy, Office of Science, Office of Basic Energy Sciences(contract no. DE-AC02-06CH11357).

References

[1] Miao J, Charalambous P, Kirz J and Sayre D 1999 *Nature* **400** 342–344
[2] Miao J, Nishino Y, Kohmura Y, Johnson B, Song C, Risbud S and Ishikawa T 2005 *Phys. Rev. Lett.* **95** 085503
[3] Robinson I and Harder R 2009 *Nature Materials* **8** 291–8
[4] Huang X, Nelson J, Kirz J, Lima E, Marchesini S, Miao H, Neiman A, Shapiro D, Steinbrener J, Stewart A, Turner J and Jacobsen C 2009 *Phys. Rev. Lett.* **103** 198101
[5] Shen Q, Bazarov I and Thibault P 2004 *J. Synchrotron Radiat.* **11** 432–8
[6] Chen B, Dilanian R A, Teichmann S, Abbey B, Peele A G, Williams G J, Hannaford P, Van Dao L, Quiney H M and Nugent K A 2009 *Phys. Rev. A* **79** 23809
[7] Abbey B, Whitehead L W, Quiney H M, Vine D J, Cadenazzi G A, Henderson C A, Nugent K A, Balaur E, Putkunz C T, Peele A G *et al.* 2011 *Nature Photonics* **5** 420–424
[8] Whitehead L W, Williams G J, Quiney H M, Vine D J, Dilanian R A, Flewett S, Nugent K A, Peele A G, Balaur E and McNulty I 2009 *Phys. Rev. Lett.* **103** 243902
[9] Clark J, Putkunz C, Curwood E, Vine D, Scholten R, McNulty I, Nugent K and Peele A 2011 *Opt. Lett.* **36** 1954–1956
[10] Williams G, Quiney H, Peele A and Nugent K 2007 *Phys. Rev. B* **75** 104102
[11] Chen B, Abbey B, Dilania R, Balaur E, van Riessen G, Junker M, Tran C, Jones M, Peele A, McNulty I, Vine D, Putkunz C, Quiney H and Nugent K 2011 *Phys. Rev. B* **86** 235401
[12] Starikov A and Wolf E 1982 *J. Opt. Soc. Am.* **72** 923–928
[13] Fienup J R 1982 *Appl. Opt.* **21** 2758–2769
[14] McNulty I, Khounsary A, Feng Y P, Qian Y, Barraza J, Benson C and Shu D 1996 *Rev. Sci. Instrum.* **67** 3372
[15] Paterson D, Allman B E, McMahon P J, Lin J, Moldovan N, Nugent K A, McNulty I, Chantler C T, Retsch C C, Irving T H K and Mancini D C 2001 *Optics Communications* **195** 79–84

Development of Scanning-Imaging X-Ray Microscope for Quantitative Three-Dimensional Phase Contrast Microimaging

Akihisa Takeuchi, Yoshio Suzuki, and Kentaro Uesugi

JASRI / SPring-8, 1-1-1 Koto, Sayo, Hyogo 679-5198, Japan

take@spring8.or.jp

Abstract. A novel x-ray microscope system has been developed for the purpose of quantitative and sensitive three-dimensional (3D) phase-contrast x-ray microimaging. The optical system is a hybrid that consists of a scanning microscope optics with a one-dimensional (1D) focusing (line-focusing) device and an imaging microscope optics with a 1D objective. These two optics are orthogonally arranged regarding their common optical axis. Each is used for forming each dimension of two-dimensional (2D) image. The same data acquisition process as that of the scanning microscope system enables quantitative and sensitive x-ray imaging such as phase contrast and absorption contrast. Because a 2D image is measured with only 1D translation scan, much shorter measurement time than that of conventional scanning optics has been realized. By combining a computed tomography (CT) technique, some 3D CT application examples are demonstrated

1. Introduction

Since phase contrast (PC) imaging in hard x-ray region shows much higher sensitivity than absorption contrast (AC) imaging especially for low-Z materials, various types of PC x-ray microscope has been developed in the last decade. In particular, many approaches implemented in imaging (full-field) x-ray microscope optics have been proposed. On the other hand, scanning x-ray microscopy has also been widely studied for differential phase contrast (DPC) imaging. One of the advantages of the imaging microscope optics is high-throughput 2D imaging. It makes easy to apply to 3D measurement by introducing tomographic method. However, they are not still at a practical level in quantitative PC 3D imaging because it is difficult to satisfy all requirements for the PC 3D imaging such as sensitivity, linearity of image contrast, and system stability. On the other hand, scanning optics satisfies all requirements above mentioned. However, 3D imaging application has not been in practical use yet because data acquisition requires a lot of time

To satisfy advantages of the both systems; high sensitivity and high contrast linearity of scanning optics and high throughput of imaging optics, a novel x-ray microimaging system called as scanning-imaging x-ray microscope (SIXM) has been developed (Fig. 1) [1]. The system consists of a scanning microscope optics with a 1D focusing (line-focusing) device and an imaging microscope optics with a 1D objective. These two optical systems are set normal to each other regarding the optical axis for obtaining the scanning microscope data at each sample height. Therefore, a translation scan of the sample along the focusing direction is only required to obtain a 2D image. DPC image and AC image are arbitrary obtained by image processing after data acquisition. Some application examples on 3D CT experiments will be presented.

Journal of Physics: Conference Series **463** (2013) 012034 doi:10.1088/1742-6596/463/1/012034

2. Optical system and experiment

The experiment has been carried out at the beamline 20XU of SPring-8. As shown in Fig. 1, the experimental setup of SIXM-CT system consisted of a 1D scanning CT system in the horizontal direction and a 1D imaging optics in the vertical direction. The system was illuminated with full-coherent beam of 8 keV x-ray. A pair of 1D Fresnel zone plates (1D-FZPs) with the same parameters was used as the 1D focusing device and as the 1D objective. The 1D-FZPs were fabricated at NTT-AT. Tantalum zone pattern with 1 μm thickness is deposited on a silicon carbide (SiC) membrane with 2 μm thickness. Outermost zone width is 100 nm, and total width of the 1D-FZP pattern is 155 μm. Focal distance at 8 keV is 100 mm. In the experiment, these 1D-FZPs were set inclined by 60 deg. in order to increase the effective zone thickness. Under this condition, measured value of diffraction efficiency was 0.279 (theoretical value is 0.316) [1]. An indirect-sensing x-ray camera consisted of a visible-light conversion unit called "Beam Monitor" (BM AA40P, Hamamatsu Photonics, K. K.) and a frame-transfer type charge coupled device (CCD) camera (C9100-02, Hamamatsu Photonics, K. K.) was used as an image detector. The Beam Monitor consists of a powder scintillator P43 (Gd_2O_2S: Tb, 10 μm thickness), and a couple of relay lenses. Typical parameters of the detector are as follows; number of the pixel is 1000 × 1000, maximum frame rate for full frame is 30 Hz, analog-digital conversion rate is 14 bit, and effective pixel size is 11.4 μm /pixel. Typical distance between the object plane and the image plane was roughly ~7 m. Therefore, typical magnification factor of the x-ray imaging microscope optics was ~70. In the raw image data obtained with the image detector, horizontal and vertical intensity distributions represent the far-field image of focusing 1D-FZP and 1D magnified image of the object, respectively. The raw image sequential dataset is recorded as translating the line probe across the sample. After the data acquisition, a 2D image is reconstructed from the raw image dataset. The reconstruction theory is just same as that of the conventional scanning optics with multi pixel detector.[3] In the case of the SIXM system, the reconstruction is processed only in the horizontal direction because the vertical position of the raw image is simply representing the magnified image of the object. Since the phase gradient can be measured only in the scanning direction, only the horizontal DPC image is obtainable (vertical DPC is not obtainable). A PC image representing phase shift distribution is reconstructed with linear integration of a horizontal DPC image. Effective pixel size in the vertical direction is determined by the magnification factor of the imaging microscope optics whereas that in the horizontal direction is the same as the scan pitch. Such asymmetric pixel ratio is numerically compensated after image reconstruction. Figure 2 shows x-ray images of a tantalum test chart (NTT-AT) obtained by the SIXM system. Spatial resolution of the system was measured to be 110 nm in the horizontal direction and approximately 400 nm in the vertical direction. More details of the SIXM system are described elsewhere. [1,2]

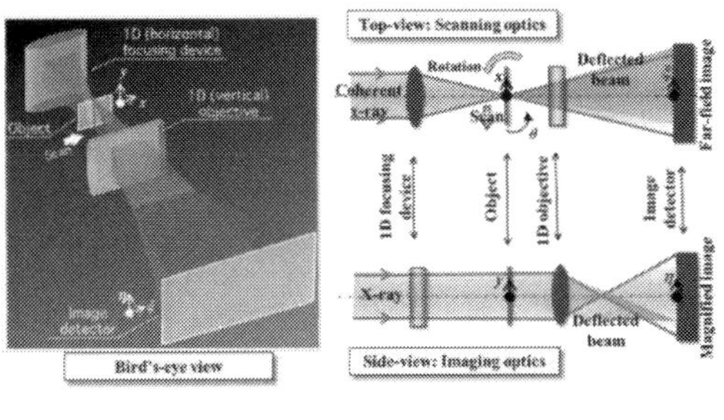

Figure 1. Conceptual drawing of SIXM optical system, left: bird's-eye view, upper-right: top-view, and lower-right: side-view.

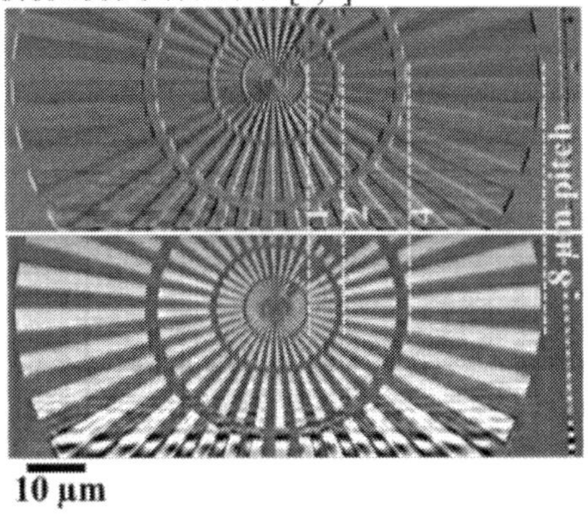

Figure 2. X-ray images of a tantalum test chart (NTT-AT), above: DPC image, and below: AC image. Pixel size 125 nm (H) x 162 nm (V), dwell time is 0.1 s, and measurement time is 80 sec.

3. 3D CT applications

3.1. Human hair

A human hair was observed as a feasibility test for the application of low-Z materials. Although internal structures of human hairs have been widely studied by using scanning electron microscope and so on, it has been impossible to observe the intact structure because samples must be cut out at the observed region and are required some preparation such as metal shadowing. Owing to use of the DPC scanning x-ray microscope, inner structures of intact human hairs was successfully observed,[3] and this method is now utilized to the developments for hair-care products. The scanning x-ray microscope, however, has never enabled to investigate 3D internal structure because of its low throughput as mentioned in chapter 1, although the 3D observation have been strongly required to understand the inner structures of human hairs completely. Although x-ray 3D observations by using several types of PC imaging microscope optics such as Zernike method, Talbot interferometer, and defocus were collaterally attempted, these attempts were hardly succeeded due to poor contrast linearity, low stability and low sensitivity of these systems.

Figure 3 shows CT images of human hair obtained with the SIXM-CT system. Measurement time was approximately 3.5 hours. Figures 3(a) and 3(b) show the virtual cross-sectional images in the transversal plane and in the sagittal plane of the sample, corresponding to the horizontal plane and the vertical plane of the system, respectively. Three-dimensional rendered image and its virtually cropped view are also shown in Figs. 3(c) and 3(d), respectively. In the transversally sliced CT image of Fig. 3(a), some defects and melanin indicated as small particles with lower and higher density than the neighbors, respectively, medulla at the center region of the hair shaft, and layered structure in the peripheral region of cuticles are clearly seen. In the sagittal cross section image of Fig. 3(b), shapes of defects and melanin are found to be extent along to the direction of hair shaft, while they are granular in the transversal direction. Therefore, the SIXM-CT system has distinctly revealed for the first time that their three-dimensional geometries are not globular shape, but prolate (rugby ball-like) shape whose long axes are along to the hair shaft.

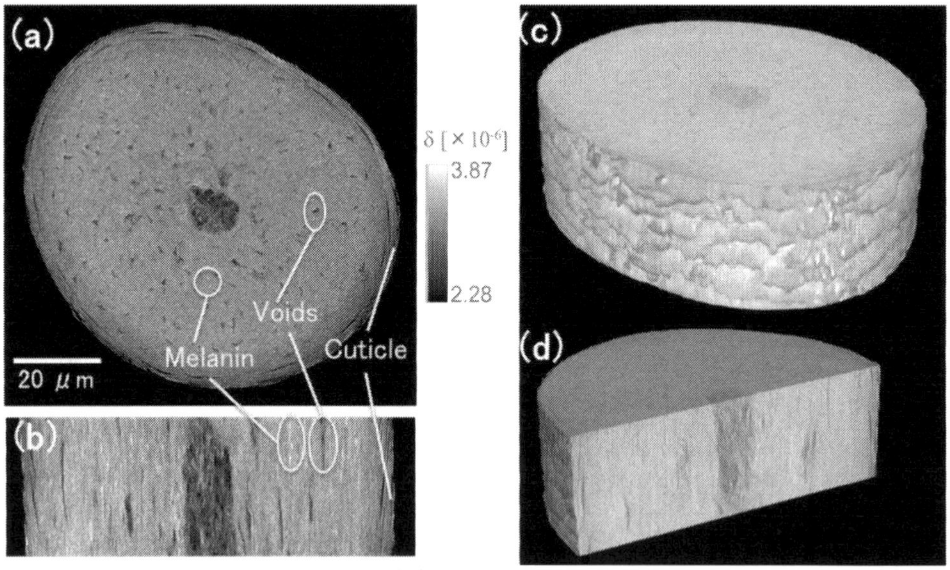

Figure 3. CT images of human hair; upper-left, horizontally-sliced CT image, lower-left: vertically-sliced CT image, and right: 3D rendered images. Scan pitch (H) is 125 nm, pixel size (V) is 323 nm, number of CT projection is 301 for 180 deg, dwell time is 20 ms, and measurement time is 210 min.

3.2. Carbonaceous chondrite (Allende meteorite)

Figure 4 shows CT images of a piece of Allende meteorite, which is classified into a carbonaceous chondrite, including organic materials, silicates and metals. PC-CT images and AC-CT images are shown in left and right half of Fig. 4, respectively. These two image contrast modes are derived from a single CT measurement data because image contrast modes can be arbitrary selected after data acquisition. Four times of CT scan were repeatedly performed with different vertical position of the sample because the field of view of the imaging microscope system (the vertical direction) was smaller than the vertical size of the sample. Measurement time for each CT scan was 2 hours. Shown images in Fig. 4 are the stitched ones of four CT image dataset in the vertical direction. Boundaries of stitched images are represented with arrowed lines in Fig. 4. Because the major component of the sample (silicates and metals) has a sufficiently high absorption for 8 keV x-ray, AC-CT image shows high image contrast as well as the PC-CT image. This method which enables 3D observation of PC and AC mode in one measurement shows feasibility of simultaneous investigation of high-Z such as metals and low-Z such as organic materials. Therefore, in the case of extraterrestrial sample, especially, the SIXM method will be one of the promising methods to search the beginning of life.

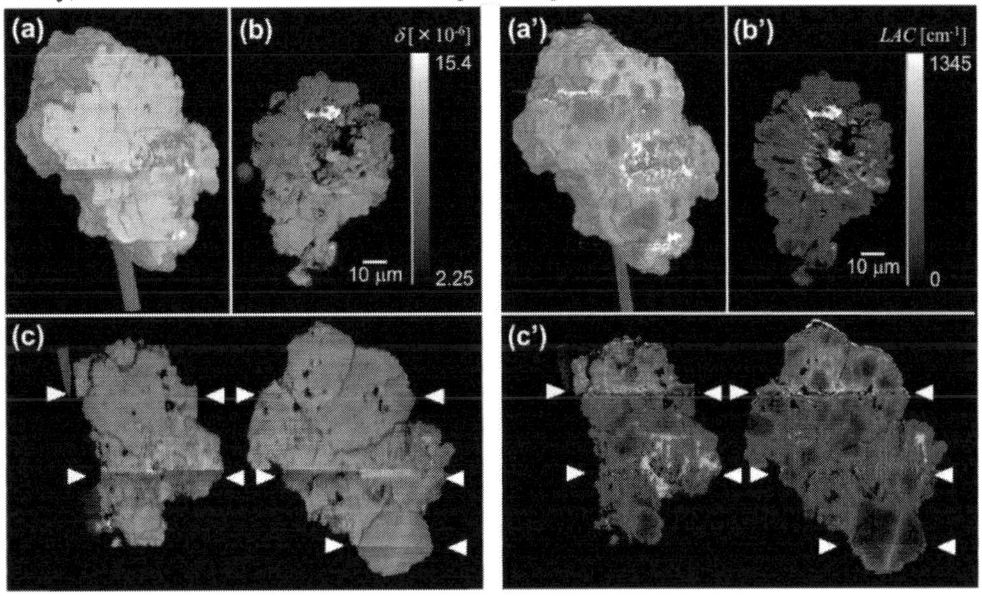

Figure 4. X-ray CT images of Allende meteorite, left: PC mode, and right: AC mode. (a) and (a') 3D rendering views, (b) and (b') virtual sectional images perpendicular to the rotation axis, (c) and (c') virtual sectional images parallel to the rotation axis. Horizontal scan pitch is 175 nm, vertical pixel size is 172 nm, number of CT projection is 501 for 180 deg, dwell time is 20 ms and measurement time is 120 min x 4.

Acknowledgements

The authors would like to appreciate Dr. T. Inoue and Dr. K. Takehara for providing of human hair sample. The authors would also like to appreciate Mr. Tomoki Fukui for their helpful technical supports. The experiment was performed under the approval of SPring-8 committee 2010B1097, 2011A1449, 2011A1453, and 2012A1239.

References

[1] Takeuchi A, Suzuki Y, and Uesugi K 2012 *Rev. Sci. Instrum.* **83** 083701.
[2] Takeuchi A, Suzuki Y, and Uesugi K, in preparation.
[3] Takeuchi A, Uesugi K and Suzuki Y 2010 *AIP Conf. Proc.* **1266** 42.

The development of synchrotron-assisted scanning probe microscopy at NSRRC

Yuet-Loy Chan[1], Xihui Liang[2], Tsung-Hsuan Wu[4], Dah-An Lu[1, 2], Meng-Fan Luo[2], Yao-Jane Hsu[1], and D H Wei[1,3,4,*]

[1] National Synchrotron Radiation Research Center, Hsinchu 30076, Taiwan
[2] Department of Physics, National Central University, Taoyuan 32001, Taiwan
[3] Graduate Program for Science and Technology of Synchrotron Light Source, National Tsing Hua University, 30013 Hsinchu, Taiwan
[4] Department of Physics, National Sun Yat-sen University, Kaohsiung 80424,Taiwan

E-mail: dhw@nsrrc.org.tw

Abstract. Synchrotron-based X-ray microspectroscopy is a technique that brings together microscopy and X-ray spectroscopy. It can be considered as an experimental approach capable of extracting X-ray spectrum from a finite area, or an alternative way of constructing images with spectroscopic contrast. The goal of this project is to integrate the functions of scanning tunnelling electron microscope (STM) with near edge X-ray absorption fine structure (NEXAFS) spectroscopy. Here, we describe our experimental setup, followed by recent results that demonstrate the feasibility of acquiring NEXAFS spectrum with a SiO_2 coated STM tip.

1. Introduction

Scanning tunneling microscopy (STM) is a technique capable of resolving the surface structure at atomic scale. However, while sensitive on current detection, the STM does not generally discriminate the kinetic energies of electrons. As a result, there is little chemical information can be extracted from the STM images. On the other hand, the synchrotron-based photoelectron spectroscopy is known to have a high chemical sensitivity through the incorporation of either an electron energy dispersive device or an energy tunable photon source. As it is not a trivial task to focus soft X-ray into nanometer scale, the photoelectron spectroscopy is hardly the technique of choice to examine a single structure prepared in nanometer dimension. Nevertheless, considering both techniques detect electrons, it seems reasonable to explore if it is possible to integrate STM with synchrotron radiation (SR) so that both element specific and spatially-resolved information can be accessed via a single hybrid instrument. The concept of integrating a scanning probe microscope with other spectroscopy techniques has been under pursuit since 1990s[1-8]. At NSRRC, the effort in constructing a synchrotron-assisted scanning probe microscope was initiated jointly in 2009 by National Central University and NSRRC. In this paper we report the status and recent progress of the SR-STM station at Taiwan Light Source (TLS).

2. Experimental setup

The SR-STM system at TLS is powered by a 3.9 m long undulator, U50, which delivers 3.6×10^{17} photons mm^{-2} $mrad^{-2}$ at 360 mA storage ring current [9]. Using a spherical grating monochromator, the beamline BL09 hosting U50 is operated within an energy window of 80 to 1200 eV. To enhance the photon brightness, a set of Kirkpatrick-Baez (KB) mirror was installed at BL05A2 branch to focus incoming photon down to a 20 μm x 70 μm spot at the sample position. The STM system used in this study consists of a UNISOKU USM 1000 system, a sample preparation chamber, and a fast load lock system. Optical alignment was performed when the system is connected to the beamline to ensure a proper X-ray illumination on STM tip/sample. In current setup, we have a 5° of X-ray incident angle measured from the sample surface. Figure 1 shows the optical layout of SR-STM at BL05A2.

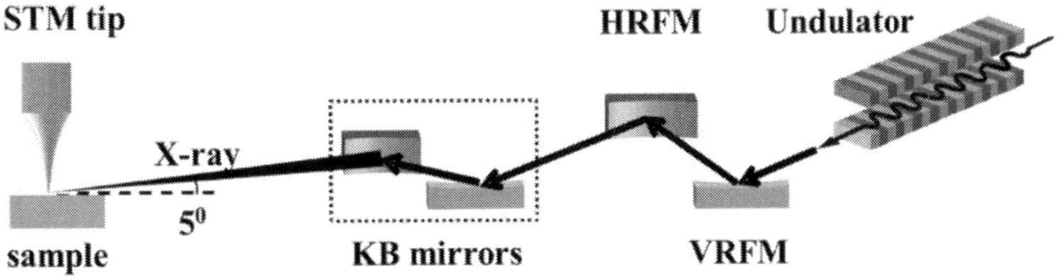

Figure 1. Schematic drawing of SR-STM setup at TLS beamline BL05A2.

3. Insulated tip preparation

One challenging issue in developing the synchrotron-assisted STM (ST-STM) is to collect only those electrons from regions of interest. However, the grazing incidence of micrometer-sized photon beam means that the area capable of emitting photo-excited electrons is in the order of millimeter. To avoid the collection of electrons emitting from regions far away from the tip, the metallic tip needs to be modified [10,11]. Our approach is to cover the tip with a thin glass layer except its apex. Following is a brief description of how the glass-covered tip was fabricated; a tungsten (W) tip of 0.5 mm in diameter was inserted into a glass tube that has inner and outer diameters of 0.75 and 1.0 mm, respectively. The glass surrounding tip apex was made thinner by pulling the glass tube at a temperature close to its melting point. A 50 ML hydrofluoric acid was use to etch away the final glass layer covering tip apex. Glass covered W-tip with less than 5 μm opening at apex was successfully fabricated. Figure 2 shows the schematic drawing of our in-house setting for tip modification. In (I), (a) is an aluminum made frame which supports of all the installation, (b) is a macro ceramic which holds up the electrodes (c), power connector (d) and filament(e), (f) is an O.D. 1mm I.D. 0.75mm capillary glass tube, (g) is an O.D. 2mm capillary glass tube and it connects with (f), (h) is a screw used to fix (f) in order to stops the its sliding , (i) is a plastic tube connected to pumping system, (j) is a ~3g metal plate for pulling the capillary glass(g). In (II), it illustrates the detail arrangement between tip and filament. (III) and (IV) are SEM images of a glass-coated W tip before and after the HF treatment.

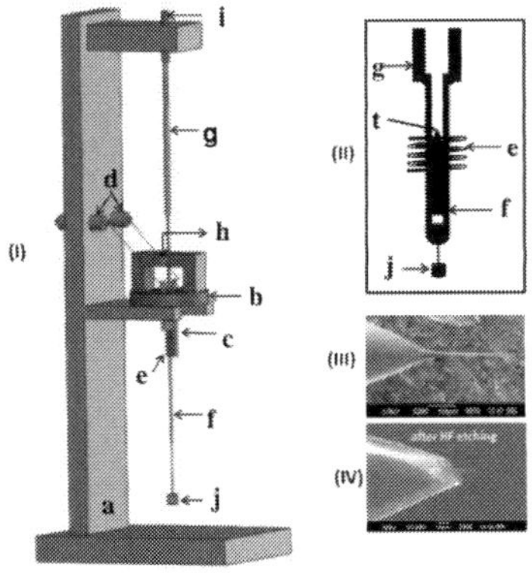

Figure 2. Schematic drawing of the apparatus that was used to fabricate tip insulation.

4. Results and discussions

Fig. 3(a) is a STM image of Cu(100)/Co (200 nm x 200 nm) recorded with a standard W-tip at constant current mode (0.2 nA). The Co islands were clearly observed before X-ray irradiation. However, upon the irradiation of energetic photon (778 eV), the image gets distorted immediately. Such a phenomenon is understandable based on the facts that X-ray illumination would cause both W-tip and surface to emit photoelectron. The tip would first experience a sudden increase in current upon the arrival of photon, followed with a response of fast retraction from the proximity of surface. What was unexpected in fig. 3(a) is the appearance of strip patterns within the period of time that X-ray irradiates. As the spatial modulation seen in image corresponds to a 60 Hz modulation in time domain, we believed the periodic patterns seen in fig. 3(a) are actually originated from the responses of STM electronics trying to accommodate photon-induced current. However, unlike the tunneling current is a quantity sensitive to tip-surface separation, the photon-induced current is not. As a result, the STM feedback system would lead to a period pattern as depict in fig. 3(a). Besides the image acquisition, a STM tip can in principle act as an electron detector. Our next attempt is thus to extract localized NEXAFS spectra by approaching a miniature detector (STM tip) to the area of interest. This is what depicted in fig. 3(b) in which NEXAFS spectra were recorded with a glass coated tip positioned a little beyond the tunneling distance but remains close to the Cu(100)/Co surface. During the spectra acquisition, the feedback loop in STM controller was turned off. Fig. 3(b) shows the cobalt *L*-edge absorption resonances recorded at four different bias (bias is applied on sample). As expected, a large negative bias would repel more electrons away from the surface to cause larger tip current. We are now working on improving the vibration isolation of our SR-STM system, and designing a chopping system to discriminate tunneling and photo-excited electrons.

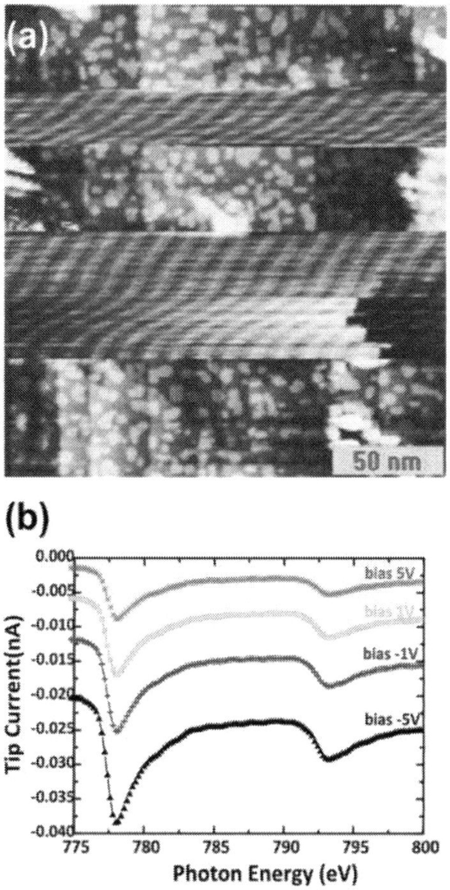

Figure 3. (a) With a standard W-tip, the STM image of Cu(100)/Co is severely perturbed upon X-ray illumination. (b) The intensities of NEXAFS spectra collected by an insulated tip are sensitive to the bias.

We would like to thank Dr. Chi-Yi Huang and Dr. Chien-I Ma for their efforts in setting up the KB mirrors system, Mr. Shang-Wei Lin for his support in beamline programming. National Synchrotron Radiation Research Center provided financial support of this work.

References:

[1] Kouichi Tsuji and Kazuaki Wagatsuma, Jpn. J. Appl. Phys., 36,1264(1997).

[2] Kouichi Tsuji, Yukio Hasegawa, Kazuaki Wagatsuma and Toshio Sakurai, Jpn. J. Appl. Phys. , 37, L1271(1998).

[3] Kouichi Tsuji, Kazuaki Wagatsuma, Kazuaki Sugiyama, Kenji Hiraga and Yoshio Waseda, Surf. Interface Anal. 27, 132 (1999).
[4] Toyoaki Eguchi, Taichi Okuda, Takeshi Matsushima, Akira Kataoka, Ayumi Harasawa, Kotone Akiyama, Toyohiko Kinoshita, Yukio Hasegawa, Masanori Kawamori, Yuichi Haruyama, and Shinji Matsui, Appl. Phys. Lett. 89, 243119 (2006).
[5] Akira Saito, Junpei Maruyama, Ken Manabe, Katsuyuki Kitamoto, Koji Takahashi, Kazuhiro Takami, Makina Yabashi, Yoshihito Tanaka, Daigo Miwa, Masashi Ishii,Yasumasa Takagi, Megumi Akai-Kasaya, Shik Shin, Tetsuya Ishikawa, Yuji Kuwaharaa, and Masakazu Aonoa, J. Synchrotron Rad. 13, 216 (2006).
[6] Akira Saito, Yasumasa Takagi, Koji Takahashi, Hiromasa Hosokawa, Kazuhisa Hanai, Takehiro Tanaka, Megumi Akai-kasaya, Yoshihito Tanaka, Shik Shin, Tetsuya Ishikawa, Yuji Kuwaharaa, and Masakazu Aono, Surf. Interface Anal. 40, 1033 (2008).
[7] Taichi Okuda, Toyoaki Eguchi, Kotone Akiyama, Ayumi Harasawa,Toyohiko Kinoshita,Yukio Hasegawa,Masanori Kawamori,Yuichi Haruyama, and Shinji Matsui, Phys. Rev. Lett. 102, 105503 (2009).
[8] V. Rose, J. W. Freeland, K. E. Gray, and S. K. Streiffer, Appl. Phys. Lett. 92, 193510 (2008).
[9] Cheng-Hao Ko, Ruth Klauser, Der-Hsin Wei, Hei-Hing Chan and T. J. Chuang, J. Synchrotron Rad. 5, 299 (1998).
[10] Ching-Yuan Chiu, Yuet-Loy Chan, Y. J. Hsu, and D. H. Wei, Appl. Phys. Lett. 92, 103101 (2008).
[11] A. Saito, K. Takahashi, Y. Takagi, K. Nakamatsu, K. Hanai, Y. Tanaka, D. Miwa, M. Akai-kasaya, S. Shin, S. Matsui, T. Ishikawa, Y. Kuwahara and M. Aono, Surf. Sci., 601, 5294 (2007).

In situ study on solidification controlled by electric current pulse in Al-Bi immiscible alloy via synchrotron radiation imaging technology

Zhu Jing[a], Wang Tongmin[a*], Cao Fei[a], Fu Yanan[b], Xie Honglan[b], Xiao Tiqiao[b]

[a] School of Materials Science and Engineering, Dalian University of Technology, Dalian 116024, P.R.China

[b] Shanghai Synchrotron Radiation Facility, Shanghai 215600, China

Email: tmwang@dlut.edu.cn

Abstract. The Bi-rich droplets motion in a solidifying Al-Bi immiscible alloy was *in situ* investigated using a synchrotron microradiography technique. The electric current pulse (ECP) was applied to control the droplets motion and distribution in the Al-matrix during solidification. Without ECP treatment, it was found that the second phase presented spherical droplets and distributed in the Al-matrix without obvious aggregation and segregation. With ECP treatment, the second phase seriously aggregated and formed lots of big clusters in the matrix. The electromagnetic pinch force induced by ECP was used to explain this phenomenon.

1. Introduction

The immiscible alloys exhibit tribological property as engine bearings [1] due to the special distribution of second phase in the matrix. The excellent tribological property depends strongly on the homogeneous distribution of the second-phase droplets in the Al-matrix. However, it is difficult to achieve ideal distribution because of the segregation and coagulation of the minority phase during solidification. In order to avoid these undesired phenomena, researchers carry out many experiments and numerical simulations to probe the aggregation mechanism of second-phase droplets. Wu et al. [2] studied the influence of phase-transport phenomena on macrosegregation and structure formation during solidification. Silva et al. [3] performed directional unsteady-state solidification experiments with different Al-Bi alloys. Wang et al. [4] described phase separation and structural evolution of undercooled monotectic alloy.

It is noted that the experimental results obtained above were static analysis using a scanning electron microscope or metallographic microscope after quenching. This will lead to some crucial missing in microstructure variation. The synchrotron radiation technology provided a new method to real time record the dynamic solidification process of metal alloys. So far, it has been successfully used to study dendrites growth [5-7].

It is well known that the electric current pulse (ECP) has been widely applied on the solidification process of metal alloys in recent years. Zhai et al. [8] researched the microstructure refinement of pure aluminum by ECP. Gao et al. [9] probed the effect of ECP on tensile strength and elongation of casting ZA27 alloy. Li et al. [10] studied the influence of ECP on the microstructure of MnBi/Bi eutectic alloy. To date, little work about the effect of ECP on second-phase droplets motion of immiscible alloys has been reported. In this paper, consequently, second-phase droplets motion of Al-Bi immiscible alloy controlled by ECP has been studied in situ by synchrotron radiation imaging technology.

2. Experimental method

The experiments were carried out on the BL13W1 beamline at SSRF. During the experiments, the incident x-ray beam was monochromatized at 18 keV and the beam size applied gave a field of view corresponding to 7×7 mm^2. A large field of view was essential for these types of studies to observe the dynamical phenomena of dendrite growth in real time and a fast readout-low noise CCD video camera was used to capture digital images simultaneously. Its pixel size was down to 3.7 μm. The sample-to-detector distance was adjustable from 15 cm to 25 cm. The exposure time per frame was about 2-7 s. The in-line X-ray phase imaging method was used in this work, as shown in figure 1.

The solidification process occurred in a self-designed furnace. It was composed of high- and low-temperature furnace rooms, each of which was made up of heating coil wrapped with heat insulation fiber-cotton. Both furnace rooms were controlled independently in the same temperature ranges by precise temperature controller. During these experiments, the sample was fixed between the high- and low-temperature furnace rooms with the irradiation of x-rays passing through. Figure 2 shows the schematic diagram of the furnace.

A binary alloy was prepared with the composition of Al-6.5wt.%Bi. A well polished thin Al-6.5wt.%Bi alloy sample was cut into 1 mm thick rectangular slices with sizes of 20×10 mm^2 and polished to a final thickness of 200 μm. The sample was set between a thin mica sheet with a thickness of 100 μm used to fix the alloy sample as a mold, and placed between two ceramic plates. Two rectangular 280 μm thick ceramic plates were enclosed with gypsum pattern all around. Two strips of nickel foil, as electrodes, were respectively set on the upper part and lower part of the sample.

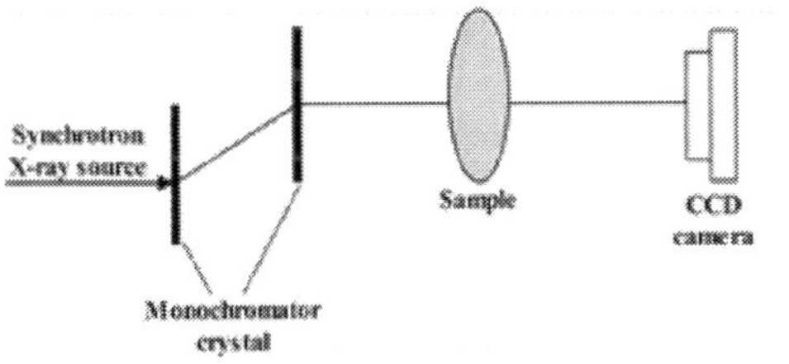

Figure 1. Schematic diagram of in-line X-ray phase-contrast imaging.

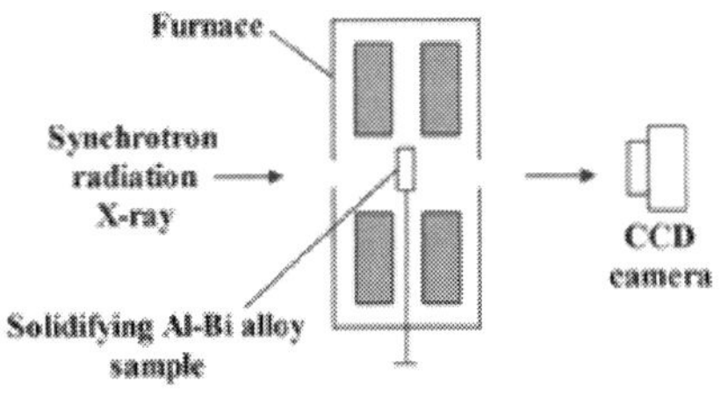

Figure 2. Schematic diagram of the furnace.

3. Results and discussion

In this experiment, when alloy was melted completely at 750 °C, both high- and low-temperature furnace rooms were cooled synchronously at the cooling rate of 5 °C/min. The morphology evolution of second-phase droplets in the Al-matrix without ECP treatment is shown in figure 3. At the initial stage of solidification, the second phase emerges in the field of vision with breezing droplets, as shown in figure 3(a). As the solidification proceeds, Bi-rich droplets separate out and tend to move up owing to the effect of Marangoni force (figure 3(b-c)). At the final stage of solidification, some Bi-rich droplets coagulate and suspend in the melt in order to balance the Marangoni force and gravity (figure 3(d)).

When ECP is applied in the whole solidification, the morphology of second-phase droplets varies obviously. At the early stage of solidification, the Bi-rich droplets nucleate slowly, as shown in figure 4(a). However, from 666 s, Bi-rich droplets grow up rapidly and aggregate seriously. The morphology evolution of second-phase droplets look like pieces of clouds rather than spherical droplets and the big clusters form at the final stage of solidification (figure 4(b-d)).

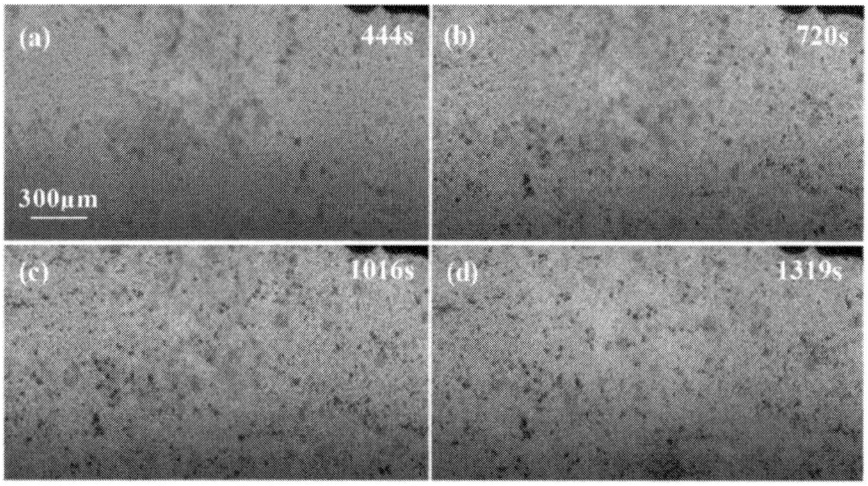

Figure 3. The morphology evolution of second-phase droplets in the Al-matrix without ECP treatment.

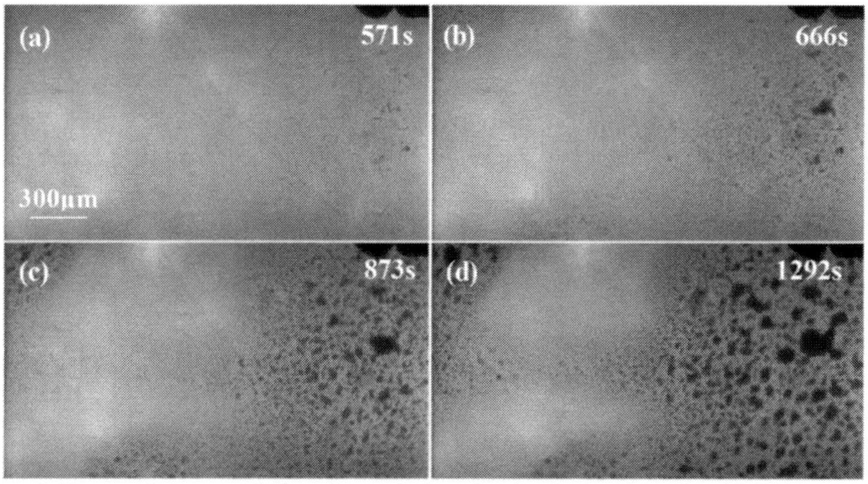

Figure 4. The morphology evolution of second-phase droplets in the Al-matrix with ECP treatment.

Figure 5 shows the formation of droplet clusters (explained by droplet A and B). When ECP is applied, the electromagnetic oscillation force with the direction parallel matrix centre exists due to the induced magnetic field. Meanwhile, another force, namely electromagnetic pinch force with the direction opposite to electromagnetic oscillation force appears. When the electric conductivity of the droplet is much smaller than the electric conductivity of the matrix, the equation can be described as following [11]:

$$F_p = -\frac{3}{4}\frac{\pi d_p^3}{6}F \tag{1}$$

Where F is electromagnetic oscillation force and d is droplet diameter. When droplet A and droplet B move, there is the motion velocity v_A and v_B acting on droplet A and droplet B, respectively. The motion velocity of the droplets, v_A and v_B, can be derived as follows [12]:

$$v_A = -\frac{F}{24\mu}d_A^2 \tag{2}$$

$$v_B = -\frac{F}{24\mu} d_B^2 \tag{3}$$

Where μ is the viscosity of the liquid metal. According to the formula (2) and (3), droplet A moves faster than droplet B. In the early stage, droplet A and droplet B move in the same direction. In the middle stage, as the direction of electromagnetic pinch force changing, droplet B moves in the opposite direction. Droplet A, however, cannot change the motion direction immediately owing to its original motion velocity. In the last stage, pieces of big clusters appear due to the collision between droplet A and droplet B.

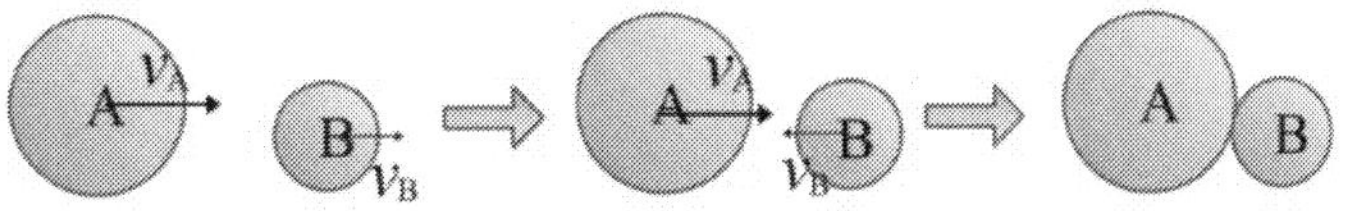

Figure 5. The collision explanation of second-phase droplets in Al-Bi immiscible alloy with ECP treatment.

4. Conclusions

The synchrotron radiation imaging technology is used successfully to *in situ* investigate the second-phase droplets motion in Al-Bi alloy under ECP. With ECP treatment, it is clear that the second droplets aggregate seriously. The electromagnetic pinch force is considered as the main factor to affect the droplets motion. The droplets aggregation is undesired, but the application of ECP is an effective method to control second-phase droplets motion in the matrix melt.

5. Acknowledgements

The authors gratefully acknowledge supports of National Natural Science Foundation of China (Nos. 50601003, 50971032, 51071035, 51274054), the Keygrant Project of Chinese Ministry of Education (No. 313011). The authors wish to thank all the staff members of the BL13W1 beamline of SSRF and 4W1A beamline of Beijing Synchrotron Radiation Facility (BSRF).

References

[1] Schaffer P L, Mathiesen R H and Arnberg L 2009 Acta Marter. **57** 2887
[2] Wu M H and Ludwig A 2003 Adv. Eng. Mater. **5** 62
[3] Silva A P, Spinelli J E, Mangelinck-Noël N and Garcia A 2010 Mater. Des. **31** 4584
[4] Wang N and Wei B 2003 Mater. Sci. Eng. A **345** 145
[5] Wang T M, Xu J J, Li J, Huang W X, Liu S C, Li T J 2010 Sci China Technol Sc **53** 1278
[6] Yasuda H, Yamamoto Y, Nakatsuka N, Yoshiya M, Nagira T, Sugiyama A, Ohnaka I, Uesugi K and Umetani K 2009 Int J. Cast Metal Res. **21** 125
[7] Li B, Brody H D and Kazimirov A 2007 Metall. Mater. Trans. A **38A** 599
[8] Li J, Ma J H, Gao Y L and Zhai Q J 2008 Mater. Sci. Eng. A **490** 452
[9] Gao M, He G H, Yang F, Guo J D, Yuan Z X and Zhou B L 2002 Mater. Sci. Eng. A **337** 110
[10] Li F, Regel L L and Wilcox W R 2001 J. Cryst. Growth **223** 251
[11] Taniguchi S J and Brimacombe J K 1994 ISIJ Int. **34** 722
[12] Zhong Y B, Ren ZM, Zhang B W, Deng K and Xu K D 2004 The Chinese Journal of Nonferrous Metals **14** 1329 (In Chinese)

Spatially resolved NEXAFS spectroscopy of siderophores in biological matrices

J Thieme[1], D Kilcoyne[2], T Tyliszczak[2], K Haselwandter[3]
[1]Brookhaven National Laboratory, NSLS-II, Upton, NY-11973, USA
[2]Lawrence Berkeley National Laboratory, ALS, 1 Cyclotron Road, Berkeley, CA-94720, USA
[3]University of Innsbruck, Department of Microbiology, Technikerstrasse 25, A-6020 Innsbruck, Austria

E-mail: jthieme@bnl.gov

Abstract. Iron is an essential nutrient for almost all forms of life. In the presence of oxygen iron is present in its ferric form which precipitates under formation of rather insoluble oxide-hydroxide polymers. Hence the bioavailability of iron is extremely low (< 10^{-17} M at pH 7 for Fe^{3+}). Under such conditions almost all microorganisms synthesize siderophores as iron chelating agents, thus solubilizing ferric iron from rather insoluble iron sources. Siderophores form soluble complexes with Fe^{3+}. The present study aims at developing a methodology that would allow for the specific detection and localization of such iron chelators in their natural environment. The applicability of spatially resolved NEXAFS spectroscopy in the soft X-ray energy (E < 1 keV) range was evaluated for localization of typical fungal hydroxamate siderophores like ferrichrome or coprogen, which can be present in various biological materials. Results obtained with the scanning transmission X-ray microscopes at beamlines 11.0.2 and 5.3.2 of the ALS have shown characteristic signatures for siderophores. Thus NEXAFS spectroscopy at the carbon K-edge, nitrogen K-edge and iron L-edge with high spatial resolution has proven to be extremely useful for their identification in their natural environment. Spectra of different siderophores as well as spectra and images of biological material containing siderophores are presented.

1. Introduction

In soil environments siderophores may function as ligand-based weathering agents leading to iron dissolution [1]. Iron is an essential nutrient for almost all forms of life. However, in an aerobic environment iron is present as Fe^{3+} and precipitates forming rather insoluble oxide-hydroxide polymers. Hence the bioavailability of iron is extremely low with a maximum concentration in the range of 10^{-17} M at pH 7 for Fe^{3+}. Under such conditions almost all microorganisms synthesize siderophores as iron chelating agents with very high stability constants, thus solubilizing ferric iron from rather insoluble iron sources. Siderophores form soluble complexes with Fe^{3+}. Most of the fungi release a variety of siderophores with different physico-chemical properties [3, 5]. Ascomycota and Basidiomycota typically produce hydroxamate-type siderophores [2]. Siderophores of fungal symbionts such as mycorrhizal fungi appear to play a role in mobilization of iron in the soil environment [4].

2. Materials and Method

Lichens are one of the most outstanding examples of a symbiosis between a fungal mycobiont and a photosynthetic algal and/or cyanobacterial photobiont partner. They are capable of colonizing bare rocks and other mineral substrates. Therefore, lichens are extremely important in soil formation during ecological successions. The fungal partner in the lichen mutualism gains important benefits through access to photosynthetic products of the alga or bacterium. The algal partner profits from the availability of a protected habitat, and greater access to inorganic nutrients. The latter immediately leads to the question whether the fungal partner synthesizes siderophores to interact with mineral surfaces. Siderophores are highly specific Fe^{3+} ligands, their formation constants often showing values $> 10^{30}$. Fungi and bacteria synthesize these organic molecules. The aim of the study presented here has been to evaluate whether NEXAFS spectroscopy can be a useful tool for the unambiguous identifcation of siderophores in their natural environment, when combined with the high spatial resolution of a scanning transmission X-ray microscope.

The two K-absorption edges of carbon and nitrogen and the L-absorption edge of iron have been used to find characteristic signatures useful for an identification. Beamlines 11.0.2 and 5.3.2 at the Advanced Light Source at the Lawrence Berkeley National Laboratory have been used for these studies. Both beamlines offer access to all three absorption edges. Beamline 11.0.2 is an undulator beamline, covering an energy range of E = 200-1900 eV with an energy resolution of $E/\Delta E < 7500$. Fresnel zone plate optics are used to create a focal spot that allows for a spatial resolution below 20 nm. Beamline 5.3.2 is a beamline using bending magnet radiation, covering an energy range of E = 250-800 eV with an energy resolution of $E/\Delta E < 5000$. Also here, Fresnel zone plate optics are used to create a focal spot that allows for a spatial resolution below 30 nm. Hence, both beamlines are very well suited for these studies. A detailed description of these beamlines can be found elsewhere [6, 7]. For the studies presented here, NEXAFS spectra of the siderophores coprogen and ferrichrome have been taken at the carbon and the nitrogen K-edge. The spectra shown in figure 4, are used to analyze the spectral signature of a lichen sample. The lichen species investigated, *Parmelia Saxatilis,* was growing on the surface of hematite rocks. Minute particles of this lichen sample have been placed on a silicon nitride (Si_3N_4) membrane with 100 nm thickness. Small parts such as isolated hyphae stuck to the surface of the membrane and allowed taking images and spectra. No further sample preparation steps have been applied, therefore the samples are considered to be in an almost unaltered natural state. X-ray images as well as image stacks have been taken from these samples. The latter has been used to extract spectra from regions of interest within the samples, using the programs AXIS2000 and StackAnalyze [8, 9]. Figure 1 shows as a typical example isolated fungal hyphae. Subsequent to taking X-ray images of these hyphae, image stacks across the three absorption edges of carbon, nitrogen and iron have been taken for spectral analysis. To identify siderophores within the organic matrix of the hyphae, images below and above the L-edge of iron have been used. As siderophores form complexes with iron, this can be used to identify regions of interest for further analysis.

3. Results and Outlook

Fungal hyphae isolated from the lichen thallus are shown in Figure 1. Figure 2 shows an example of an iron containing area found in a small part of the fungal hyphae imaged in figure 1. Figure 3 shows the corresponding spectra at the carbon and the nitrogen K-edge. Figure 4 compares spectra, taken at the carbon K-edge and the nitrogen K-edge, of two siderophores and a sample of *Parmelia saxatilis*. These experiments demonstrate that it is possible to obtain specific signatures for different fungal siderophores using NEXAFS spectroscopy at the absorption edges of carbon, nitrogen and iron. The carbon K-edge spectrum of *Parmelia saxatilis* shows three distinct peaks, at the energy positions E = 285.2 eV, E = 286.8 eV, and E = 288.5 eV. Two peaks are explainable by the spectra of the siderophores coprogen and ferrichrome. Coprogen shows two peaks at almost the same positions, at E = 285 eV, and E = 288.2 eV. Ferrichrome has a prominent peak at an X-ray energy of E = 288.2 eV. Comparing these positions with values from a large number of publications of spectra from simpler molecules, these peaks arise from the aromatic C=C double bond around E = 285 eV and the C=O double bond around E = 288.5 eV. The third peak in the spectrum of the lichen species at E = 286.8 eV might arise from a phenolic C=C double bond, showing the difference of aromatic and phenolic C=C. Here, more data are necessary and future experiments are planned to address this uncertainty. At the nitrogen K-edge, the spectrum of *Parmelia saxatilis* could be explained mainly by coprogen, although the mismatch in peak heights indicates missing components. Ferrichrome shows a prominent peak at E = 401.5 eV with a smaller peak at the higher energy of E = 402.6 eV. The coprogen spectrum has a small peak at E = 399.1 eV and a prominent peak at E = 401.5 eV. *Parmelia saxatilis*, spectra (Figures 3 and 4), show four clearly visible absorption peaks in the spectrum, at the energies E = 398.8 eV, E = 399.9 eV, E = 400.9 eV, and E = 401.5 eV. All these peaks represent 1s - π^* resonances, the σ^* resonances are found at higher energies [10, 11] The peaks at E = 401.5 eV and E = 402.6 eV originate for the amide groups within the siderophores, whereas the peaks at lower energies stem from quinoid structures within the molecules. These assignments fit well with the known structure of these molecules. Still, further measurements will be necessary to enable a more in-depth analysis.

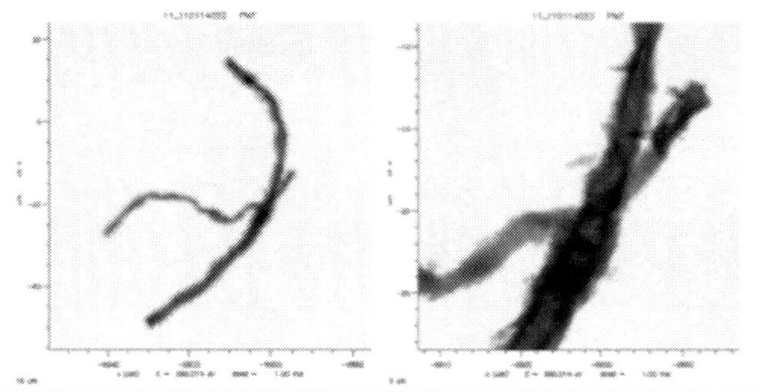

Figure 1. STXM images of fungal hyphae isolated from the lichen *Parmelia saxatilis*, growing on hematite and imaged on a Si_3N_4 membrane. Left: Image size 80x80 μm^2 at 160x160 pxl, X-ray energy E = 690 eV, dwell time per pixel t = 1 ms. Right: Image size 20x20 μm^2 at 200x200 pxl, X-ray energy E = 690 eV, dwell time per pixel t = 1ms.

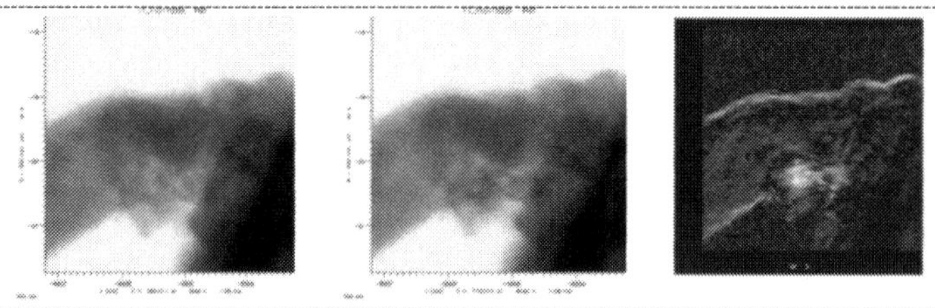

Figure 2. STXM images of a detail of the hyphae shown in Figure 1, taken from a stack of images across the Fe L-absorption edge. Image size 3 x 3 μm^2 at 100 x 100 pxl^2, dwell time per pixel t = 1ms. Left: X-ray energy E = 690 eV; middle: E = 710 eV; right: division of left by middle image.

The results of the first experiments presented here demonstrate that siderophores can be distinguished from each other and can be found in a lichen thallus, hitherto unknown to contain siderophores. The next step will be to search for a unambiguous identification of siderophores in lichens. Spectral signatures of more siderophores, such as ferricrocin, fusigen, rhizoferrin or enterobactin are needed for this purpose. In addition, more lichen species will have to be studied to to strengthen the claim that NEXAFS spectroscopy with high spatial resolution is a very useful tool for these studies. The ultimate goal will be to detect and localize siderophores in lichen thalli and to identify the species of siderophores therein. Such an approach will facilitate studies at the mineral / lichen interface, helping to better understand processes underlying e.g. the weathering of rocks and soil formation.

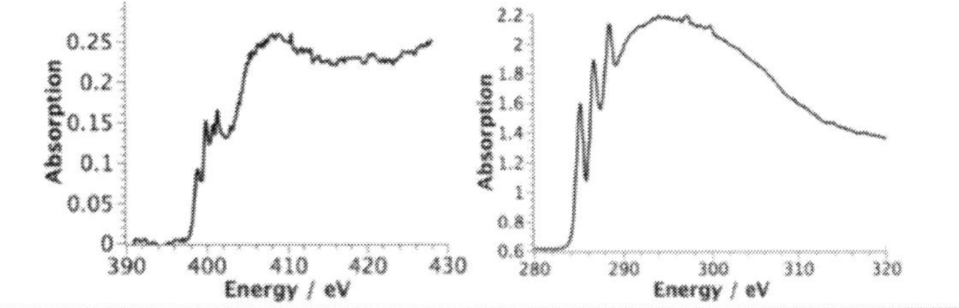

Figure 3. NEXAFS spectra extracted from image stacks taken across the K-absorption edges of nitrogen (left) and carbon (right). The region of interest imaged using such stacks is shown as the iron containing area (bright spot) in Figure 2 (right).

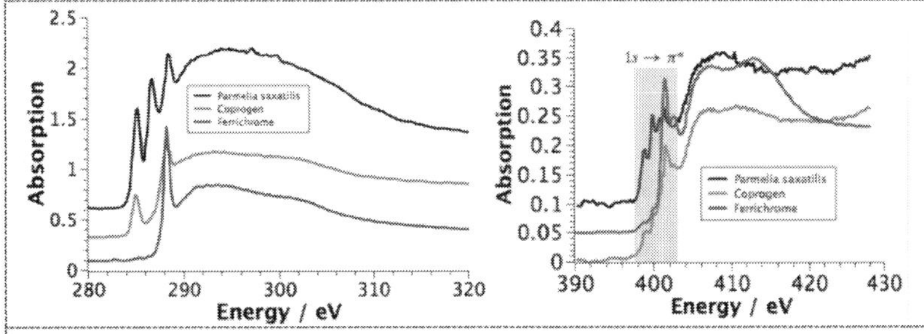

Figure 4. Comparison of NEXAFS spectra taken from the two siderophores coprogen and ferrichrome and from the lichen species *Parmelia saxatilis*. A detailed analysis is discussed in the text.

Acknowledgments
Data were acquired at beamlines 5.3.2.2 and 11.0.2 at the ALS, which is supported by the Director of the Office of Science, U.S. Department of Energy under Contract No. DE-AC02-05CH11231.

References

[1] Hoffland E, Kuyper TW, Wallander H, Plassard C, Gorbushina AA, Haselwandter K, Holmström S, Landeweert R, Lundström US, Rosling A, Sen R, Smits MM, van Hees PAW, van Breemen N 2004 *Front Ecol Environ* **2** 258-264
[2] Winkelmann G 2007 *Biometals* **20** 379-392
[3] Renshaw JC, Robson GD, Trinci APJ, Wiebe MG, Livens FR, Collison D, Taylor RJ 2002 *Mycol Res* **106** 1123-1142
[4] Haselwandter K 2008 *Mineralogical Magazine* **72** 61-64
[5] Haselwandter K, Häninger, G, Ganzera, M 2011 *Biometals* **24**, 153-157
[6] Kilcoyne ALD, Tyliszczak T, Steele WF, Fakra S, Hitchcock P, Franck K, Anderson E, Harteneck B, Rightor EG, Mitchell GE, Hitchcock AP, Yang L, Warwick T, Ade H, 2003 *J Synchrotron Rad* **10**, 125–136
[7] Warwick T, Ade H, Kilcoyne ALD, Kritscher M, Tylisczcak T, Fakra S, Hitchcock A, Hitchcock P, Padmore H 2002 *J Synchrotron Rad* **9(4)** 254–257
[8] http://unicorn.mcmaster.ca/aXis2000.html
[9] Jacobsen C, Flynn G, Wirick S, Zimba C 2000 *J of Microscopy* **197(2)** 173-184
[10] Vairavamurthy A, Wang S 2002 *Environ Sci Technol* **36**, 3050-305
[11] Leinweber P, Kruse J, Walley F, Gillespie A, Eckhardt KU, Blyth RIR, Regier TZ 2010 *Developments in Soil Science* **34**, 255-288

Multi-contrast computed laminography at ANKA light source

Y Cheng[1,2], V Altapova[3], L Helfen[1,4], F Xu[1], T dos Santos Rolo[1], P Vagovič[1], M Fiederle[2] and T Baumbach[1,3]

[1] Institute for Photon Science and Synchrotron Radiation /ANKA light Source, Karlsruhe Institute of Technology (KIT), D-76344 Karlsruhe, Germany
[2] Freiburger Materialforschungszentrum (FMF), University of Freiburg, D-79104 Freiburg, Germany
[3] Laboratory for Applications of Synchrotron Radiation (LAS), Karlsruhe Institute of Technology (KIT), Postfach 6980, D-76128 Karlsruhe, Germany
[4] European Synchrotron Radiation Facility (ESRF), F-38043 Grenoble, France

E-mail: yin.cheng@kit.edu

Abstract. X-ray computed laminography has been developed as a non-destructive imaging technique for inspecting laterally extended objects. Benefiting from a parallel-beam geometry, high photon flux of synchrotron sources and modern high-resolution detector systems, synchrotron radiation computed laminography (SRCL) results in a powerful three-dimensional microscopy technique. SRCL can be combined with different contrast modes, such as absorption, phase and dark-field contrasts, in order to provide complementary information for the same specimen. Here we show the development of SRCL at the TopoTomo beamline of the ANKA light source. A novel instrumentation design is reported and compared to the existing one. For this design, experimental results from different contrast modalities are shown.

1. Introduction

Computed laminography using synchrotron radiation has been developed as a non-destructive three-dimensional (3-D) imaging method for planar specimens [1]. Compared with its laboratory counterparts, a parallel-beam geometry and high flux enables fast and monochromatic imaging with high resolution. In comparison to computed tomography (CT), SRCL is based on the inclination of the tomographic rotation axis with respect to the incident X-ray beam by a defined angle, the so-called laminographic angle θ. In other words, $\theta = 90°$ and $\theta = 0°$ are the limiting cases of CT and radiography, respectively.

SRCL provides a possibility to zoom into regions of interest (ROIs) of macroscopically large specimens with microscopic resolution without affecting their integrity. Originally it has been motivated by and developed for non-destructive microsystem device inspection [2, 3]. Recently, the technique has been extended to applications in materials science [4, 5], paleontology [6] and cultural heritage [7]. Nowadays by exploiting the high degree of coherence of the synchrotron beam, applications of SRCL have been further broadened by their combination with phase-contrast imaging methods and microscopy techniques [8, 9, 10].

In this paper, we will report on the development of SRCL at ANKA's TopoTomo beamline.

2. Principles and Instrumentation Design

The TopoTomo beamline at ANKA is a micro-imaging station dedicated to topography and tomography with a maximum beam cross section of 5 mm × 15 mm (V×H, FWHM) at the sample position 30 m downstream of the bending magnet source [11]. For typical imaging applications, a filtered white beam or a double-multilayer monochromator provides x-ray energies of 5 ∼ 30 keV. The combination of different scintillator crystals and microscope optics yields a large flexibility in spatial resolution values (pixel sizes from 0.36 to 6.1 μm).

A schematic drawing in Figure 1(a) depicts the basic principle of SRCL. In order to realize the SRCL geometry, an accurate inclination of the sample rotary stage with respect to the X-ray beam has to be performed. To guarantee a large freedom of the laminographic angle θ, two designs have been proposed (see Figure 1(b),(c)). The Design I (Figure 1(b)) is a novel design which utilizes a horizontal rotation by using an additional rotary stage to enable a full circular tilting (theoretically) of the sample stage. The Design II (Figure 1(c)) realizes the tilting of the sample stage by simply using a goniometer stage together with a fixing angle wedge (details of Design II has been demonstrated elsewhere [12]). Both designs have advantages and drawbacks. Comparing the degrees of freedom (DOF) for both designs, the novel Design I has a flexible inclination angle range from -180° to +180°, which allows one to perform radiography ($\theta = 0°$) if there is an aperture through the sample rotary stage. The Design II only allows an accessible inclination angle range of 0° to +32° (limited by the travel range of the goniometer).

Since both designs can complement each other and fit to different requirements or conditions, both have been implemented at the TopoTomo beamline. Figure 1(d), (e) show experimental implementations that correspond to Design I (b) and Design II (c), respectively.

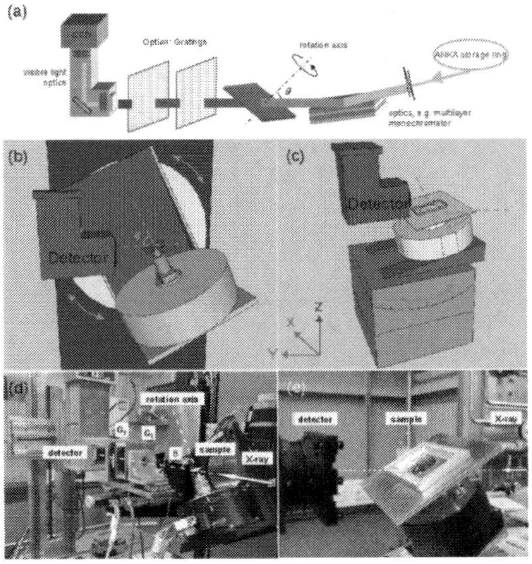

Figure 1. (a) Principle of CL at synchrotron beamlines with parallel-beam geometry. (b)(c) 3-D drawings of the different proposed designs for the SRCL sample manipulator. (d)(e) Experimental setup corresponding to Designs I (b) and II (c) respectively. Both are implemented at the TopoTomo beamline.

3. Multi-contrast SRCL

Similar to CT, SRCL can be combined with diverse contrast modes, including inline phase-contrast and Talbot interferometry in addition to the traditional absorption contrast. In the following, various contrast modalities for SRCL at the TopoTomo beamline will be illustrated by an example.

3.1. Absorption and inline phase-contrast modalities

A wavefront of spatially coherent hard X-rays becomes distorted when it penetrates an object. It undergoes not only attenuation (related to the imaginary part β of the complex refractive index) but also a phase shift, caused by the non-uniform distribution of the refractive index decrement δ in the sample. While the attenuation results in an intensity modulation behind the object, the phase distortion develops into intensity modulations only as the wave continues to propagate (provided that there is sufficient partial coherence). In the hard X-ray region, δ is orders of magnitude higher than β for light materials. Therefore, by changing the sample to detector distance we can achieve absorption laminography for attenuating materials or inline phase-contrast laminography for weakly absorbing materials.

To demonstrate the resolution achievable via SRCL at TopoTomo, we imaged a test pattern (Xradia Inc., Concord, CA) in absorption mode (Figure 2). An axis inclination angle of around 30° ($\theta \approx 60°$) and a monochromatic beam of 17 keV X-ray energy were used. Volumes were reconstructed from 1599 projections (4 s exposure time) using filtered backprojection [12].

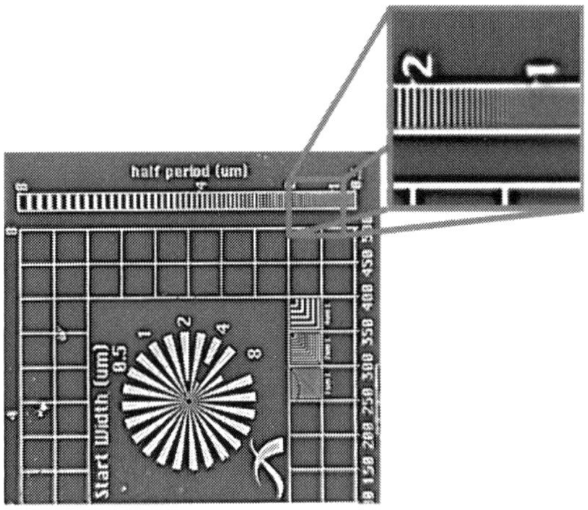

Figure 2. Reconstructed slice of Xradia resolution test pattern (X500-200-2). A zoom-in region shows that the in-plane spatial resolution allows one to distinguish features smaller than 1.5 μm size. The effective pixel size was 0.36 μm.

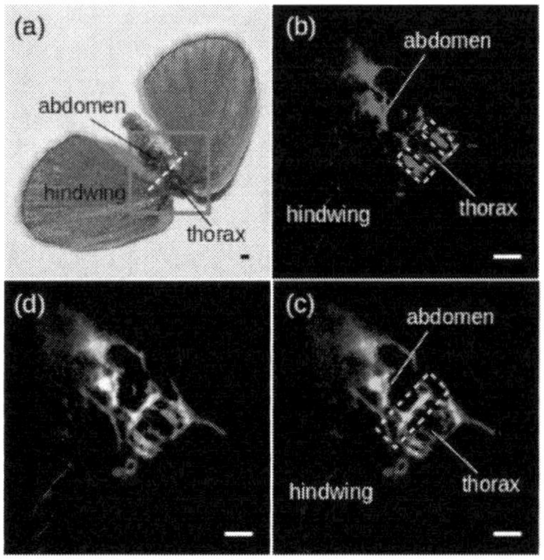

Figure 3. Images of a butterfly (*Pararge aegeria*). (a) the photograph. (b) slice of reconstructed transmission image. (c) slice of the dark-field image. (d) a signal map combining absorption (red) and dark-field (green). The effective pixel size is 6.1 μm. Scale bars: 1 mm.

3.2. Grating interferometry based modality

The grating interferometry (GI) technique is based on the Talbot effect and consists of a phase grating (G_1) and an amplitude grating (G_2), for details of this method see Refs [13, 14]. It has received great attention as a highly sensitive phase-contrast technique, providing simultaneously absorption, phase and dark-field contrast modes. Subsequently, GI-based CT has been conducted mostly for biological applications [15]. The principle of GI-based SRCL [16, 17] has already been shown in Figure 1(d). The pair of gratings is positioned downstream of the sample. We use the phase-stepping method to retrieve the absorption, differential phase and dark-field signals of the sample.

Figure 3 shows slices through the reconstructed volume of a butterfly (red box in Figure 3(a)) scanned by GI-based SRCL. The X-ray energy was chosen to 18 keV monochromatic beam, the laminographic angle was $\theta \approx 60°$. As before, 1599 projections have been taken with 4 s exposure time per projection. Phase-stepping has been conducted by moving the phase grating G_1 over 1 period using 4 steps.

From the comparison of the absorption image (Figure 3(b)) and dark-field image (Figure 3(c)) (the differential phase-contrast image not shown here), we see that at the interface of the thorax and abdomen (cyan box in Figure 3(c)), where only thin nerves, ducts or vessels are passing, shows a very strong scattering signal rather than absorption. While at the wing bases where the two wings attach to the thorax on each side (cyan box in Figure 3(b)), it shows strong absorption due to relatively heavy chitinous compositions rather than showing scattering contrast. These complementary information are visualised in a combined signal map showing in Figure 3(d).

4. Conclusion

SRCL has been successfully implemented at the TopoTomo beamline of the ANKA light source, KIT, Germany. Equipped with a double multilayer monochromator, a pink beam with a spectral bandwidth of $\triangle E/E=10^{-2}$ can be obtained to perform μm-scale inline phase-contrast laminography with energies up to 30 keV. By combining CL with grating interferometry, high-sensitivity dark-field contrast of flat specimens can be achieved. These flexible combinations allow a complementary investigation of laterally extended objects. Multi-contrast SRCL is under operation and opens up new possibilities for 3D imaging in life and material sciences.

Acknowledgments

Many thanks go to Harald Schade, Stefan Uhlemann, Thorsten Müller, Jan Psariotis and Gernot Buth for beamline support. Arndt Last and the use of KNMF Facility are acknowledged for precisely fixing the resolution pattern onto its holder. Yin Cheng would like to thank Yang Yang and acknowledges financial support by the CSC-Helmholtz Scholarship Program.

References

[1] Helfen L, Baumbach T, Mikulík P *et al.* 2005 *Appl. Phys. Lett.* **86** 071915
[2] Helfen L, Myagotin A, Pernot P, DiMichiel M, Mikulík P, Berthold A and Baumbach T 2006 *Nucl. Inst. Meth. A* **563** 163–166
[3] Tian T, Xu F, Han J K, Choi D, Cheng Y, Helfen L, Di Michiel M, Baumbach T and Tu K N 2011 *Appl. Phys. Lett.* **99** 082114
[4] Moffat A J, Wright P, Helfen L *et al.* 2010 *Scr. Mater.* **62** 97
[5] Morgeneyer T, Helfen L, Sinclair I, Proudhon H, Xu F and Baumbach T 2011 *Scripta Mater.* **65** 1010 – 1013
[6] Houssaye A, Xu F, Helfen L, De Buffrnil V, Baumbach T and Tafforeau P 2011 *J. Vertebr. Paleontol.* **31** 2–7
[7] Krug K, Porra L, Coan P, Wallert A, Dik J, Coerdt A, Bravin A, Elyyan M, Reischig P, Helfen L and Baumbach T 2008 *J. Synchrotron Rad.* **15** 55–61
[8] Helfen L, Baumbach T, Cloetens P and Baruchel J 2009 *Appl. Phys. Lett.* **94** 104103
[9] Xu F, Helfen L, Moffat A J, Johnson G, Sinclair I and Baumbach T 2010 *J. Synchrotron Rad.* **17** 222–226
[10] Xu F, Helfen L, Suhonen H, Elgrabli D, Bayat S, Reischig P, Baumbach T and Cloetens P 2012 *PLoS ONE* **7** e50124
[11] Rack A *et al.* 2009 *Nucl. Instrum. Meth. B* **267** 1978 – 1988
[12] Helfen L, Myagotin A, Mikulík P, Pernot P, Voropaev A, Elyyan M, Di Michiel M, Baruchel J and Baumbach T 2011 *Rev. Sci. Instrum.* **82** 063702
[13] Momose A, Kawamoto S, Koyama I, Hamaishi Y, Takai K and Suzuki Y 2003 *Jpn. J. Appl. Phys.* **42** L866–L868
[14] Weitkamp T, Diaz A, David C, Pfeiffer F, Stampanoni M, Cloetens P and Ziegler E *Opt. Express* 6296–6304
[15] Momose A, Yashiro W, Takeda Y, Suzuki Y and Hattori T 2006 *Jpn. J. Appl. Phys.* **45** 5254–5262
[16] Harasse S, Hirayama N, Yashiro W and Momose A 2010 *Proc. SPIE* **7804** 780411
[17] Altapova V, Helfen L, Myagotin A, Hänschke D, Moosmann J, Gunneweg J and Baumbach T 2012 *Opt. Express* **20** 6496–6508

Journal of Physics: Conference Series **463** (2013) 012039 doi:10.1088/1742-6596/463/1/012039

Development of high-accuracy X-ray ptychography apparatus

Akihiro Suzuki[1], Yasunori Senba[2], Haruhiko Ohashi[2], Yoshiki Kohmura[3], Kazuto Yamauchi[1], Tetsuya Ishikawa[3], Yukio Takahashi[1]

[1]Graduate school of Engineering, Osaka University, 2-1 Yamada-Oka, Suita, Osaka 565-0871, Japan
[2] Japan Synchrotron Radiation Research Institute, SPring-8, 1-1-1 Kouto, Sayo, Hyogo, 679-5198, Japan
[3] RIKEN SPring-8 Center, 1-1-1 Kouto, Sayo, Hyogo, 679-5148, Japan

E-mail: suzuki@up.prec.eng.osaka-u.ac.jp

Abstract. X-ray ptychography allows us to observe the internal structures of non-isolated objects without lenses. We developed a high-accuracy X-ray ptychography apparatus equipped with Kirkpatrick-Baez mirrors. The full width at half-maximum of the focal profiles was ~600 nm for the vertical and horizontal directions. The estimated flux was $\sim 3\times 10^7$ photons/s. We also developed a constant-temperature system to suppress sample and/or beam drift during measurements. The temperature change of the apparatus was controlled to less than 0.04 °C over 10 hours. The total amount of drift over 10 hours was suppressed to less than 500 nm. By using both the constant-temperature system and a drift compensation method, the reconstructed image of the test object was markedly improved.

1. Introduction

Coherent X-ray diffraction imaging [1] (CXDI) is a lensless imaging technique based on coherent X-ray diffraction measurements and an oversampling method with iterative algorithms. CXDI allows us to three-dimensionally observe thick objects [2] and also provides us with unique structural information, i.e., electron density distribution, not obtainable by electron microscopy. The original geometry of CXDI, in which the sample is illuminated with an X-ray plane wave, is limited to an isolated object. X-ray ptychography [3] is an experimental approach to overcoming this important limitation, wherein a probe is scanned and coherent diffraction patterns are collected so that the illumination area overlaps with the adjacent position. Because X-ray ptychography is very sensitive to drift between the sample and the illumination optics [4], the spatial resolution of the reconstructed image is degraded by the positioning error of the X-ray beam. Therefore, exact control of the X-ray beam is important for high-resolution X-ray ptychography. We developed a high-accuracy X-ray ptychography apparatus equipped with Kirkpatrick-Baez (KB) mirrors and a constant-temperature system.

2. Experimental

An experiment on coherent X-ray diffraction was performed at BL29XUL in SPring-8. An X-ray beam was two-dimensionally focused onto a spot of 600 nm size using KB mirrors at X-ray energy of 11.8 keV. The flux of the focused X-ray beam was estimated to be $\sim 3\times 10^7$ photons/s. To suppress the thermal expansion of the optics and their supports during the measurement, we developed a constant-temperature system. We maintained the temperature in the experimental hutch at 26.0 °C using a feedback

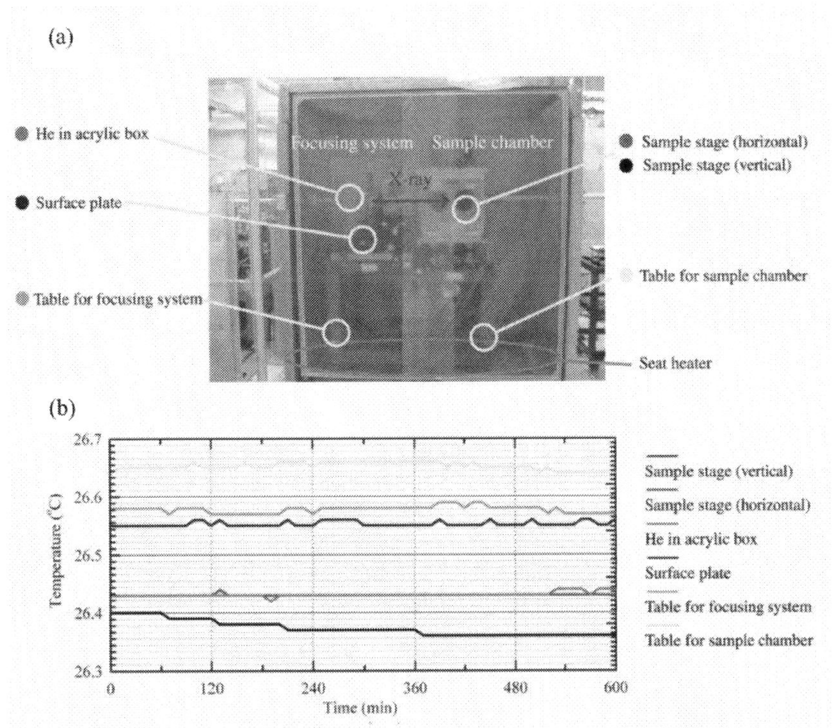

Figure 1. (a) Inside of constant-temperature room. (b) Temperature of apparatus during measurement.

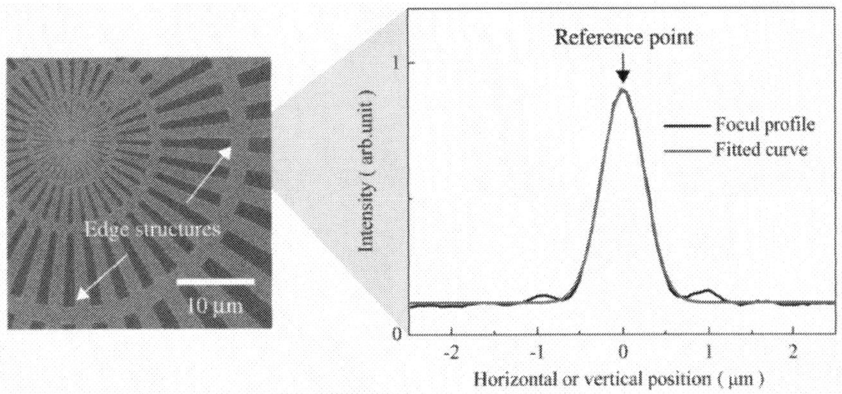

Figure 2. Schematic view of drift measurement technique using sample structures. The scattering intensities from the edge structure were measured by an in-vacuum front-illuminated charge-coupled device detector when the focused beam with a step size of 25 nm and an X-ray exposure time of 1.5 s at each position crossed the structures.

system consisting of sheet heaters , a resistance temperature detector, and a proportional- integral-derivative controller. In addition, both the focusing system and the sample chamber were set up in a constant-temperature room with insulation panels. The temperature of the constant-temperature room was maintained at 26.4 °C in the same manner as that of the experimental hutch. Figure 1(a) shows the inside of the constant-temperature room. Yellow circles indicate the measurement points of temperature. Figure 1(b) shows the temperature of the apparatus over 10 hours. The temperature change of the apparatus was controlled to less than 0.04 °C. Before installing the constant-temperature system, the temperature change of the apparatus was more than 0.5 °C. Figure 2 shows the schematic view of the drift measurement technique using sample structures. To precisely measure the amount of drift, the dark-field knife-edge scanning [5] was performed at the edge structure of the Siemens star test pattern, which consists of a 200-nm-thick nanostructured tantalum layer on a SiC membrane and contains the finest lines and spaces of 50 nm. The intensity profiles were fitted using a Gaussian function, and then the peak-top position of the fitting curve was defined as the reference point. The drift was measured using the reference point. The measurements were performed at 10 min intervals in the vertical and horizontal directions. The total measurement time was 10 hours.

3. Results and discussion

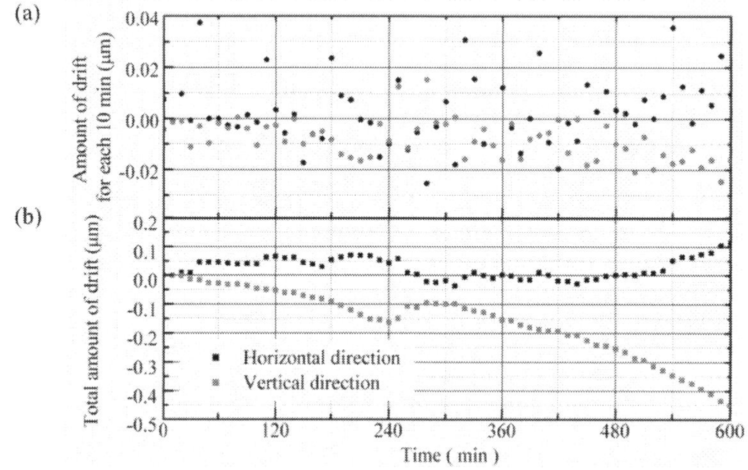

Figure 3. Amount of drift occurring with constant-temperature system. (a) Drift for every 10 minutes. (b) Accumulated drift value.

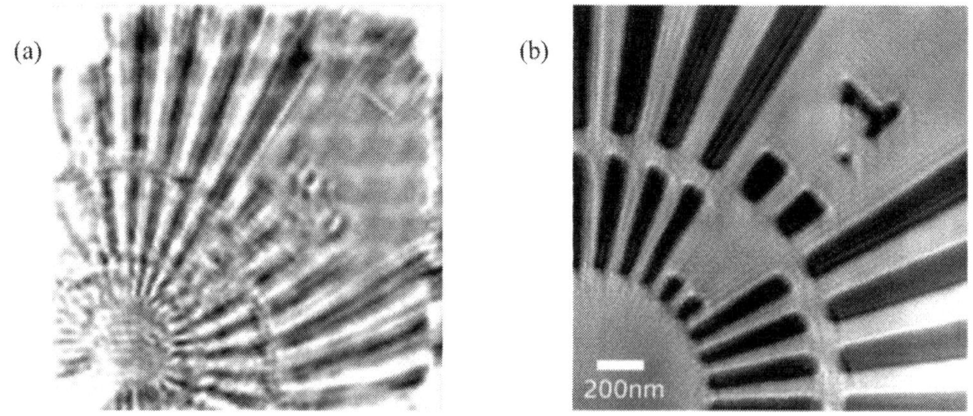

Figure 4. X-ray ptychographic images of Siemens star test pattern obtained (a) without and (b) with constant-temperature system and drift compensation method.

Figures 3(a) and (b) show the amount of drift for every 10 minutes and the total amount of drift, respectively. The amounts of drift for every 10 minutes in the vertical and horizontal directions were controlled to less than 25 nm and 40 nm, respectively. The total amounts of drift over 10 hours in the vertical and horizontal directions were 500 nm and 150 nm, respectively. Without the constant temperature system, there was an approximately 5 μm positioning error over 10 hours owing to drift. Figure 4(a) shows the X-ray ptychographic image obtained without the constant-temperature system or the drift compensation method [6]. The sample image could not be reconstructed owing to drift. On the other hand, with both the constant temperature system and the drift compensation method, the 50 nm lines and spaces on the sample were sharply resolved [see Figure 4 (b)]. The artifacts due to the drift could not be seen, indicating that the constant-temperature system was effective.

4. Conclusion

We developed a high-accuracy X-ray ptychography apparatus equipped with a KB mirror focusing system and a constant-temperature system. The temperature fluctuation was controlled to less than 0.04 °C when using the constant-temperature system. The total amount of drift over 10 hours was held below 500 nm. By combining the constant-temperature system with a drift compensation method, we succeeded in preventing the degradation of the spatial resolution of the reconstructed image owing to drift.

Acknowledgements

This work was supported by a Grants-in-Aid for Scientific Research (Grants No. 24651137 and No. 23102504), the Global COE Program "Center of Excellence for Atomically Controlled Fabrication Technology", and X-ray Free Electron Laser Priority Strategy Program from the Ministry of Education, Culture, Sports, Science and Technology.

References

[1] Chapman H N, and Nugent K A 2010 *Nature Photon.* **90** 833

[2] Takahashi Y, Zettsu N, Nishino Y, Tsutsumi R, Matsubara E, Ishikawa T and yamauchi K 2010 *Nano Lett.* **10** 1922

[3] Rodenburg J M, Hurst A C, Cullis A G, Dobson B R, Pfeiffer F, Bunk O, David C, Jefimovs K and Johnson I 2007 *Phys.Rev.Lett.* **98** 034801

[4] Schropp A, Boye P, Feldkamp J M, Hoppe R, Patommel J, Samberg D, Stephan S, Giewekemeyer K, Wilke R N, Salditt T, Gulden J, Mancuso A P, Vartanyants I A, Weckert E, Schöer S, Burghammer M and Schroer C G 2010 *App. Phys. Lett.* **96** 091102

[5] Suzuki Y, Takeuchi A, Takano H and Takenaka H 2005 *Jpn. J. App. Phys.* **44** 1994

[6] Takahashi Y, Suzuki A, Zettsu N, Kohmura Y, Senba Y, Ohashi H, Yamauchi K and Ishikawa T 2011 *Phys. Rev. B* **83** 214109

X-ray ptychography with highly-curved wavefront

S. Wang[1], D. Shapiro[2], K. Kaznatcheev[1],
[1]NSLSII, Brookhaven National Laboratory, Upton NY 11973, USA
[2]ALS, Lawerence Berkeley National Laboratory, Berkeley CA 94720, USA

E-mail: shellywang@bnl.gov

Abstract. The recent development of scanning coherent x-ay diffraction microscopy (also known as ptychography) eliminates several constraints exerted by coherent imaging. In particular, an illuminating wave (the probe) can have an arbitrary shape, as a diffraction data redundancy due to multiple measurements at overlapping neighboring probe positions permits its independent reconstruction along with the scattering potential (the object wave). A priori knowledge, such as a finite sample support, is reduced to a recording of sequential probe positions and a plausible guess to be used as a starting estimate for iterative phase retrieval. Using a focusing probe, such as one produced by a zone plate, we investigate the effectiveness of the reconstruction algorithm and find that it is significantly less successful at reconstructing wavefronts with large curvature (extended phase variation) than the wavefronts with almost flat phase structure. Our simulations show that when the actual probe has large phase variation, the amount of overlap required for a successful reconstruction of both object and probe depends upon the phase difference between the actual probe and the probe used as a starting estimate for the reconstruction. We quantitatively define the circumstances for successful reconstruction of an object placed away from focus. We use an experimental dataset measured with a moderate amount of overlap to show that a successful reconstruction of the test sample can be done using a curved probe as an initial guess.

1. Introduction.

For a successful demonstration of real-space reconstruction of a non-periodic object based on a far field diffraction pattern [1], x-ray coherent diffractive imaging (CDI) emerges as an alternative to x-ray microscopy. CDI was successfully applied to 2D [2] and 3D [3,4] fabricated nanostructures and a specimen as complex as a yeast cell [5,6,7,8]. A five orders of magnitude increase in average brightness of novel x-ray free-electron lasers further fuels the coherent scattering technique development [9,10], raising the hope that the x-ray coherent scattering will be capable of imaging complex specimens at a spatial resolution below 10nm, or beyond the quality limit given by x-ray optics [11].

As only scattering intensity is measured, the scattering wave phase has to be reconstructed via a computational procedure that at each iterative step equalizes the square modulus of the scattering wave Fourier transform to a measured diffraction intensity while phase is left to be complied with other experimental constraints [12]. As a result, many CDI experimental setups use pinhole and object reconstruction assuming plane-wave illumination. This is not ideal for soft x-rays, since pinholes scatter very strongly when used alone and, if the pinhole is not very close to the sample, the scattering pattern is large and difficult to properly record. Earlier attempts to use focusing light demonstrate an improved iterative algorithm convergence, but they still rely on a priori knowledge of the probe-wave derived from experimental geometry [13,14]. The non-uniform optic illumination and optics fabrication errors or aberrations limit the object reconstruction accuracy, as only the probe- and object-wave product is measured. In scanning x-ray diffraction microscopy, the measurements are performed at overlapped adjacent object areas, so collected data have a redundancy. A critical advance provided by Thibault et al. was the probe retrieval step as a means of verifying the probe structure in parallel with the object reconstruction [15,16]. As a result, both self-consistent solutions of the probe- and object-waves are derived.

2. Simulations.

Numerical simulations were performed using diffraction data generated from the USAF 1951 resolution test pattern and probes with a Gaussian intensity profile and a quadratic phase term. For the area outside the three-line patch the object-wave amplitude is set to unity, at the line it is set to zero, and the phase jump is set to π. The probes, shown to the left in Figure 1, have different phase terms: flat (probe 0 with no phase variation), curved with a maximum phase variation of 100 π (probe 1), and far out-of-focus probe with a maximum phase of 200 π (probe 2). Notation "probe x-y" means that the data are generated with probe x but probe y is used as an initial guess. These probes were sampled on a grid with 128x128 pixels, so the maximum phase jump per pixel was kept well below π in all cases. A circular scan pattern of a total of 61 positions was chosen to avoid artificial square-grid aliasing, and the dose delivered to the object was varied by adjusting the spacing between grid rings. Typically, the ptychographic datasets are characterized by the probe overlap as set by the ratio of the distance between neighboring probe positions to the width of the probe [17]. However, when the probe has a complicated intensity distribution, as it does for the out-of-focus sample position of the zone plate probe, the overlap is poorly defined. Here, we define the overlap as the dose per pixel averaged over the illumination area, assuming the normalized Gaussian intensity distribution.

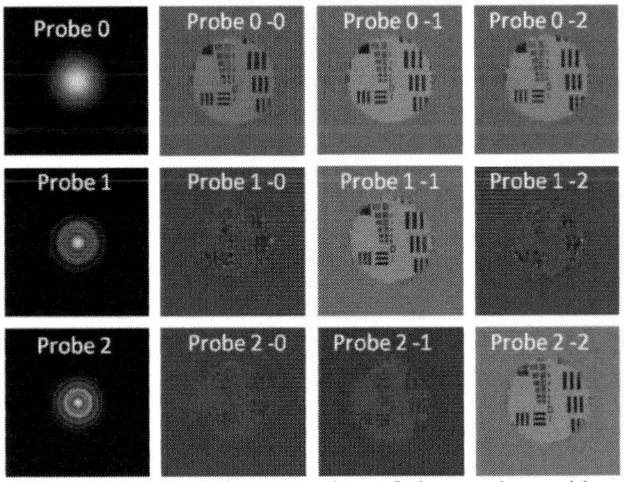

Figure 1. Reconstruction images at average dose of 0.7. Results were all obtained after 500 ePIE iteration steps. The renderings of the complex-valued reconstructions use brightness to represent the magnitude (it also applies to Figure 3 (d) and (e)). The background outside of the illumination area is darker for the unsuccessful images because of the presence of bright pixels.

Since the original demonstration of the ptychographic approach [18], several versions of phase retrieval procedures have been developed [19, 20]. We chose to follow the extended ptychographic reconstruction engine (ePIE) algorithm and use the same notation as in [21]. An update parameter β, used to seed the probe at each step, was set to 0.75 and has not been changed during the reconstruction. The object update function α was equal to 0.001. The initial guess probe is set to have the same intensity as the probe used to generate the scattering data but an incorrect phase term in a form of quadratic exponent. The reconstruction error, as a normalized difference between the square of the simulated diffraction intensity and modulus of the scattering wave Fourier transform (omitting constant phase offset) is computed after every tenth iteration and is used to determine an algorithm convergence (Figure 2).

Figure 1 shows the results of object reconstruction after 500 ePIE iterations for the dataset with 0.7 overlap. Images at the cross illustrate the result of object reconstruction where the scattering pattern is generated with one probe function but at a different initial guess. Good reconstruction is observed across the diagonal line, where the departure of initial guess and the resulting probe is small. Data generated by the flat probe are also insensitive to initial probe-wave guess. In contrast, the data sets with large phase variation require a good estimate of initial probe. Otherwise, the algorithm converges to a wrong result.

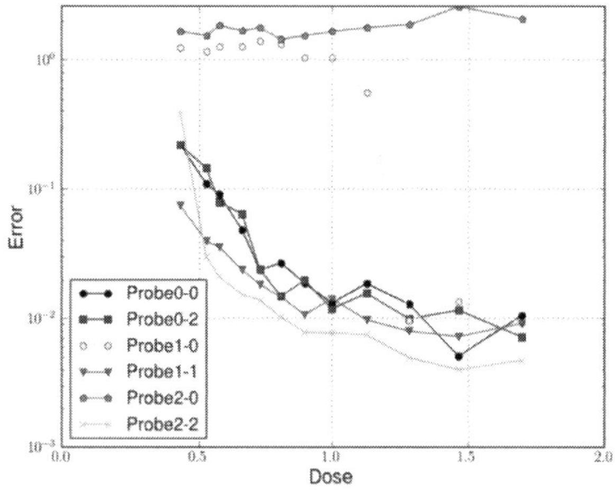

Figure 2. Error (log scale) of simulation data reconstruction after ePIE 500 iterations evolves as function of average dose for different combinations of actual probe and starting probe.

As both good initial guess and the overlap function define a degree of algorithm convergence, Figure 2 shows the result of reconstruction (represented by the error function) as a function of probe overlap (dose). For clarity, we omit some results but left one defining the trend. Our simulation shows that there is an optimum overlap (dose) range, 0.5-1.8, for a successful reconstruction to take place. At a dose below 0.5 even a good initial guess of illumination function may result in large reconstruction errors. For high dose (above 1.80, not shown), the reconstruction based on the ePIE algorithm also fails—possibly due to poor convergence. As in the case of large overlap, it might be more appropriate to use the hybrid input-output algorithm as discussed in [17].

3. Experiment.

The experimental data set was acquired using the Scanning Transmission X-ray Microscope at the Canadian Light Source ID10 beamline. The test sample of lithographically fabricated gold nano-structure consists of elbow-shaped 150nm wide lines deposited on a 100nm thick Si_3N_4 window. The sample was placed approximately 30μm downstream of the zone plate focus. At 710eV energy, for the zone plate of 240μm diameter with 25nm outermost zone, the width of the x-ray spot size was just over 2 μm. The sample was scanned on a 5x5 grid at 560 nm stepsize. Andor DX camera was located 35mm downstream of the sample. The scattering images were cropped to the central part (256x256 pixels, pixel size of 13μm) of a CCD image, providing a 12 nm pixel size of reconstructed image. The calculated dose is 0.85, which, as suggested by the simulation, is above the critical dose needed for good convergence. Figure 3 shows one of the scattering images (Figure 3 (b)) as well as two alternative initial guess probe-waves (Figure 3(a), for large phase variation, but constant intensity, consistent with 30μm out-of-focus zone plate beam, and Figure 3(c), with constant phase across ~2μm size pupil, as in pinhole probe). Only out-of-focus probe-waves result in a successful reconstruction. Note that the reconstructed probe-wave has a more donut-like shape consistent with zone plate pupil function, and it is different from the initial guess. The estimate of the gold thickness (80nm), based on tabulated x-ray cross-sections and the observed phase jump across the elbow line, is also consistent with actual gold plated thickness (~100nm).

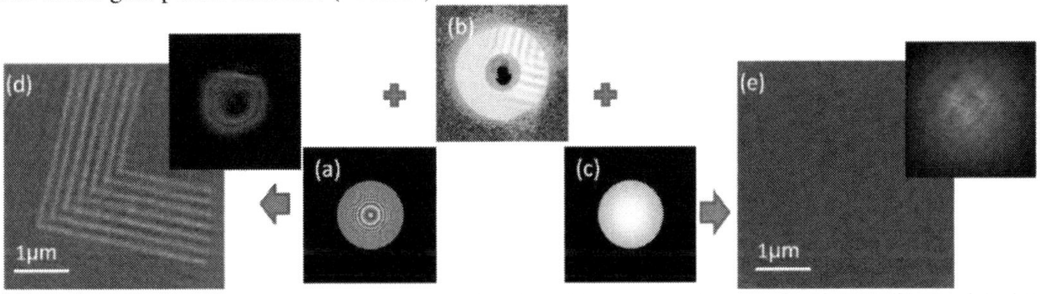

Figure 3. (a) the starting zone plate type curved probe; (b) an example scattering image (log scale); (c) the starting pin hole type probe; (d) successful reconstruction image (amplitude) and resultant probe when using probe in (a); (e) failed reconstruction image and resultant probe when using probe in (c).

4. Conclusion.

Using numerical simulation we study how ptychographical reconstruction error varies with the illumination dose for near-to-focus probe-waves. We find that the ePIE algorithm requires a more accurate initial guess of the probe when the actual probe has a large phase variation or it requires significantly higher overlap. Inaccurate guesses or insufficient overlap lead to reconstruction failure. We then demonstrate experimentally the ability to reconstruct ptychographic data with the test sample spaced away from the zone plate focus. Even a Gaussian-shaped initial probe with close to real geometry quadratic phase term was sufficient to lead to a successful test object reconstruction.

Acknowledgement.
The current work is based on BNL research development grant 11-030. The National Synchrotron Light Source II is supported by US DOE contract DE-AC02-98CH10886.

References

[1] Miao J, Charalambous P, Kirz J and Sayre D1999 *Nature* **400** 342
[2] Chapman H et al. 2006 *Nat. Phys.* **2** 839
[3] Chapman H et al. 2006 *J.Opt.Soc.Am.* A **23** 1179
[4] Barty A et al. 2008 *Phys.Rev.Lett.* **101** 055501
[5] Shapiro D et al. 2005 *PANS* **102** 15343
[6] Nelson J et al. 2010 *PNAS* **107** 7235
[7] Giewekemeyer K et al. 2010 *PNAS* **107** 529
[8] Huang X et al. 2009 *Phys. Rew. Lett.* **103** 198101
[9] Martin A V et al. 2012 *Opt. Express* **20** 13501
[10] Martin A V et al. 2012 *Opt. Express* **20** 16650
[11] Boutet S and Williams G L 2010 *New J. Phys.* **12** 035024
[12] Fienup J R 1982 *Appl. Opt.* **21** 2758
[13] Quiney H M, Peele A G, Cai Z, Paterson D and Nugent K A 2006 *Nat. Phys.* **2** 101
[14] Williams G J et al. 2010 *New J. Phys.* **12** 035020
[15] Thibault P, Dierolf M, Menzel A, Bunk O, David C, and Pfeiffer F 2008 *Science* **321** 379
[16] Rodenburg J M, Dierolf M, Kynde S, Johnson I, Marti O and Pfeiffer F 2007 *Phy. Rev.Lett.* **98** 034801.
[17] Bunk O, Dierolf M, Kynde S, Johnson I, Marti O and Pfeiffer F 2008 *Ultramicroscopy* **108** 481
[18] Rodenburg J M 2004 *App. Phys. Letters* **85** 4795
[19] Marchesini S 2007 *Rev. Sci. Instrum.* **78** 011301
[20] Yang C et al. 2011 arXiv:1105.5628
[21] Maiden A M and Rodenburg J M 2009 *Ultramicroscopy* **109** 1256

Low-dose phase-contrast X-ray imaging: a comparison of two methods

T Zhou, U Lundström, D H Larsson, H M Hertz and A Burvall

Biomedical and X-Ray Physics, Department of Applied Physics, Royal Institute of Technology/AlbaNova, 10691 Stockholm, Sweden

Email: tunhe@kth.se

Abstract. Propagation- and grating-based X-ray phase-contrast imaging methods are compared theoretically. As a prospective application of phase-contrast methods in medical or small animal imaging, carbon dioxide (CO_2) angiography is the simulated task. The required dose for the observable blood vessel is compared through simulation. The result indicates that the propagation-based method requires lower dose in this application.

1. Introduction

In recent years X-ray phase-contrast imaging [1] has been developed as an alternative to conventional X-ray radiography for materials with low absorption coefficients. For small features, phase-contrast imaging reduces the dose compared to absorption-contrast imaging [2]. The grating-based phase-contrast method [3] and propagation-based phase-contrast method [4] are currently of most interest for laboratory systems. As illustrated in figure 1, the propagation-based method detects edge-enhanced images due to refraction or diffraction at the object. The grating-based method uses a Talbot interferometer previously applied in visible light [5] and more recently extended to the X-ray regime [6], as shown in figure 2. The phase shift caused by the object is transformed into a transverse shift of the interference pattern. By transversely scanning one of the two gratings, the shifted fringes, which are too fine to be resolved, are transformed into intensity variations and then detected. The purpose of this study is to compare the two methods from a dose perspective which is highly relevant for future medical applications.

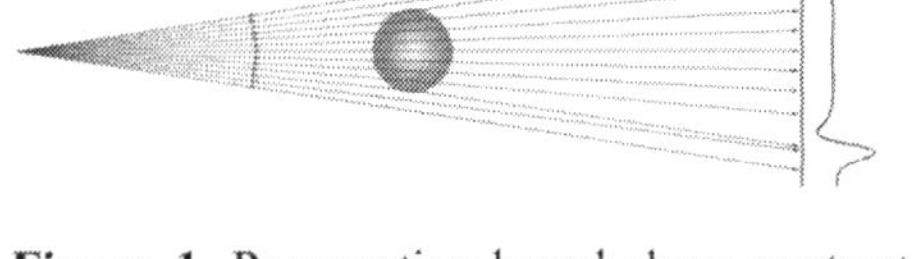

Figure 1. Propagation-based phase-contrast method.

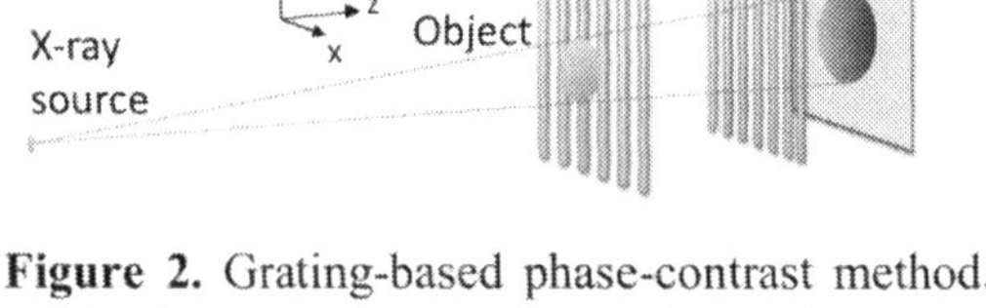

Figure 2. Grating-based phase-contrast method. G1 is the phase grating, and G2 the absorption grating. No source grating is used due to small source size.

2. Simulation and comparison

To theoretically compare the two methods, both the grating- and the propagation-based [2] methods are modeled numerically using a wave-propagation approach. The intensity at the detector is calculated by numerical evaluation of the Fresnel diffraction integral [7] using a Fourier method. For the grating-based method, one image is simulated for each step in the phase scan. From those images the refraction angle of the object can be reconstructed [8]. The simulation processes for both methods have basically the same structure except that two gratings are included in the grating-based method as illustrated in figure 3. As our main interest is in differential phase, so far the objects are defined as ideal and scattering is not taken into account in simulations.

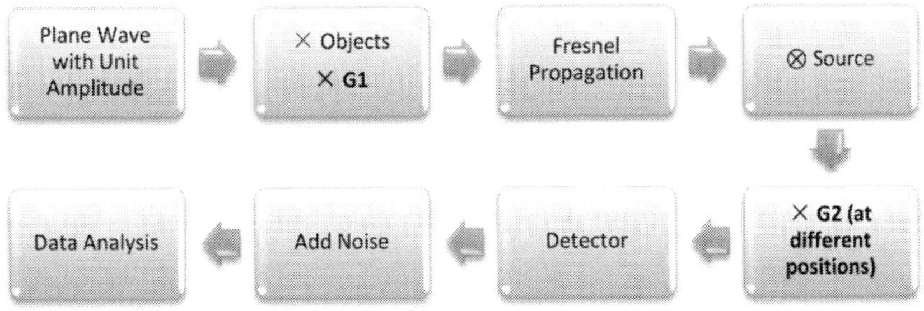

Figure 3. Flow chart of the simulation program structure. ⊗ represents convolution and × multiplication.

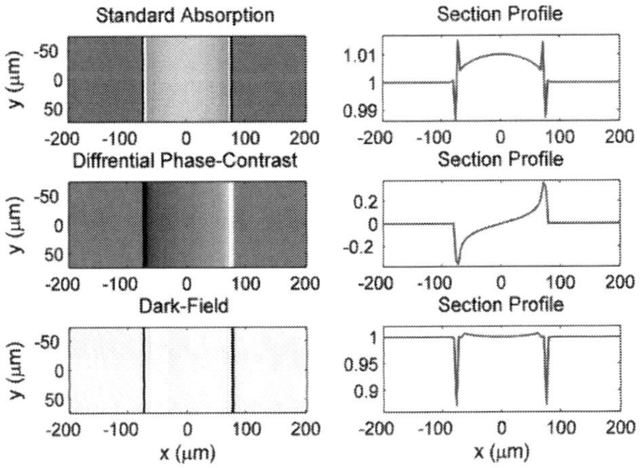

Figure 4. The three images of the grating-based method (standard absorption, differential phase-contrast and dark-field image) reconstructed from simulated images. The object is a cylindrical CO_2-filled blood vessel with 150 μm diameter and length within a 2-cm-thick soft tissue.

Carbon dioxide angiography is of interest as it is a useful diagnostic tool with varieties of clinical applications. Therefore the object is assumed to be a CO_2-filled blood vessel inside a 2-cm-thick block of soft tissue, which is an approximation of a small-animal sample. For the purpose of comparing the detectability of small objects of these two methods, the blood vessel is simplified as CO_2 cylinder with varying diameter. Meanwhile, in order to avoid influence of the vessel length on visibility, each cylinder is specified to have the same length as its diameter. Figure 4 shows an example of a simulated blood vessel in the three reconstructed images of the grating-based method.

To quantitively evaluate the observability of the image, the signal-to-noise ratio (SNR) is calculated. For an ideal observer, a so-called signal-known-exactly/background-known-exactly (SKE/BKE) image with uncorrelated Gaussian noise has [9]

$$\mathrm{SNR}^2 = \frac{1}{\sigma^2}\sum_m \left|\Delta s_m\right|^2, \tag{1}$$

where Δs_m is the difference of pixel m with and without the object and σ^2 is the variance of the reference image, as the variances with or without object are very similar. The sum is taken over all pixels of the image. An object is considered observable when $\mathrm{SNR}^2 \geq 25$ according to Rose's criterion [10].

During the scan process of the grating-based method, the intensity of each pixel varies with grating shift x_g and can be expressed as

$$I_m(x_g) = I_{0,m}\left(1 + v_m \cos\left(\frac{2\pi}{p_2}x_g + \varphi_m\right)\right), \tag{2}$$

where $I_{0,m}$ is the mean transmitted intensity which corresponds to the absorption image at the same distance, v_m is the visibility, and φ_m is a phase parameter which is proportional to the differential phase shift of the sample. SNR^2 of the three parameters from fitting result are calculated by using equation (1) for all pixels. The phase parameter φ_m, compared to $I_{0,m}$ and v_m, has the highest SNR^2 as shown in figure 5. Therefore the differential phase-contrast image is used for comparison with the propagation-based method, rather than the other two parameters here.

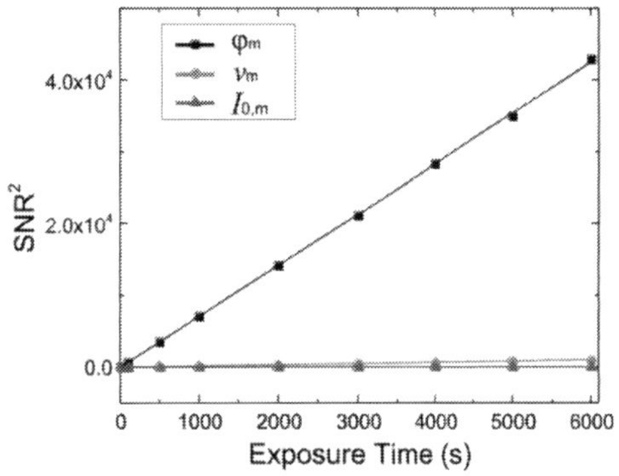

Figure 5. Example of the SNR^2 comparison of the three parameters of the grating-based method, resulting from the fitting process.

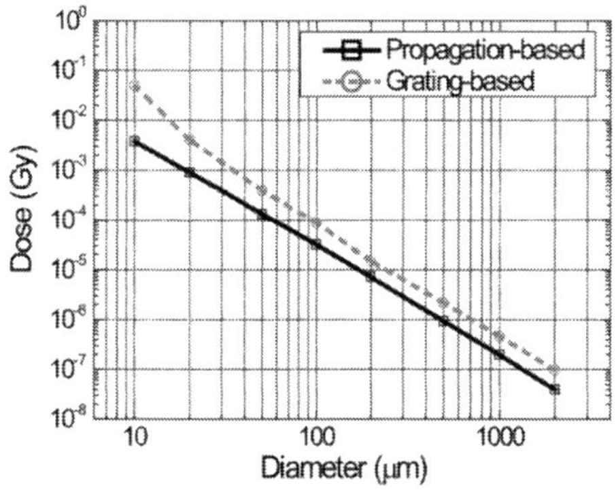

Figure 6. Comparison of dose as a function of vessel diameter between the grating-based and the propagation-based methods using a 20 keV monochromatic 8 μm FWHM X-ray source ($\mathrm{SNR}^2 = 25$).

3. Results and outlook

In the comparison the source is simulated as a 20 keV monochromatic X-ray source with an 8 μm FWHM Gaussian distribution, which is a simplified model of the Galinstan liquid-metal-jet source at Biomedical and X-Ray Physics, KTH, similar to the source presented by Larsson et al. 2011 [11]. The period of the phase grating p_1 is 4 μm, and the period of the absorption grating p_2 is 2 μm under plane wave, corresponding to a true period of 2.24 μm under spherical wave in this geometry. The source-to-object distance is 1.5 m for both methods. The object-to-detector distance for the propagation-based method is 1.5 m and for the grating-based method 0.18 m, which is the 5th Talbot distance in this situation. The detector resolution is 9 μm for the propagation-based method. For simplicity, the detector resolution for the grating-based method is set to four times p_2, which is 8.96 μm, very close to 9 μm.

The result is given as the dose delivered to the sample when $SNR^2 = 25$ as illustrated in figure 6. It indicates that the doses of the two methods have almost the same decreasing trend as the diameter of the vessel increases. The propagation-based method requires slightly lower doses in this application, in particular for small vessels when the pixel size is comparable with the vessel diameter, and the phase gradient signal is not distributed on a sufficient number of pixels.

Further work includes simulations of polychromatic radiation, using the full spectrum of the Galinstan liquid-metal-jet source.

References

[1] Momose A 2003 *Opt. Express* **11** 2303-14
[2] Lundström U, Larsson D H, Burvall A, Takman P A C, Scott L, Brismar H and Hertz H M 2012 *Phys. Med. Biol.* **57** 2603-17
[3] Pfeiffer F, Weitkamp T, Bunk O and David C 2006 *Nat. Phys.* **2** 258-61
[4] Wilkins S W, Gureyev T E, Gao D, Pogany A and Stevenson A W 1996 *Nature* **384** 335-8
[5] Silva D E 1972 *Appl. Opt.* **11** 2613-24
[6] David C, Nohammer B, Solak H H and Ziegler E 2002 *Appl. Phys. Lett.* **81** 3287-9
[7] Goodman J W 1996 *Introduction to Fourier optics* 2nd ed (New York: McGraw-Hill)
[8] Weitkamp T, Diaz A, David C, Pfeiffer F, Stampanoni M, Cloetens P and Ziegler E 2005 *Opt. Express* **13** 6296-304
[9] Barrett H H and Myers K J 2004 *Foundations of image science* (Hoboken, NJ: Wiley-Interscience) equation (13.120)
[10] Rose A 1946 *J. Soc. Motion Pic. Eng.* **47** 273-94
[11] Larsson D H, Takman P A C, Lundström U, Burvall A and Hertz H M 2011 *Rev. Sci. Instrum.* **82** 123701

X-ray microscopy using two phase contrast imaging techniques: two dimensional grating interferometry and speckle tracking

Hongchang Wang[1], Sebastien Berujon[1,2], Ian Pape[1] and Kawal Sawhney[1]

[1] Diamond Light Source, Didcot, Oxfordshire, OX11 0DE, UK
[2] European Synchrotron Radiation Facility, BP-220, F-38043 Grenoble, France

E-mail: Hongchang.Wang@diamond.ac.uk

Abstract. Two phase contrast imaging techniques, namely two dimensional grating interferometry and X-ray speckle tracking (XST), have been combined with the use of a Fresnel Zone Plate (FZP) for application to X-ray microscopy. Both techniques allows the phase shift introduced by a sample on a hard X-ray beam in two dimensions, to be recovered with a high sensitivity and low requirements on transverse and longitudinal coherence. Sub-micron phase imaging of carbon fibres was achieved using the two methods thanks to the high magnification ratio of the FZP. Advantages, drawbacks and differences between these two techniques for X-ray microscopy are discussed.

1. Introduction

In the last decade, X-Ray phase contrast imaging using grating interferometry (GI), has been well developed and is widely used[1, 2, 3]. The X-ray source has been extended from high brilliant synchrotron sources to low brilliance, lab-based X-ray sources by adding a source mask grating [4]. In addition to the phase stepping method, Moiré fringe analysis has also been used since only a single exposure is required [5, 6]. Furthermore, this device is largely achromatic[3], and can be used with broad energy bandwidth X-rays [7]. Based on a different concept, the X-ray speckle tracking (XST) technique has recently been extended into the hard X-ray domain [8, 9]. The principle of this technique is to follow the propagation direction of near field-speckle produced by a random object inserted into the beam, and then reconstruct the wavefront from the calculated trajectory of the X-rays. Although, both techniques are compact and robust with sub-microradian angular sensitivity, the spatial resolution is limited either by lateral shear for GI, or the detector pixel size for XST [3, 10, 11]. In this paper, we demonstrate that the spatial resolution for the two techniques can be further enhanced by combining with a Fresnel Zone Plate (FZP) [10, 12]. Two dimensional (2D) GI has been employed to simultaneously recover beam phase in both the horizontal and vertical directions [6, 13], and the results were compared with those obtained from XST.

2. Experiment

All measurements were performed at B16, the Test beamline on a bending magnet (BM) source at Diamond Light Source[5]. As presented in the sketch in Fig. 1a, the FZP and an order sorting aperture

(OSA) were mounted on dedicated motorized towers. The aperture of the FZP was 200μm, and its outermost zone width was 100nm. The energy of the X-ray beam was 8.2 keV for grating interferometry, selected using a silicon double-crystal monochromator. The 2D grating shearing interferometer and 2D CCD X-ray detector with an effective pixel size of S_p=6.4 μm were mounted on three different motorized towers of a versatile optics test bench, and the distance (d_m) between the phase grating (G_1) and the absorption grating (G_2) was set to the fifth order Talbot distance of 61mm. The pitch of the first grating p_1=3.576μm was less than twice the pitch of the second grating p_2=2μm in order to match the beam's divergence. The specimen was composed of a bunch of carbon fibres and placed about 10mm upstream of the FZP. The focal length of FZP is 132mm at 8.2keV, and the distance between G_1 and FZP is 945mm. Hence, the corresponding magnification ratio M is 7.2. The setup of x-ray speckle tracking technique is illustrated in Fig. 1b, wherein a filter membrane with pore size of 3μm was mounted downstream of focal plane. In this case, the photon energy was set to E=11 keV, and the focal length of FZP is 178mm. Here, the carbon fibres were placed L_1=29mm downstream of the focal plane, and the distance between the specimen and membrane is L_2=826mm. The CCD was installed on a separated tower with L_3=3310mm, and the magnification ratio $M = (L_1 + L_2 + L_3)/ L_1$ was 143.7.

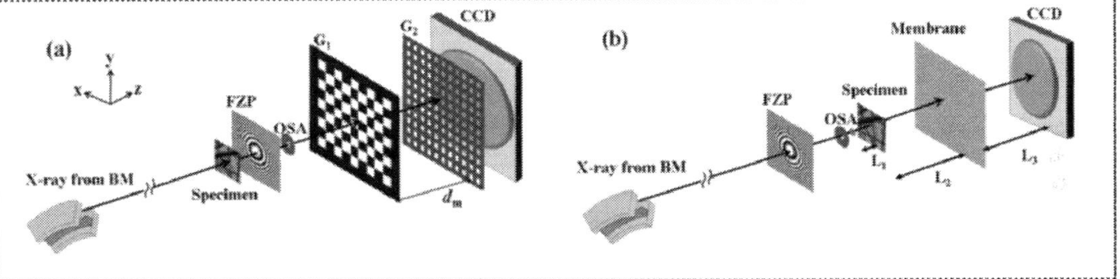

Figure 1. Setup of the (a) grating interferometer and (b) x-ray speckle tracking combined with the use of FZP on bending magnet (BM) beamline B16

3. 2D grating interferometer

For the 2D GI measurement, a raster scan of 8×8 points of the first grating G_1 was made over one period of the grating in each direction using a piezo motors [10]. The phase stepping scan was performed with and without sample in the beam such that the flat field could be removed. From a Fourier analysis of the signal in each pixel, the horizontal and vertical phase gradient image, absorption and dark-field images of the sample can be recovered simultaneously. Fig. 2 shows the results of a bunch of carbon fibres using the 2D GI. It can be seen that visibility of the vertical features is low in the horizontal phase gradient image (Fig. 2a) and vice versa. The conventional x-ray absorption image is shown in Fig. 2d, whereas Fig. 2e contains the dark-field image. Once the two transverse phase gradients have been calculated, the wavefront phase (Fig. 2c) induced by the carbon fibres can be reconstructed by solving a Poisson equation with a pseudo inversion matrix algorithm[14]. As expected, only the edges of carbon fibres are clearly visible in the conventional absorption image. The corresponding phase image shown in Fig. 2c reveals complementary details of the inside parts of carbon fibres. The expected phase shift for the carbon fibre at 8.2 keV with a diameter of 11μm is 1.5 radian, which is very close to the values calculated in the reconstruction and demonstrates the quantitative accuracy of this method [10]. Here, the spatial resolution is limited by the detector pixel size rather than the shear of the first grating [10]. Hence, the effective spatial resolution is defined as S_p/M=0.84μm, where M is the magnification ratio.

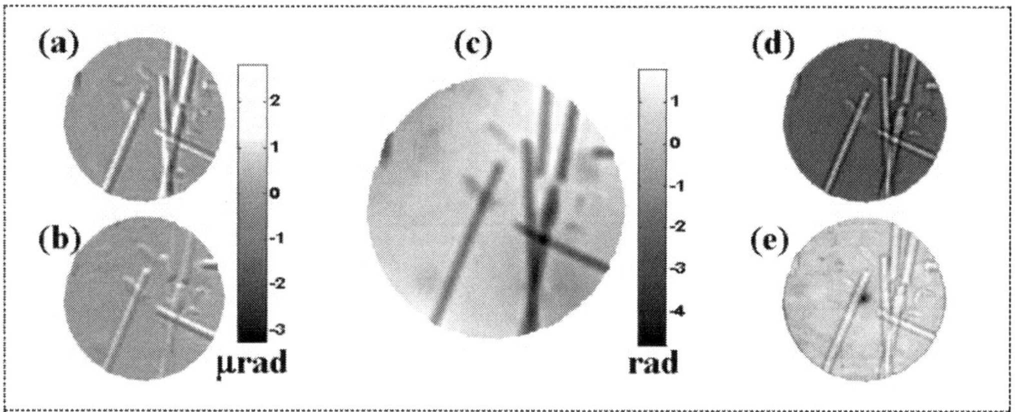

Figure 2. (a) Horizontal and (b) Vertical phase gradient. (c) Wavefront reconstruction from the two directional gradient (a) and (b). (d) Absorption image and (e) Dark-field image.

4. X-ray Speckle Tracking Technique

X-ray phase microscopy using speckle tracking was performed in a differential configuration [9]. The speckle image without and with carbon fibres present in the X-beam are shown in Fig. 3a and Fig. 3b, respectively. Movement of the speckles can be tracked with accuracy better than 0.05 pixel using a digital image correlation algorithm [9, 11]. Displacement vectors, calculated from the speckle image, give the two directional phase gradients (Fig. 3c and 3d) induced by the carbon fibres. Using the same reconstruction method [14], the phase calculated from the horizontal and vertical phase gradient is shown in Fig.3e. The imaginary part of refractive index of the carbon fibre is 3.77×10^{-6} at 11 keV, and the expected phase shift for the above carbon fibre is 1.7 rad. It shows that the reconstructed phase agrees well with the theoretical calculation. Moreover, the spatial resolution for the XST technique can reach 45nm due to the large magnification ratio achieved by placing the sample close to the focal plane, as illustrated in Fig. 1b.

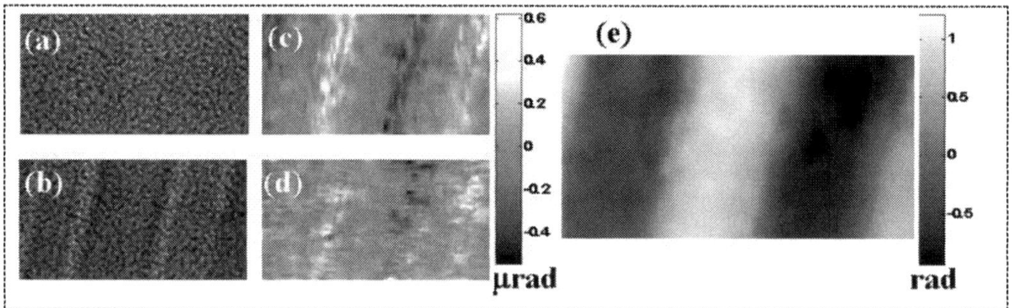

Figure 3. Speckle image (a) without and (b) with carbon fibers in the beam. (c) Horizontal and (d) Vertical phase gradient. (e) Reconstructed phase image from the two directional gradients (c) and (d).

5. Summary

The 2D grating interferometer and X-ray speckle tracking technique have been established and further developed. The 2D x-ray sub-micrometer phase contrast imaging was performed by combining the FZP with the two techniques. Compared with XST technique, the two grating pitches are required to match the beam divergence and the experimental setup is relatively complicated. In addition, the total transmission of the 2D GI is several times lower than the membrane used in XST technique. XST technique only requires collecting two images for phase imaging, but at least dozens of images are required for GI technique to perform the phase stepping scan. However, it is also possible to perform a single shot for GI technique using the Fourier analysis of moiré fringe.[6] Furthermore, the dark-field image can be retrieved simultaneous with phase gradient image for GI technique. Both the XST and GI are compact and robust, and they are thereby complementary to other phase retrieval methods for x-ray phase microscopy.

Acknowledgments

This work was carried out with the support of the Diamond Light Source Ltd UK. The authors wish to thank Christian David, Simon Rutishauser and Joan Vila-Comamala for fabricating the gratings and FZP. We also acknowledge Andrew Malandain for his technical assistance and Simon Alcock for his help in the correction of this manuscript.

References

[1] David C, Nöhammer B, Solak H, Ziegler E 2002 Differential x-ray phase contrast imaging using a shearing interferometer. *Appl. Phys. Lett.* **81** 3287
[2] Momose A 2003 Phase-sensitive imaging and phase tomography using X-ray interferometers. *Opt. Express* **11** 2303
[3] Weitkamp T, Diaz A, David C, Pfeiffer F, Stampanoni M, Cloetens P 2005 X-ray phase imaging with a grating interferometer *Opt. Express* **13** 6296
[4] Pfeiffer F, Weitkamp T, Bunk O, David C 2006 Phase retrieval and differential phase-contrast imaging with low-brilliance X-ray sources. *Nat. Phys.* **2** 258
[5] Wang H, Sawhney K, Berujon S, Ziegler E, Rutishauser S, David C 2011 X-ray wavefront characterization using a rotating shearing interferometer technique *Opt. Express* **19** 16550
[6] Itoh H, Nagai K, Sato G, Yamaguchi K, Nakamura T, Kondoh T 2011 Two-dimensional grating-based X-ray phase-contrast imaging using Fourier transform phase retrieval *Opt. Express* **19** 3339
[7] Momose A, Yashiro W, Maikusa H, Takeda Y 2009 High-speed X-ray phase imaging and X-ray phase tomography with Talbot interferometer and white synchrotron radiation *Opt. Express* **17** 12540
[8] Morgan K, Paganin D, Siu K 2012 X-ray phase imaging with a paper analyzer. *Appl. Phys. Lett.* **100** 124102
[9] Bérujon S, Ziegler E, Cerbino R, Peverini L 2012 Two-Dimensional X-Ray Beam Phase Sensing *Phys. Rev. Lett.* **108** 158102
[10] Berujon S, Wang H, Pape I, Sawhney K, Rutishauser S, David C 2012 X-ray submicrometer phase contrast imaging with a Fresnel zone plate and a two dimensional grating interferometer. *Opt. Lett.* **37** 1622
[11] Berujon S, Wang H, Pape I, Sawhney K 2012 X-Ray phase microscopy using the Speckle Tracking technique paper in submission.
[12] Atsushi M 2009 X-ray phase tomography with a Talbot interferometer in combination with an X-ray imaging microscope. *Journal of Physics: Conference Series* **186** 012044
[13] Zanette I, Weitkamp T, Donath T, Rutishauser S, David C 2010 Two-dimensional x-ray grating interferometer *Phys. Rev. Lett.* **105** 248102.
[14] T. Simchony RC, M. Shao 1990 Direct Analytical Methods for Solving Poisson Equations in Computer Vision Problems. *IEEE Transactions on Pattern Analysis and Machine Intelligence* **12** 435

Damage study of optical substrates using 1-μm-focusing beam of hard X-ray free-electron laser

T Koyama[1], H Yumoto[1], Y Senba[1], K Tono[1], T Sato[2], T Togashi[1], Y Inubushi[2], J Kim[3], T Kimura[4], S Matsuyama[3], H Mimura[5], M Yabashi[2], K Yamauchi[3], H Ohashi[1], T Ishikawa[2]

[1]JASRI/SPring-8, 1-1-1 Kouto, Sayo-cho, Sayo-gun, Hyogo 679-5198, Japan
[2]RIKEN SPring-8 center, 1-1-1 Kouto, Sayo-cho, Sayo-gun, Hyogo 679-5148, Japan
[3]Department of Precision Science & Technology, Faculty of Engineering, Osaka University, 2-1 Yamada-oka, Suita, Osaka 565-0871, Japan
[4]Research Institute for Electronic Science, Hokkaido University, Kita 21 Nishi 10 Kita-ku, Sapporo 001-0021, Japan
[5]Graduate School of Engineering, The University of Tokyo, 7-3-1, Hongo, Bunkyo-ku, Tokyo 113-8656, Japan

E-mail: koyama@spring8.or.jp

Abstract. We evaluated the ablation threshold of silicon and synthetic fused silica, which are widely used as optical substrates such as those in X-ray mirrors. A focusing XFEL beam with a beam size of approximately 1 μm at a photon energy of 10 keV was used. We confirmed that the ablation thresholds of these materials, which were 0.8 $\mu J/\mu m^2$ for the silicon and 4 $\mu J/\mu m^2$ for the synthetic fused silica, approximately agreed with the melting dose.

1. Introduction

X-ray free-electron lasers (XFELs), such as the Linac Coherent Light Source (LCLS) [1] and the SPring-8 Ångstrom Compact free electron LAser (SACLA) [2], deliver intense ultra-short pulses and fully transverse coherent photons in the hard X-ray region. Such intense beams possibly induce damage in optical elements, which can be a serious problem with degraded beam quality. Deformation of the surface of optical elements, such as total reflection mirrors that function as focusing optics and/or low-pass filters and crystals of monochromators, causes particularly disturbing wavefronts of XFEL beams. It is important to obtain information on the damage properties of silicon and amorphous SiO_2, which are used as the substrates of total reflection mirrors.

Damage by FEL irradiation is very actively being investigated in the XUV region [3-5]. Irradiation tests in the hard X-ray region have also been reported to evaluate focusing beam properties [6] since the advent of XFEL facilities.

In this work, we investigated the ablation threshold of optical substrates by using focused an XFEL beam, which had sufficient power density to study ablation phenomena, to obtain knowledge on the damage properties of X-ray optics. The samples used in this work were silicon and synthetic fused silica (amorphous SiO_2), which are widely used substrates in X-ray mirrors.

2. Performance of focusing beam

The experiments were performed at the beamline BL3 of SACLA. Fully transverse coherent X-rays were generated from a 400-m long linear accelerator and a 120-m long undulator section. The SACLA source was operated at a mean pulse energy of 130 μJ, a pulse duration of 20 fs, and a pulse repetition rate of 10 Hz. An X-ray photon energy of 10 keV was chosen. Unwanted contributions of light from higher-order harmonics and gamma-rays were largely suppressed by passing it through a double-mirror system under a grazing incidence of 2 mrad in an optics hutch. The X-ray beam was collimated both vertically and horizontally by a four quadrant slit to a size that was slightly smaller than the focusing mirror aperture to prevent the mirror substrate edges from being irradiated. The focusing mirror consisted of two carbon coated elliptical mirrors aligned in a Kirkpatrick–Baez configuration. The mirror parameters are summarized in Table 1. The focusing mirrors were developed in collaboration with Osaka University, SPring-8, and JTEC Corp. High quality long X-ray mirrors (400 mm in length) were fabricated and their diffraction limited performance was confirmed [7, 8]. The focusing mirror system was located 115 m downstream from the exit of the undulator. The size of the incident beam in front of the focusing mirror was approximately 200 μm in diameter at full-width at half maximum (FWHM). The spatial acceptance of the focusing mirror was 600 × 600 μm^2, and reflectivity through the two mirrors was 97%. A knife-edge scanning method was applied to measure the focusing beam profile. A 200-μm diameter Au wire was used as the knife-edge. The beam intensity was reduced with a silicon attenuator during the scan to prevent the knife-edge surface from being deformed. The knife-edge was scanned step by step and 20 XFEL pulses were averaged in each step to reduce shot to shot fluctuations in the normalized intensity. The beam size was measured as 0.95 × 1.2 μm^2 at FWHM, as shown in figure 1. The pulse energy at the focal spot was available up to 100 μJ.

Table 1. Focusing mirror parameters.

Surface profile	Elliptical cylinder
Substrate material	Synthetic fused silica
Mirror substrate size	420 × 50 × 50 mm^3
Surface coating	Carbon 50 nm
Graz. incidence angle	(H) 1.5 mrad, (V) 1.55 mrad
Spatial acceptance	(H) 615 μm, (V) 632 μm
Focal length	(H) 1.55 m, (V) 2.00 m

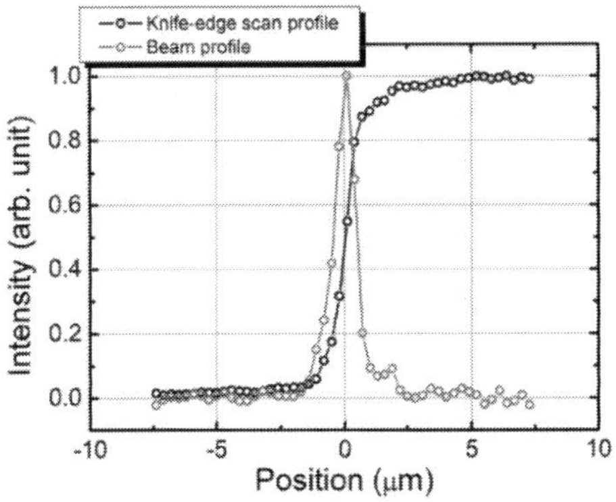

Figure 1. Typical beam profile measured with knife-edge scanning method in the horizontal direction.

3. Evaluation of ablation threshold

A series of focused single pulses irradiated the samples. Silicon and synthetic fused silica were used as the samples, which are widely used as optical substrates such as those in X-ray mirrors. The surface roughness of the silicon and the synthetic fused silica were 0.2 nm (rms) and 0.5 nm (rms). The pulse energy was controlled with silicon attenuators of various thicknesses placed in front of the focusing mirror. The random shot-to-shot fluctuations in the XFEL pulse energy were monitored by using a scattering based gas monitor [9]. A pulse energy ranging from 0.01 to 100 μJ was applied in this experiment. The ablation thresholds of the samples were evaluated by measuring the imprint diameters of the irradiated area [10] by means of scanning probe microscopy (SPM) and scanning electron microscopy (SEM).

Figure 2 shows imprint SEM images of irradiated samples of (a–c) silicon and (d–f) synthetic fused silica. The imprint diameters increased with increased fluence. Molten materials were ejected from the craters.

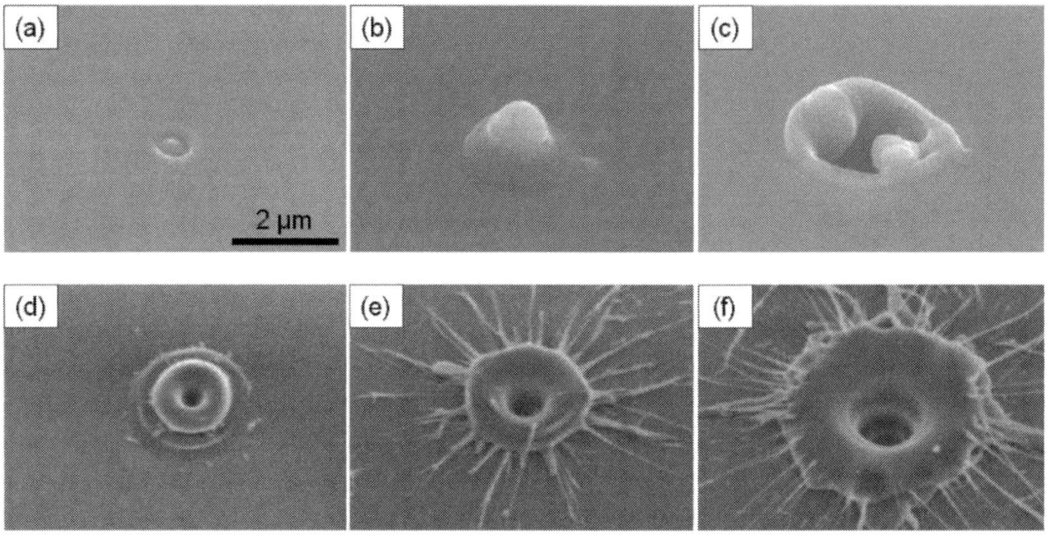

Figure 2. (a–c) Imprint SEM images of silicon under angle of 45° at fluence of 2.7 μJ/μm^2, 4.3 μJ/μm^2, and 6.7 μJ/μm^2. Imprint diameters were 1.1 μm, 2.4 μm, and 3.1 μm. (d–e) Imprint SEM images of synthetic fused silica under angle of 30° at fluence of 6.7 μJ/μm^2, 8.3 μJ/μm^2, and 12 μJ/μm^2. Imprint diameters were 1.7 μm, 2.4 μm, and 3.8 μm.

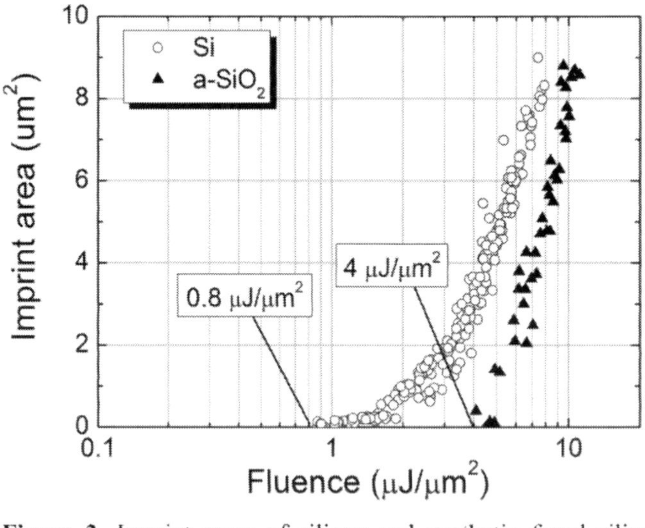

Figure 3. Imprint areas of silicon and synthetic fused silica (amorphous SiO_2) are plotted as a function of fluence. Imprint areas were measured by SPM.

The imprint areas are plotted as a function of fluence in figure 3. The ablation threshold of silicon was evaluated to be 0.8 μJ/μm^2, which was converted to the dose of a single atom [11, 12] as 0.8 eV/atom. This value was within the range of the calculated melting dose of 0.4–0.9 eV/atom. The melting dose of silicon was calculated from the thermodynamic properties [13], which took into

Journal of Physics: Conference Series **463** (2013) 012043 doi:10.1088/1742-6596/463/1/012043

consideration the temperature dependent heat capacity and the latent heat of melting. The ablation threshold of synthetic fused silica was evaluated to be 4 μJ/μm^2, which was converted to the dose of a single atom as 1.5 eV/atom. This value was close to the calculated melting dose of 1.1 eV/atom, which assumed a softening point of 1600°C.

We confirmed from these results that the ablation thresholds of silicon and synthetic fused silica approximately agreed with the melting dose in the experiment using a focusing XFEL beam at a photon energy of 10 keV.

4. Summary

We evaluated the ablation threshold of silicon and synthetic fused silica, which are widely used as optical substrates such as those in X-ray mirrors. We confirmed that the ablation thresholds of silicon and synthetic fused silica approximately agreed with the melting dose.

Acknowledgments

This work was performed at the BL3 of SACLA with the approval of the Japan Synchrotron Radiation Research Institute (JASRI) (Proposal No. 2012A8056).

References

[1] Emma P *et al.* 2010 *Nature Photon.* **4** 641
[2] Ishikawa T *et al.* 2012 *Nature Photon.* **6** 540
[3] Sobierajski R, Klinger D, Jurek M, Pelka J B, Juha L, Chalupský J, Cihelka J, Hakova V, Vysin L, Jastrow U, Stojanovic N, Toleikis S, Wabnitz H, Krzywinski J, Hau-Reige S and London R 2009 *Proc. SPIE* **7361** 736107
[4] Stojanovic N, Linde D, Sokolowski-Tinten K, Zastrau U, Perner F, Förster E, Sobierajski R, Nietubyc R, Jurek M, Klinger D, Pelka J, Krzywinski J, Juha L, Cihelka J, Velyhan A, Koptyaev S, Hajkova V, Chalupsky J, Kuba J, Tschentscher T, Toleikis S, Düsterer S and Redlin H 2006 *Appl. Phys. Lett.* **89** 241909
[5] Hau-Riege S P, London R A, Bionta R M, Soufli R, Ryutov D, Shirk M, Baker S L, Smith P M and Nataraj P 2008 *Appl. Phys. Lett.* **93** 201105
[6] David C, Gorelick S, Rutishauser S, Krzywinski J, Vila-Comamala J, Guzenko V A, Bunk O, Färm E, Ritala M, Cammarata M, Fritz D M, Barrett R, Samoylova L, Grünert J and Sinn H 2011 *Sci. Rep.* **1** 57
[7] Kimura T, Ohashi H, Mimura H, Yamakawa D, Yumoto H, Matsuyama S, Tsumura T, Okada H, Masunaga T, Senba Y, Goto S, Ishikawa T and Yamauchi K 2010 *Nucl. Instrum. Methods* **A616** 229
[8] Mimura H, Morita S, Kimura T, Yamakawa D, Lin W, Uehara Y, Matsuyama S, Yumoto H, Ohashi H, Tamasaku K, Nishino Y, Yabashi M, Ishikawa T, Ohmori H and Yamauchi K 2008 *Rev. Sci. Instrum.* **79** 083104
[9] Tono K, Kudo T, Yabashi M, Tachibana T, Feng Y, Fritz D, Hastings J and Ishikawa T 2011 *Rev. Sci. Instrum.* **82** 023108
[10] Liu J M 1982 *Opt. Lett.* **7** 196
[11] Bionta R M 2000 LCLS Technical Note No. LCLS-TN-00-3
[12] Yabashi M, Higashiya A, Tamasaku K, Kimura H, Kudo T, Ohashi H, Takahashi S, Goto S and Ishikawa T 2007 *Proc. SPIE* **6586** 658605
[13] NIST Chemistry WebBook, http://webbook.nist.gov/chemistry/

High Resolution Higher Energy X-ray Microscope for Mesoscopic Materials

I. Snigireva[1] **and A. Snigirev**
European Synchrotron Radiation Facility (ESRF), 38043 Grenoble, France.

E-mail: snigireva@esrf.fr

Abstract. We developed a novel X-ray microscopy technique to study mesoscopically structured materials, employing compound refractive lenses. The easily seen advantage of lens-based methodology is the possibility to retrieve high resolution diffraction pattern and real-space images in the same experimental setup. Methodologically the proposed approach is similar to the studies of crystals by high resolution transmission electron microscopy. The proposed microscope was applied for studying of mesoscopic materials such as natural and synthetic opals, inverted photonic crystals.

1. Introduction

Mesoscopic structures - photonic crystals have attracted much attention in the past decade for their ability in manipulation of the light. Photonic crystals are the structures in which the refractive index varies periodically in space on the length scale of light. They possess a photonic band analogue of the electron band gap of semiconductors. High quality photonic crystals have promising applications in the area of photonics. Colloidal crystals obtained by convective assembly technique can be used as a template for making inverted photonic crystals with a desired refractive index. In order to obtain crystals with a full photonic band gap the structure needs to be controlled since the photonic band gap is highly sensitive to defects like stacking faults, dislocations and other deformations. To achieve perfect crystals, detailed knowledge about the crystal growth mechanism and resulting defect structures is required.

A major obstacle in studying the defect structure is the lack of appropriate methods that allow seeing the internal structure of photonic crystal in three dimensions. The application of TEM in the study of these materials is limited since the technique's probing depth is a few hundred nanometers. SEM is restricted to obtaining structural information from the surface. The only possibility for studying internal crystal structure by SEM is through physically cutting crystals apart and studying their cut edges. This can modify the crystal structure in the cutting process. Different optical methods, such as laser diffraction, confocal optical microscopy and optical spectroscopy are widely used for the in-situ characterization of immersed, fluorescent, and refractive-index matched colloidal crystals with the particle diameters in the order of a micrometer. However numerous inverse opals, fabricated from highly-absorbing materials, including carbon, transition metal oxides and metals, are dry and in contact with air, which implies that the structures are not refractive-index matched. One way to overcome this is the infiltration of crystal with a refractive index-matching fluid before imaging the structure. But capillary forces acting on the crystal during this process may change the structure,

[1] To whom any correspondence should be addressed.

resulting in unreliable characterization. Recently X-ray high resolution diffraction based on Fourier transform by compound refractive lenses allows accessing the internal structure of the colloidal crystals [1-4]. X-rays can be used for practically all materials and they do not require index matching or fluorescent labelling. Their shorter wavelength lift restrictions related to diffraction limitations. In present work we demonstrate a novel X-ray microscopy technique to study mesoscopically structured materials, employing compound refractive lenses, where we can get diffraction and imaging in the same experimental setup [5]. The proposed microscope was applied for studying of natural and synthetic opals, inverted photonic crystals and colloidal goethite board-like particles.

2. Concept

In transmission electron microscopy (TEM) switching from diffraction to imaging mode is achieved by changing the lenses strength in order to bring different parts of the wave front to the detector area. General consideration in both cases of X-rays and electrons is conveniently based on the Abbe formulations of the imaging process. The significant advantage of using lens-based methodology is the possibility to retrieve the diffraction pattern and real-space images in the same experimental setup. This concept, well-known in TEM, is the key ingredient of our approach. In the present work we demonstrate the transferability of the optical scheme of a TEM to x-ray scattering from mesoscopic structures, applying the concept of X-ray refractive optics as a Fourier transformer [4]. Similar to the studies of "traditional" crystals by high resolution transmission electron microscopy (HRTEM), the sample has to be tilted so that a low-index direction is exactly perpendicular to the X-ray beam. All lattice planes about parallel to the X-ray beam will be close enough to the Bragg position and will diffract the primary beam. The diffraction pattern is the Fourier transform of the periodic (complex) refractive index in two dimensions, and if in the objective lens diffracted beams and the primary beam are brought together again, their interference provides a back-transformation and leads to an enlarged picture of the periodic structure – in its 2D projection.

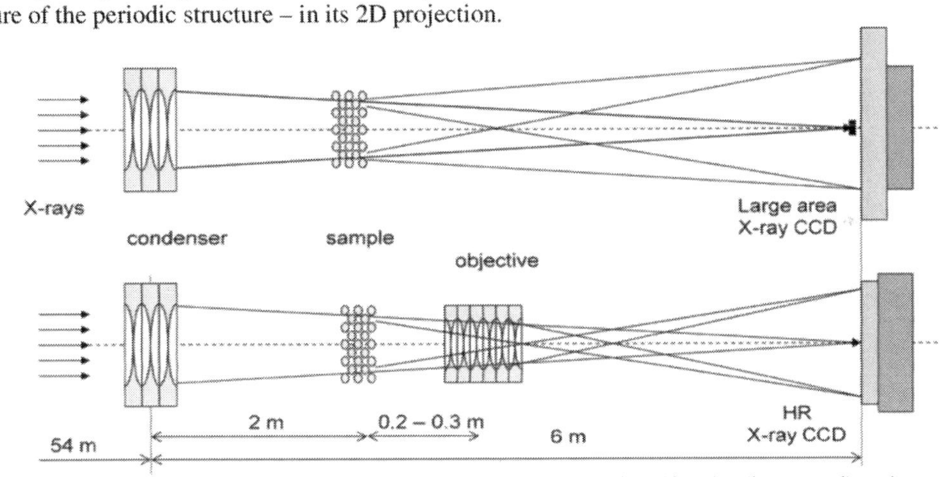

Fig. 1. Conceptual layout of the microscope. Diffraction (upper panel) and imaging (lower panel) modes

The microscope operates under a coherent illumination where a diffraction pattern of the specimen is formed in the back focal plane of the condenser and an inverted two-dimensional image of the object is formed by objective lens in the image plane [5]. The diffraction mode is used to investigate the crystal structure over the macroscopic distances and to orient the crystals parallel to the low index direction to perform high resolution imaging on the local scale. The structural image formation relies on phase contrast since it is made by interference of several diffracted beams. Only by using coherent illumination we can obtain contrast variations in the image of the specimen showing only variations in refractive index. The illumination coherence or spatial coherence length is determined by the effective source size and it is in the order of 100 - 200 μm. To ensure a reasonable contrast it should be a few orders of magnitude larger than the amount of fine detail to be imaged. The coherence in terms of the

Journal of Physics: Conference Series **463** (2013) 012044 doi:10.1088/1742-6596/463/1/012044

angular source size determines the lens angular resolution and it is in the order of 10^{-6} rad. High angular resolution is needed for diffraction studies because for the mesoscopic materials the scattering angles are extremely small.

3. Experiment

The microscope was realized at the MOTB ID06 beamline using X-rays from 10 to 30 keV. It consists of condenser – used for coherent illumination of the sample in imaging mode and as Fourier transformer in diffraction mode, the objective lens and two CCD detectors. The condenser is comprised of 19 Be parabolic lenses with 300 micrometer radius of parabola apex producing focal length ~6 m. Condenser lenses were placed at a distance of 54 m from the undulator source. The collimating slits in front of the CRL limited the beam down to 200 μm in horizontal and vertical directions, thus fitting the effective aperture of the CRL. The objective lens assembly of 45-62 individual Be parabolic lenses with 50 micrometer radius of parabola apex was located at 2.3 m from the condenser. The sample was placed on the translation/rotation stage with the translation direction along the beam axis. Two X-ray 2D cameras were used: (i) large area Photonics science CCD detector with 18 μm resolution (9 μm pixel size) for diffraction experiments and (ii) high resolution Sensicam CCD detector with 1.3 μm resolution (0.645 μm pixel size). In the diffraction mode the direct focused beam was intercepted in front of the camera by a tungsten beamstop. Switching from the diffraction mode to the imaging mode was achieved by moving the objective lens into the beam, and the choice of proper camera (Fig. 1). Depending on the objective lens configuration the image distance (sample-to-objective lens distance) varied from 28 cm to 18 cm. Thus, the magnification factor between 10 and 30 was achieved. At maximum magnification a resolution of 100 nm was achieved.

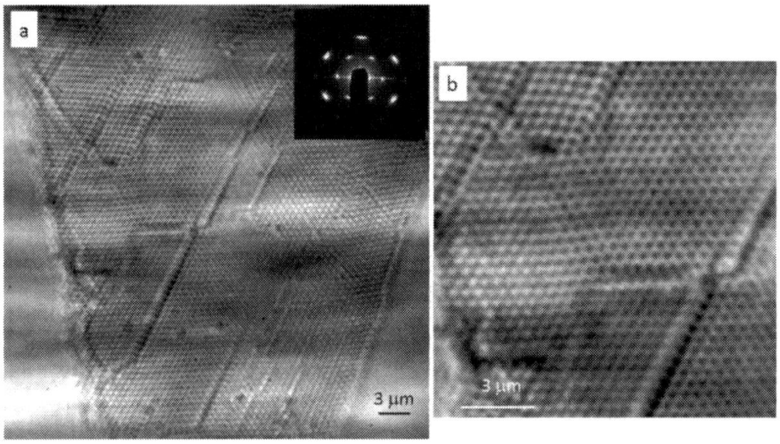

Fig.2 (a) X-ray image together with corresponding diffraction pattern of Co inverted photonic crystal; orientation is <001> zone perpendicular to the crystal surface. (b) Enlarged high resolution image of Co inverted crystal.

The microscope was applied for structural characterization of mesoscopic materials such as natural and synthetic opals, metal inverted photonic crystals. As an example we present here the microscopy study of cobalt inverted photonic crystals. Metallic inverse opal was prepared by electrodeposition of Co inside the voids of artificial opals. Artificial opals were fabricated using the novel synthetic approach based on electric-field –assisted vertical deposition of polystyrene microspheres onto conducting support. Monodisperse polystyrene microspheres with average diameter of 450 nm were synthesized by emulsifier-free emulsion of polymerization of styrene. The field of view in our experiments was varied from 50 × 70 μm^2 to 30 × 50 μm^2, depending of magnification. The sensitivity to the focal distance is quite high; 200 μm defocusing induces substantial blurring. Figure 2 shows an X-ray high resolution image of cobalt inverted photonic crystal recorded in <001> zone. A hexagonal type of arrangements is nicely visible.

The proposed microscope can be applied not only for photonic crystals, but for other mesoscopic materials, for example to study the behaviour of dispersion of colloidal goethite board-like particles under the magnetic field [6]. This system is opaque and hard to index match, therefore it unsuitable for studies with conventional microscopy methods. The particles possess a permanent magnetic moment along their long axis, together with an induced magnetic moment along the shortest particle axis. When a small magnetic field of about 200 mT is applied, the particles orient along the field and the smectic phase becomes highly oriented. The period between the vertical lines is about 380 nm, which is in agreement with the average smectic spacing obtained from diffraction (Fig. 3a). When the magnetic field was increased above 280 mT, the particles reoriented from parallel to perpendicular to the field. This is clearly seen in Fig. 3b, where the layers now possessed undulations as the long-range order was lost and the particles actually seem to have partly rearranged into a columnar phase.

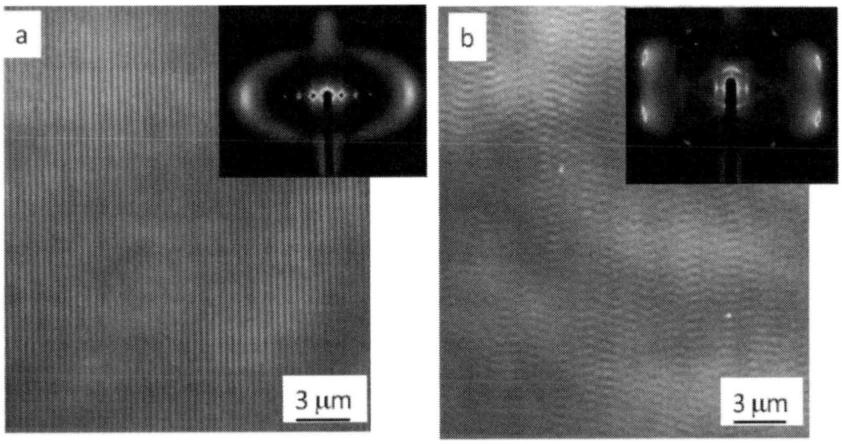

Fig.2 X-ray images together corresponding diffraction patterns of (a) highly ordered smectic phase of goethite rods in magnetic field of 200 mT and (b) reoriented smectic phase in the increased magnetic field 280 mT.

4. Conclusion.

The proposed approach provides the ultimate tool for the efficient studies of real structure of mesoscopic materials. Short acquisition times with modern area detectors allow the method to be extended to time-resolved studies and combined 3-D real/reciprocal space mapping. As immediate practical application of the microscope is the in-situ characterization of the real crystal structure during photonic crystal growth.

Developed X-ray microscope is conceptually similar to TEM; therefore a lot of approaches of TEM and HRTEM can be directly transferred to the domain of X-rays, including bright- and dark-field imaging, defocus series and exit-wave reconstruction for structure refinement, etc. Easy tunability in principle allows a multiscale investigation of materials.

References

[1] A. Snigirev, V. Kohn, I. Snigireva, B. Lengeler, *Nature* **384**, 49-51 (1996).
[2] V. Kohn, I. Snigireva, and A. Snigirev, *Opt. Comm.* **216**, 247-260 (2003).
[3] B. Lengeler, C. G. Schroer, M. Richwin, J. Tummler, M. Drakopoulos, A. Snigirev, I. Snigireva, *Appl. Phys. Lett.* **74**, 3924-3926 (1999).
[4] M. Drakopoulos, A. Snigirev, I. Snigireva, and J. Schilling, Appl. Phys. Lett. **86**, 014102 (2005).
[5] A. Bosak, I. Snigireva, K. Napolskii, A. Snigirev, *Adv. Mater.*, **22**, 3256-3259 (2010).
[6] E. van den Pol, A. V. Petukhov, D. V. Byelov, D. M. E. Thies-Weesie, A. Snigirev, I. Snigireva and G. J. Vroege, Soft matter, **6**, 4895-4899 (2010).

Applications of synchrotron X-ray micro-tomography on non-destructive 3D studies of diesel nozzle internal micro-structure

Li Zhilong, Wu Zhijun, Huang Weidi, Gong Huifeng, Zhu Minjing, Gao Ya

School of Automotive Studies, Tongji University, Shanghai, China, 200092

zjwu@tongji.edu.cn

Abstract: In diesel engines, the fuel spray breakup and atomization process have tremendous dependence on the nozzle internal geometries, then the performance, fuel consumption and emission of the engines will be affected. Based on the synchrotron X-ray micro-tomography, the three-dimensional models of the diesel nozzles were established. Then the internal geometries of the nozzle orifices, such as the diameter, the length and the inlet chamfer radius of the orifice, can be measured, making it possible to analyze the impact of different nozzle orifice processing methods. Meanwhile, the nozzle digital models were successfully applied to simulate the internal flow of high pressure fuel inside the nozzle; the effects of nozzle geometries on internal flow characteristics have been analyzed. The results show that these works are significant to the promoting of diesel nozzle processing technologies, and the revealing of high pressure fuel spray breakup mechanism.

1. Introduction

The performance, efficiency and pollutant emissions of diesel engines have tremendous dependence on the quality of fuel sprays [1,2]. Modern diesel engines employ fuel injectors with micro-orifices (typically less than 200-μm in diameter) to prepare fuel sprays and to acquire optimized liquid fuel breakup and atomization under very high injection pressure ranging from 150 to 250-MPa. At such high-injection pressures, the flow inside the nozzle is extremely sensitive to the fine internal geometries [3-5], such as the diameter, the length and the inlet chamfer radius of the orifice. When increasing the injection pressure, diesel nozzle geometry, even on μm length scale, is considered to be a key issue for the improving of fuel spray and atomization to meet the new stringent regulation [6,7].

However, as the access to the internal structure is very limited in these micro nozzle orifices, the researchers all over the world have proposed quite a few methods to obtain the internal structures of diesel nozzles, including optical diagnostic methods [8], sectioning methods [9,10], micro probe measuring methods [11,12], silicone molding method [13], commercial X-ray method [14], and synchrotron X-ray micro-tomography method [15]. However, these methods are either destructive or inaccurate and inconvenient, except for the synchrotron X-ray micro-tomography method.

By using synchrotron X-ray provided the Shanghai Synchrotron Radiation Facility (SSRF), three-dimensional models of the diesel nozzles were established. Then the characteristics of different orifice processing methods are compared based on the internal geometries acquired by measuring the 3D models. In addition, the real internal fuel flow inside the nozzle orifice has been simulated based on the corresponding internal structures.

2. Experimental Setup

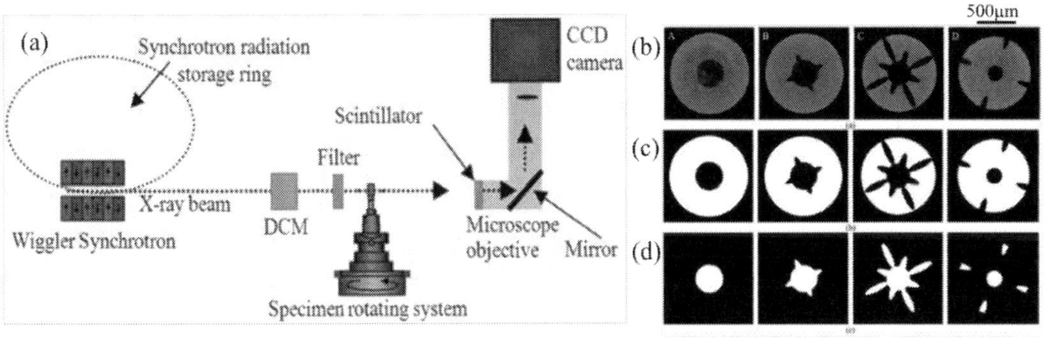

Figure 1. (a) A sketch of the synchrotron X-ray micro tomography setup at the beam line BL13W1 of the SSRF, (b) original slices, (c) binary slices, (d) binary orifice slices

In this study, the internal structures of diesel nozzles were revealed on the X-ray Imaging and Biomedical Applications Beam Line (BL13W1) of the SSRF. The detailed features of this beam line have been described by Li [16]. The experimental setup is sketched in figure 1a.

The diesel nozzle is held on the specimen rotating platform. The synchrotron X-ray penetrates its tip and irradiates the scintillator to form absorption images in a charge-coupled device (CCD) camera. The pixel size of the camera is 7.4μm×7.4μm corresponding to an actual spatial resolution of 3.7μm×3.7μm with a 2x objective lens. The X-ray photon energy of this beam line was tuned to 50 keV and the exposure time was set to 8s. During the computed tomography (CT) scanning process, the specimen rotates 180° to obtain sufficient absorption images to generate slices. In accordance with the nozzle tip size and spatial resolution, a total of 720 absorption images were captured, and the slices were then reconstructed. Fig. 1b shows one original slice. Using customized digital image processing method, the original slices were converted to binary slices and corresponding orifice slices to enhance the signal to noise ratio [15] (figures 1c and 1d). By stacking these reversed slices, the 3D digital models of the diesel nozzle were constructed directly, which can be used to illustrate the detailed internal surfaces and structures of the orifices and to measure their geometrical dimensions directly.

3. Results and Discussions

3.1. Orifice internal parameters measurement and orifice processing methods comparison

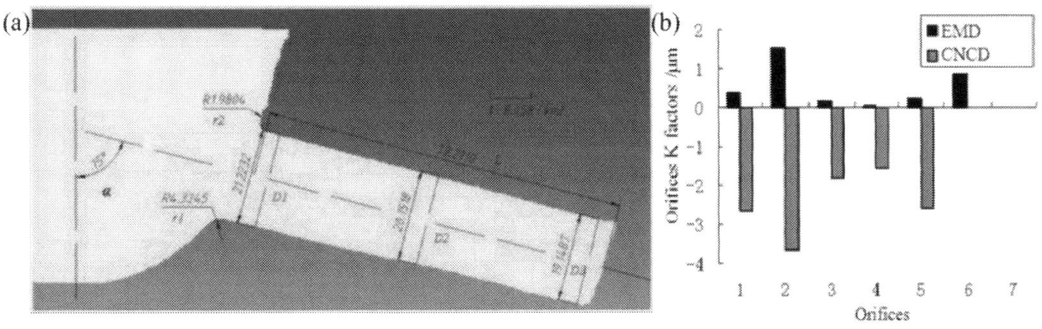

Figure 2. (a) shows the measurement of internal structure parameters, (b) is the nozzle orifice *K* factors comparison of different orifice processing methods

By cutting the 3D digital models, the internal geometry parameters, the orifice cone angle α, the orifice length L, the inlet diameter $D1$, the middle diameter $D2$, the outlet diameter $D3$, the top and bottom chamfer radii $r1$ and $r2$ can be measured, as shown in figure 2a.

Based on these parameters, the most popular orifice processing methods, computerized numerical control drilling (CNCD) and electrical discharge machining (EDM) drilling were compared and analyzed. One important parameter, K factor ($K=(D1-D3)/10$) are compared in figure 2b, the results show the K factors of the CNCD nozzle are all negative, however that of the EDM nozzle are all positive, which is good for the enhancement of the flow coefficient and the life of the nozzle.

Meanwhile, the effect of different hydro erosive grinding (HEG) time can also be researched, as shown in figure 3. By cutting the orifice from different angle (figure 3a), the parameters at different sections can be obtained. The comparison of the chamfer radii under different HEG time in figure 3b indicates that, longer HEG time can obviously smooth the orifice inlet.

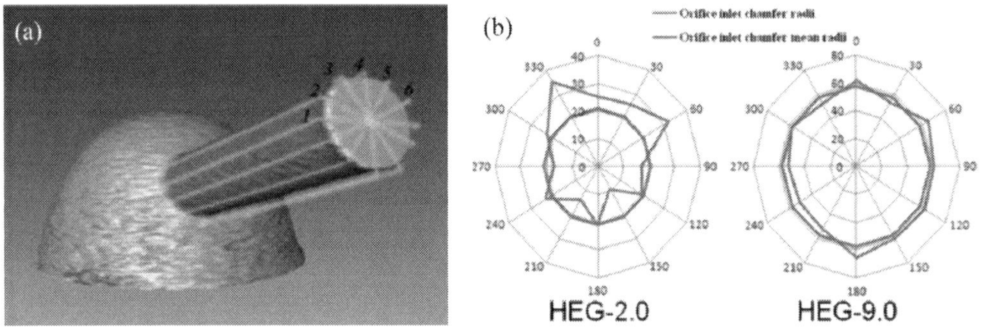

Figure 3. (a) shows different measurement sections along the orifice, (b) shows the inlet chamfer radii comparison under different hydro erosive grinding time

3.2. Fuel internal flow simulation based on orifice 3D digital models

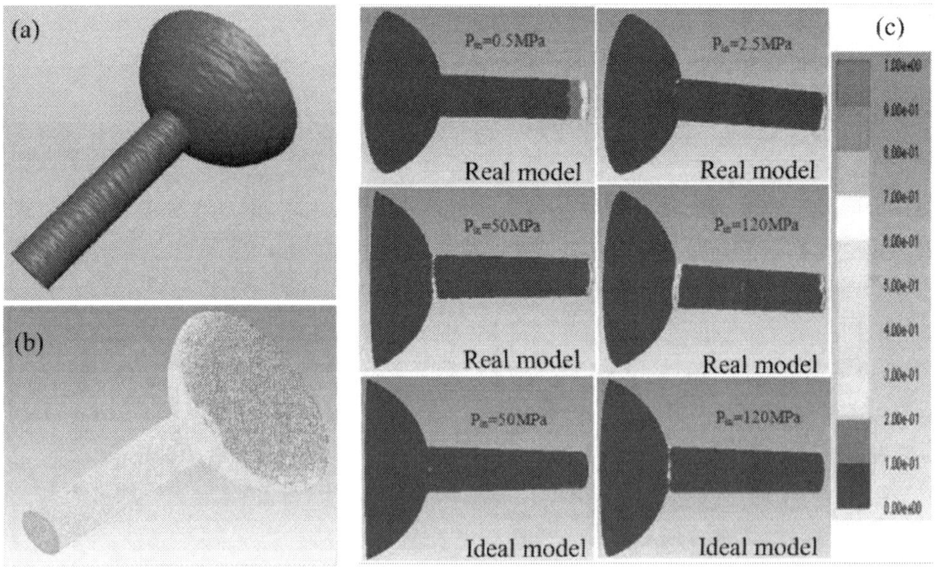

Figure 4. (a) is the real 3D digital model of a diesel nozzle, (b) shows the obtained computation gird according to (a), (c) shows the simulation results based on real model and the ideal model

The simulation of the high pressure fuel flows can acquire higher accuracy based on the orifice real 3D digital models reconstructed by the synchrotron X-ray micro-tomography, compared to the previous simulations base on the ideal 3D digital models generated directly in CAD software. The results and comparisons in figure 4 shows that, the inner flow characteristics are quite different in real and ideal models, especially the cavitation near the inlet area, which means that real model simulation can reveal the inner flows more closely to the real flow.

4. Conclusion

By using high energy X-ray micro CT technique, the internal geometries and the parameter dimensions can be observed and measured easily and accurately.

Through the analysis of nozzle inner geometry parameters, the nozzle orifice processing methods can be compared, which is significant to the quality control of diesel nozzle manufacture.

Simulation of orifice inner flow was operated successfully based on the real 3D models acquired in this study, and it represented huge difference compared to the ideal models under the same conditions.

Acknowledgments

We are grateful to Tiqiao Xiao, Honglan Xie, Yalan Fu, Guohao Du, Biao Deng, Yanlin Xue of the BL13W1 in the SSRF; Jin Wang, Yuan Gao of the Argonne National Laboratory; Rufu Zhuang, Xiangben Wang of the Wuxi Fuel Injection Equipment Research Institute.

This project was supported by the National Natural Science Foundation of China (50946052, 51076118), the New Century Excellent Talents (NCET-10-0605), Shanghai Rising-Star Program (11QH1402500), and the Fundamental Research Funds for the Central Universities.

References

[1] M. A. Patterson and R. D. Reitz, SAE Trans.**107**, 27 (1998).
[2] S. Strauss and Z. Yangbing, SAE Trans.**113** (4), 2146 (2004).
[3] Z., Im, K., Wang, Y., Fezzaa, K. et al., SAE Technical Paper 2010-01-0877, (2010)
[4] Liu, Z., Im, K., Wang, Y., Fezzaa, K. et al., ILASS Americas 2010-169, (2010).
[5] Lai, M., Zheng, Y., Xie, X., Moon, S. et al., SAE Int. J. Engines 4(1):703-719, (2011).
[6] S. Som, A. I. Ramirez, D. E. Longman, et al., Fuel **90** (3), 1267 (2011).
[7] R. Payri, F. J. Salvador, J. Gimeno, et al., Appl. Therm. Eng.**29** (10), 2051 (2009).
[8] W.L. Wang, Hu, S.J., 2004. Nanotechnology and Precision Engineering, **1**:65-70, (2004).
[9] Jung, D., Wang, W.L., Knafl, A., et al., Int. J. of Automo. Techn., **9**(1):9-15, (2008).
[10] Kostas, J., Honnerya, D., Soriaa, J., Fuel, **88**(11):2225-2237, (2009).
[11] Kao, C.C., Shih, A.J., Meas. Sci. and Techn., **18**:3603- 3611. (2007).
[12] Kim, B., Masuzawa, T., Bourouina, T., Meas. Sci. and Techno., **10**(8):697-705, (1999).
[13] Macián, V., Bermúdez, V., Payri, R., et al., Exp. Tech., **27**(2):39-43, (2003).
[14] Delphine, L., Samuel, L., Grégory, B., et al., DIR 2007, Lyon, France, (2007).
[15] Li, Z.L., Wu, Z.J., Gao, Y., et al, J. of Jilin University (Eng. and Techno. Edition), **41**(1): 128-132 (in Chinese), (2011).
[16] Li, X.H., Yu, H.X., He, J.H., Mod. Phys., **22**(3):14-19 (in Chinese), (2010).

Cross-correlation analysis of x-ray scattering from oxygen clusters

R P Kurta[1], Y Chesnokov[2], E Weckert[1] and I A Vartanyants[1,3]

[1]Deutsches Elektronen-Synchrotron DESY, Notkestraße 85, D-22607 Hamburg, Germany
[2]Moscow Institute of Physics and Technology, Institutskii per. 9, 141700 Dolgoprudny, Russia
[3]National Research Nuclear University, "MEPhI", 115409 Moscow, Russia
E-mail: ivan.vartaniants@desy.de

Abstract. Development of methods for the analysis of the structure of non-crystalline materials at the nanoscale is an important challenge for materials research with x-rays. Here, we present the results of modelling of x-ray scattering from a disordered system composed of oxygen tetrahedral pentamers. We apply x-ray cross-correlation analysis to the simulated x-ray data and show that the intensity cross-correlation functions can extract information about the local structure of the system.

1. Introduction

The standard analysis of x-ray scattering data from disordered systems such as liquids, colloids and molecules in solution, or atomic clusters in gas phase did not yet go far beyond the study of radial distribution functions extractable from the average intensity profile. Using such an approach no directional information about molecular or atomic arrangements in the system can be directly accessed. In contrast to this, it has recently been demonstrated [1–6] that the x-ray cross-correlation analysis (XCCA) of diffraction patterns from disordered systems can provide direct information on the local symmetry and the structure of these systems. This approach is based on the intensity cross-correlation functions (CCFs) introduced by Kam [7].

Application of XCCA to the scattering data from a two-dimensional (2D) disordered system of particles demonstrated a successful reconstruction of the structure of an individual particle [4, 6]. At the same time, for a three-dimensional (3D) disordered system the same problem has been solved only if additional *a priori* knowledge was used [5]. Therefore, development of direct, self-consistent approaches for the recovery of the local structure of a 3D disordered system remains a challenge.

As a step towards the solution of this problem we present the results of simulations of x-ray scattering experiments from disordered systems composed of oxygen tetrahedral pentamers (figure 1). The structure of the oxygen pentamer corresponds to the tetrahedral arrangement of water molecules in a so-called Walrafen pentamer [8], that is one of the favorable models of water that reproduces the results of x-ray and neutron scattering experiments [8-12]. We perform the XCCA of the obtained diffraction patterns and show that the Fourier spectra of the CCFs contain information on local structure of a disordered system that is not accessible in the standard small-angle x-ray scattering (SAXS) experiment.

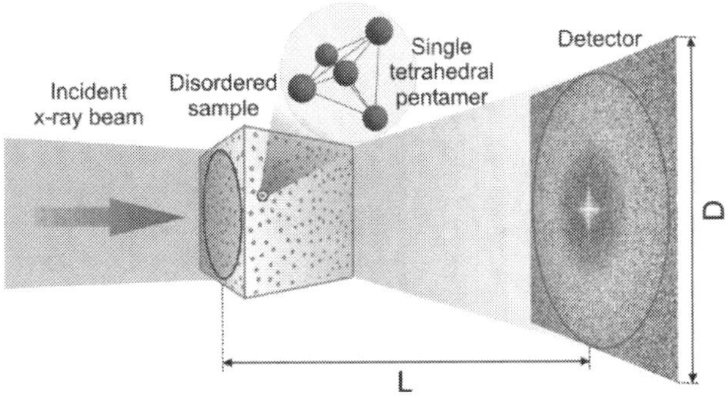

Figure 1. Geometry of the diffraction experiment in transmission geometry. The incident x-ray beam illuminates a disordered sample composed of oxygen tetrahedral pentamers.

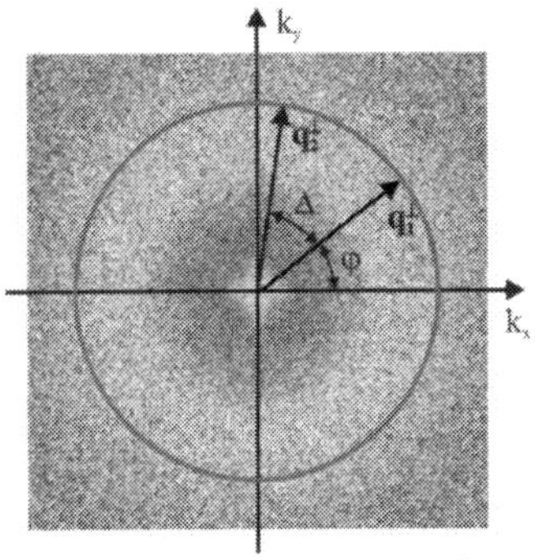

Figure 2. Typical diffraction pattern and definition of the momentum transfer vectors.

2. X-ray cross-correlation analysis

We consider here a scattering experiment on a three-dimensional (3D) disordered system composed of N oxygen pentamers (figure 1). In our model we assume a uniform distribution of the orientations of the clusters. The interference of the scattered amplitudes from different clusters in the system is neglected corresponding to the dilute limit approximation. In this case the total scattered intensity $I(\mathbf{q})$ at the momentum transfer vector $\mathbf{q}$ from N clusters can be considered as a sum of intensities $I_i(\mathbf{q})$ scattered from individual clusters, $I(\mathbf{q}) = \sum_{i=1}^{N} I_i(\mathbf{q})$. It has been demonstrated [2,3,6], that if the number N of clusters in a system is small, one can apply XCCA to extract information about the structure of a single cluster using an average of CCFs over a large number M of diffraction patterns.

For convenience we use the designation $I(\mathbf{q}) \equiv I(|\mathbf{q}^\perp|, \varphi; q^z)$, where the momentum transfer vector $\mathbf{q}$ is defined in terms of its components $\mathbf{q}^\perp$ and q^z that are perpendicular and parallel to the direction of the incident x-ray beam, and φ is an azimuthal angle of $\mathbf{q}^\perp$ (figure 2). The two-point CCF is defined at each $|\mathbf{q}_1^\perp| = |\mathbf{q}_2^\perp|$ as follows [1-7]

$$C_q(\Delta) = \left\langle I(|\mathbf{q}^\perp|, \varphi; q^z) I(|\mathbf{q}^\perp|, \varphi+\Delta; q^z) \right\rangle_\varphi - \left\langle I(|\mathbf{q}^\perp|, \varphi; q^z) \right\rangle_\varphi^2, \tag{1}$$

where Δ is the angular coordinate, and $\left\langle I(|\mathbf{q}^\perp|, \varphi; q^z) \right\rangle_\varphi = 1/(2\pi) \int_0^{2\pi} I(|\mathbf{q}^\perp|, \varphi; q^z) d\varphi$ defines the angular average. The Fourier series of the scattered intensity $I(|\mathbf{q}^\perp|, \varphi; q^z)$ and of the CCF $C_q(\Delta)$ are defined as $I_q^n = \frac{1}{2\pi} \int_0^{2\pi} I(|\mathbf{q}^\perp|, \varphi; q^z) \exp(-in\varphi) d\varphi$ and $C_q^n = \frac{1}{2\pi} \int_0^{2\pi} C_q(\Delta) \exp(-in\Delta) d\Delta$, respectively. The Fourier components C_q^n and I_q^n are related according to the expression [2-4]

$$C_q^n = \left| I_q^n \right|^2. \tag{2}$$

In the case of a dilute sample, the Fourier components of intensity can be written as [2,3]

$$I_q^n = |f(q)|^2 \sum_{k=1}^{N} \sum_{m,l}^{N_O} \exp(-i q^z z_k^{ml}) J_n(|\mathbf{q}^\perp| \cdot |\mathbf{r}_k^{lm\perp}|) \exp(-in \phi_{\mathbf{r}_k^{lm\perp}}), \tag{3}$$

where $f(q)$ is the atomic form factor of oxygen, the summation over index k is performed over N clusters in the sample, the summation over indexes l and m is performed over N_O oxygen atoms in the k -th cluster, $\mathbf{r}_k^{lm} = (|\mathbf{r}_k^{lm\perp}|, \phi_{r_k^{lm\perp}}; z_k^{lm})$ is the vector connecting two oxygen atoms in the k -th cluster with its components that are perpendicular and parallel to the direction of the incident x-ray beam similar to a vector $\mathbf{q}$, and $J_n(x)$ is the Bessel function of the first kind of integer order n.

3. Results

We simulated an x-ray scattering experiment (see figure 1) with the following parameters: x-ray wavelength λ=1.0 Å, detector size D = 24 mm (with a pixel size of p = 80 μm), sample-detector distance L = 25 mm. The samples were composed of different numbers N of oxygen clusters. The regular pentamer (see figure 1) was defined by a O-O distance of 2.82 Å between the central oxygen atom and each atom in a vertex, and a tetrahedral angle 109.47° typical for water [9].

We demonstrate results of the calculations of the Fourier spectra $\langle C_q^n \rangle_M = 1/M \sum_{m=1}^{m} (C_q^n)^m$ averaged over a sufficiently large number of realizations of a system, where the spectrum for each m-th realization was directly determined using equations (2) and (3). Such spectra statistically converge to their average values for a sufficiently large number M enabling a reliable determination of the local structural contribution of a single cluster from a large x-ray dataset [2,3].

The Fourier spectra $\langle C_q^n \rangle_M$ averaged over M=10^4 diffraction patterns for the case of N=10 clusters in the system are presented in figure 3. Calculations were performed for a regular pentamer for the case of a the flat (a) and a curved (b) Ewald sphere. Calculations for the case of a flat Ewald sphere were performed by setting the z-component of the momentum transfer vector to $q^z = 0$. The results shown in figure 3(c) demonstrate the statistical convergence of the spectra after averaging over a sufficiently large number M of diffraction patterns. The spectra $\langle C_q^n \rangle_M$ calculated for a single cluster N=1 in the system, and for a system containing N=100 clusters (not presented here), converge to the same functional dependence on q. Therefore, the results do not depend on the number N of particles in the system, but are just scaled by N.

The effect of a distortion of the regular pentamer on the Fourier spectra of the CCF is demonstrated in figure 4. The difference spectra $\langle C_q^n \rangle_M^{\text{diff}} = \langle C_q^n \rangle_M^{\text{dist}} - \langle C_q^n \rangle_M^{\text{reg}}$ were calculated for the case when a single atom in a cluster was shifted by 5% [figure 4 (a), (b)] and by 10%, [figure 4 (c), (d)] and when all atoms were randomly displaced by 5% of the O-O distance [figure 4(e),(f)]. Here, $\langle C_q^n \rangle_M^{\text{reg}}$ corresponds to the spectra calculated for the regular pentamer, and $\langle C_q^n \rangle_M^{\text{dist}}$ to the distorted one.

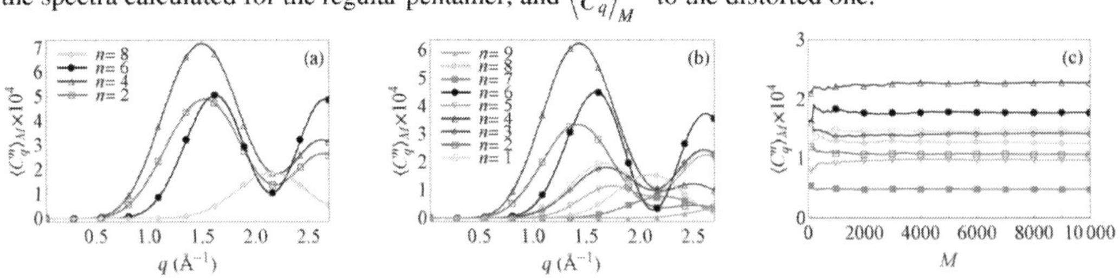

Figure 3. (a),(b) Fourier spectra $\langle C_q^n \rangle_M$ calculated for N=10 and averaged over M=10^4 diffraction patterns as a function of q for 1≤n≤9 for a flat (a) and a curved (b) Ewald sphere. In the case of a flat Ewald sphere only Fourier components with even n-values have nonzero values. (c) Convergence of $\langle C_q^n \rangle_M$ as a function of M calculated for the case of a curved Ewald sphere at a single value of q=1.9 Å^{-1} for 1≤n≤9.

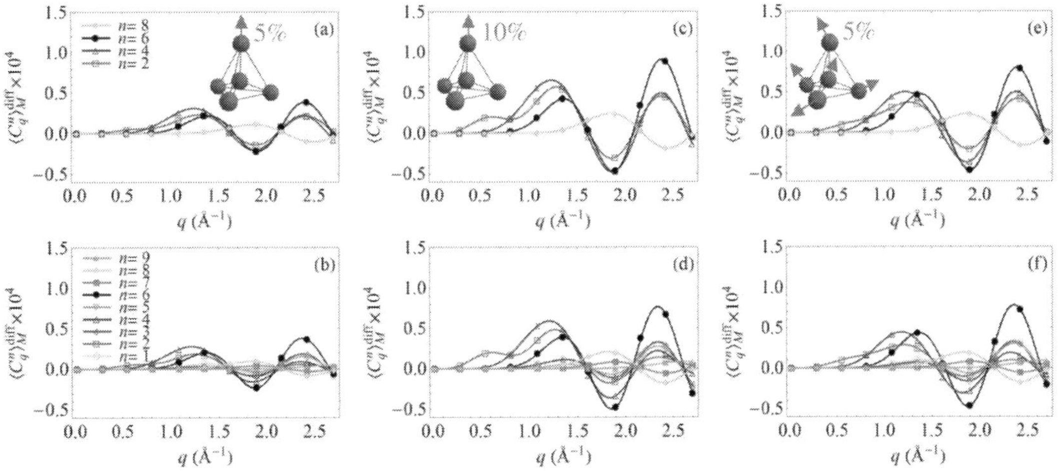

Figure 4. Difference spectra $\langle C_q^n \rangle_M^{\text{diff}}$ (see text) calculated for the different types of distortions of the regular oxygen pentamer for N=10, M=10^4 for the case of a flat (a), (c), (e) and a curved (b), (d), (f) Ewald sphere. Three cases were considered: (a), (b) a single atom is shifted by 5% and (c), (d) by 10% of the O-O distance (2.82 Å), (e), (f) all atoms are randomly displaced by 5% of the O-O distance.

4. Conclusions

Results presented here demonstrate the ability of XCCA to obtain information about the oxygen clusters with tetrahedral arrangement of oxygen atoms typical for water. The average Fourier spectra of the CCF can be used to identify the presence of this symmetry in liquids, particularly in water. Our results show, that small distortions of the oxygen pentamer do not lead to significant changes of the CCFs. This may help to identify an average local structure of water using experimentally determined CCFs. In our future research we are planning to investigate the influence of correlation effects [2-3] on the average spectra of the CCFs for dense systems such as water.

Acknowledgements

We acknowledge discussions with A. Nilsson that stimulated our research in this direction and we are grateful to M. Altarelli for a careful reading of the manuscript.

References

[1] Wochner P *et al.* 2009 *PNAS* **106** 11511
[2] Altarelli M, Kurta R P and Vartanyants I A 2010 *Phys. Rev.* B **82** 104207
[3] Kurta R P, Altarelli M, Weckert E and Vartanyants I A 2012 *Phys. Rev.* B **85** 184204
[4] Saldin D K *et al.* 2011 *Phys. Rev. Let.* **106** 115501
[5] Saldin D K, Poon H C, Schwander P, Uddin M and Schmidt M 2011 *Optics Express* **19** 17318
[6] Kurta R P, Dronyak R, Altarelli M, Weckert E and Vartanyants I A 2012 (submitted)
[7] Kam Z 1977 *Macromolecules* **10** 927
[8] Walrafen G E 1964 *J. Chem. Phys.* **40**, 3249
[9] Canpolat M *et al.* 1998 *Chem. Phys. Lett.* **294** 9-12
[10] Head-Gordon T and Johnson M E 2006 *PNAS* **103 7973-77**
[11] Ludwig R 2001 *Angew. Chem. Int. Ed. Engl.* **40** 1808
[12] Wernet Ph *et al.* 2004 *Science* **304** 995

Chemical Shift Images of Organelles in Leydig cells of Mice Testes

T. Ejima[1], Y. Neichi[1], M. Yanagihara[1], M. Kado[2], M. Ishino[2], K. Yasuda[3], and S. Tamotsu[3]

[1]Institute of Multidisciplinary Research for Advanced Materials, Tohoku University: 2-1-1, Katahira, Aoba-ku, Sendai, 980-8577 JAPAN.

[2]Kansai Photon Science Institution, Japan Atomic Energy Agency: 8-1-7, Umemidai, Kizugawa, Kyoto, 619-0215 JAPAN.

[3]Research Group of Biological Sciences, Faculty Division of Natural Sciences, NaraWomen's University: Kitauoyahigashimachi, Nara, 630-8506 JAPAN.

ejima@tagen.tohoku.ac.jp

Abstract. Soft X-ray transmission images of Leydig cells of mice testes changing incident wavelength were observed with the use of a contact microscope. After normalization of transmission images, absorbance images were obtained and compared with a visible differential interference image. Some organelles were identified by the image comparison, and absorption spectra of the organelles were obtained from the absorbance images. The absorption spectra show that peak structures are different depending on the observed organelles. The structures and the positions of organelles were clearly identified at C-K absorption.

1. Introduction

Metabolism in a bio-cell is a sequence process: **a.** an organic material ingested from outside of a bio-cell is transformed to the new one/ones changing its chemical-bonding state/states. **b.** Through the material transformation, energy is simultaneously produced to keep the vital activity of the bio-cell. **c.** The energy is consumed in/out of the bio-cell and used for the other transformations [1]. This process indicates all component materials in a bio-cell are the combination of some low-molecular-weight organic materials. For example, DNA and RNA are composed of 5 kinds of nucleotides, and the proteins in bio-cells are made of 20 kinds of amino acids [1]. These organic materials have a tendency that some of them concentrate in specific organelles. In cell nuclei, they are composed of DNA, RNA and specific proteins for keeping their functions [2]. In mitochondria, enzymes and proteins to produce the energy for metabolism concentrate in them. [3]. A membrane is composed of phosphorylated lipid. Necessary proteins and organic materials penetrate into and unnecessary substances are blocked out of the membrane.

These organic materials have different chemical bonding states, which can be identified by binding energy at an absorption edge of the component atom. The binding energies can be measured by X-ray absorption spectroscopy (XAS) [5]. Absorption spectrum of an organic material has a specific peak position with a spectral shape depending on its chemical bonding states. Therefore, interpreting the spectral shape is inseparable from understanding the chemical bonding state of the material.

Based on the chemical bonding states of nucleotides and amino acids, the XAS spectra measured at C-K edges were roughly classified into 3 groups [5-7]: **1.** Acyclic compounds composed of major atoms, such as H, C, N, and O. These compounds have a sharp peak originated from C π^* orbital observed at around 289 eV with a broad peak originated from C σ^* orbital at the higher binding energy side of the C π^* peak. **2.** Acyclic compounds including all component atoms except the group 1. These compounds show C π^* peak at 289 eV with C σ^* peaks. The C σ^* peaks are often observed at the lower binding energy side of the C π^* peak. **3.** Cyclic compounds. The peak of C π^* orbital in cyclic structure is observed at around 285 eV. Peaks originated from the side chain compounds are in almost the same energy positions as those in group 1 and 2.

According to this classification, XAS spectra of organelles can be interpreted as the combination of spectra in these 3 groups [8]. When a XAS spectrum of an organelle is measured, the spectral shape will be interpreted by the component organic materials or the chemical-bonding state of the organelle. In this study, we choose contact microscopy method to obtain absorption spectra of organelles through absorption images and select Leydig cells of mice testes as a measurement sample. The results of absorption measurements will be reported in this paper.

2. Experimental Conditions

Leydig cells were cultured on a SiN membrane and fixed by 4% paraformaldehyde. The Leydig cells preserved in the cultural fluid was packed into a cell holder. The cell holder can keep the liquid sample between the SiN membrane and a scintillator plate in a vacuum chamber. The scintillator plate used in the measurements was Ce:LYSO, which converts a shadow image under the sample to a visible fluorescent image. The converted image can be observed by a visible microscope. Thus, the Leydig cells can be observed changing the incident wavelength [9]. Observation was made at BL27SU, SPring-8, Japan [10]. Range of incident wavelength was from 4.77 to 2.07 nm (260 to 600 eV), and wavelength resolution $\lambda/\Delta\lambda$ of the incident light was 2000. Exposure time was 0.8 - 180 sec, which changed depending on the light intensity. The spatial resolution was estimated from the edge response. The edge was the boundary of the SiN membrane between the frame of the membrane and the SiN membrane (inset of Fig. 1). The value of the spatial resolution, 1.1μm, was estimated from the peak width, which was obtained by a numerical differentiation of the edge response at line A-A' in Fig. 1. After the soft X-ray observation, visible transmittance- and visible differential-interference- images were taken at the same position and compared with the soft X-ray ones.

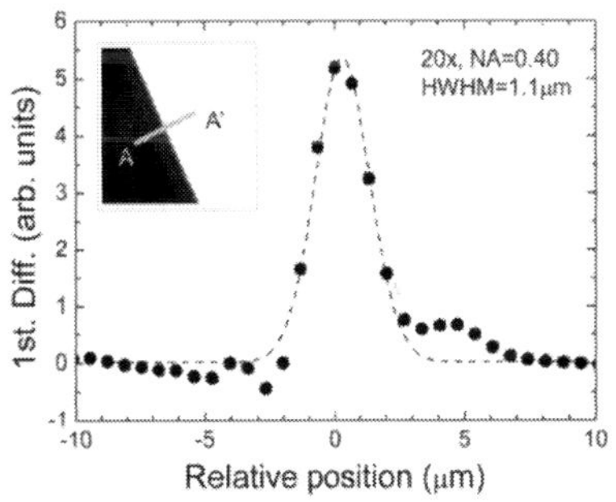

Figure 1: Edge response result to obtain the spatial resolution of the microscope. Inset: the edge used in the measurement.

3. Results

3.1. Soft X-ray Images and Spectra

At first, the background images were subtracted from the measured soft X-ray images, and the images were normalized by both the incident light images, from which the background images were also subtracted, and the exposure time to obtain transmittance images. Absorbance images were calculated from the obtained transmittance images under the assumption that there is no interference in the sample. Thus the absorbance images were compared with the visible differential interference image (Fig. 2(a)). Detailed comparison between these images indicates that absorption intensity in each organelle is different. According to the different values of the absorption intensity in organelles, regions of interest (ROIs) in each bio-cell were divided into three parts as designated in Fig. 2(d): the central region should correspond to cell nucleus (ROI A), surrounding regions around the central area are expected to be small and different organelles (ROI B), and the external part of ROI B is considered to be cell membrane under the cultural conditions (ROI C). The area outside each bio-cell is filled with the cultural fluid (ROI D).

Spectra in all ROIs were obtained from the absorption images. Spectrum of each ROI was normalized by the spectrum of ROI D on the assumption that incident light passes through the cultural fluid in all ROIs. The normalized spectra at C-K absorption edge are represented in Fig. 3. In the spectra of ROI A and B, a broad peak is observed at around 283 eV and 286 eV, respectively. In the spectrum of ROI C, no peak is observed. At N- and O-K absorption edges, absorption spectra were obtained in the same manner as at C-K edge. At N-K edge, no peak structure was observed, because the absorption of SiN membrane was stronger than the signal of the bio-cells. At O-K edge, 4 peaks at 526, 529, 532, and 536 eV were observed in all ROIs, and the spectral shape of each ROI is similar to each other. Because the wavelength interval was too wide in the measurement, further details of O-K edge results will be discussed in the future study.

3.2. Chemical Shift Images

Spectral shapes of ROI A and B show a broad peak structure at C-K edge (Fig. 3), and are reproduced well by Lorentzian curves (solid curves in Fig. 3). Fitting results show that the peak positions of ROI A and B are 283.3 eV and 285.4 eV, respectively. Peak widths of ROI A and B are 8.7 eV and 6.6 eV, respectively, in FWHM. Peak intensity of ROI A doubles that of ROI B.

The images taken at 283 eV and 285 eV were selected as the peak position images, and the background image at 275 eV was subtracted from these peak images. These image results in Fig 2 (b) and Fig 2 (c) showed the strong absorption by cell nucleus and by small organelles, respectively.

4. Discussion

To interpret the broad peak observed at 283.3eV in ROI A, the spectral shapes of the organic materials, which were mentioned in Introduction, are assumed. Nucleotides have a cyclic structure and the XAS peak originated from this structure is observed at around 285 eV. Then the broad peak observed at 283.3eV in ROI A will be originated from nucleotides. Other organic materials with XAS peaks at the higher binding energy side are the following amino acids: Phenylalanine, Tyrosine, Tryptophan, and Histidine. Most of these amino acids have an acyclic structure and XAS peaks originated from this structure are observed at around 289 eV. Then the broad peak in ROI B observed at the higher energy side of the peak in ROI A will be originated from amino acids. This speculation coincides with component materials of the organelle, which are assigned as nuclei in ROI A, as mitochondria in ROI B, and as cell membrane in ROI C [5-7].

Peak width of the C π^* peak in nucleotides is sharp and the value is less than 1 eV. If the broad peak structures observed in ROI A and B are originated from the C π^* peak, each spectral shape is an overlap of the sharp peaks at the different energy positions. Judging from the observed broad peak

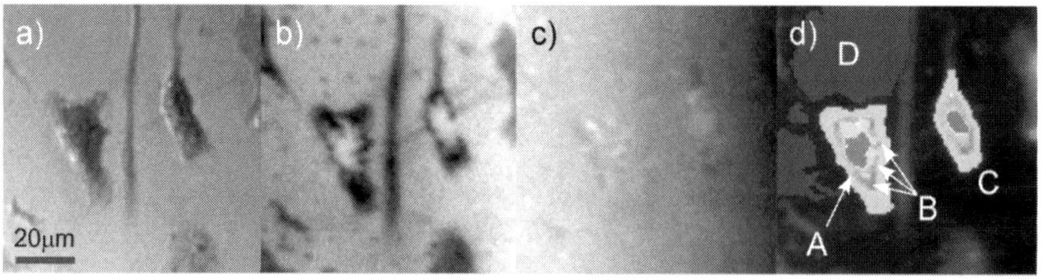

Figure 2: a) Visible differential interference image, b) SX image at 283 eV, c) SX image at 285 eV, d) ROIs are identified by colour: red is ROI A, cyan, ROI B, green, ROI C, and blue, ROI D.

structures, chemical bonding states in each organelle will be complex.

Peak positions in each ROI are about 2 eV lower than those in nucleotides and amino acids. This peak shift is caused by the sample conditions. The values in ROI A and B were measured using a liquid sample, whereas the values in these organic materials were measured in solid samples such as a thin film (in Ref. 5 and 6) and a crystalline powder (in Ref. 7). Therefore, chemical shift will occur depending on the electronic structure of the measured samples.

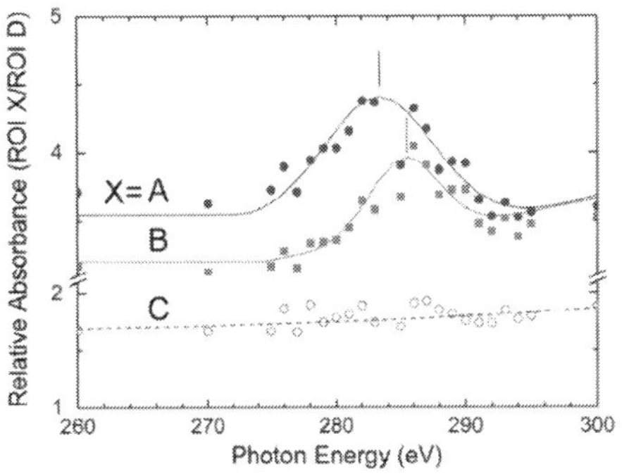

Figure 3: Normalized XAS spectra of each ROI obtained from SX images (circles). Solid and broken curves are fitting results.

5. Summary

Normalized XAS spectra of organelles in Leydig cells of mice testes were obtained from the transmission images. The different peak positions observed at C-K edge are interpreted by XAS spectra of nucleotides and amino acids which are the component organic materials of bio-cells. And the background-subtracted images at the peak positions show both structures and positions of the organelles. Identification of a small organelle or a specific protein is usually made by fluorescence proteins with the use of visible light microscopes. Present results show the identification of the organelles will be made by the binding energy difference at an absorbance edge without any modification of the organelles. For this purpose, sample conditions should be clarified to observe small organelles in a bio-cell, and laboratory-type SX microscope should be developed to change incident wavelength without any restriction of the light source.

References

[1] for example, B. Alberts, *et al.*, *Molecular biology of the cell*, (Garland Science, Taylor & Francis Group, New York, 2010) Chap. 2.
[2] for example, B. Alberts, *et al.*, *Molecular biology of the cell*, (Garland Science, Taylor & Francis Group, New York, 2010) Chap. 4.
[3] for example, B. Alberts, *et al.*, *Molecular biology of the cell*, (Garland Science, Taylor & Francis Group, New York, 2010) Chap. 6.
[4] for example, B. Alberts, *et al.*, *Molecular biology of the cell*, (Garland Science, Taylor & Francis Group, New York, 2010) Chap. 10.
[5] J. Boese, A. Osanna, C. Jacobsen, and J. Kirz, J. Elec. Spectro. Rel. Phenom. **85**, (1997) 9.
[6] K. Kaznacheyev, A. Osanna, C. Jacobsen, O. Plashkevych, O. Vahtras, H. Ågren, V. Carravetta, and A. P. Hitchcock, J. Phys. Chem. A **106**, (2002) 3153.
[7] Y. Zubavichus, A. Shaporenko, V. Korolkov, M. Grunze, and M. Zharnikov, J. Phys. Chem. B **112**, (2008) 13711.
[8] J. S-Ornstein, A. P. Hitchcock, D. H. Cruz, P. Henklein, J. Overhage, K. Hilpert, J. D. Hale, and R. E. W. Hancock, J. Phys. Chem. B **111**, (2007) 7691.
[9] T. Ejima, *et al.*, in preparation.
[10] http://www.spring8.or.jp/wkg/BL27SU/instrument/lang-en/INS-0000000287

Data-constrained characterization of sandstone microstructures with multi-energy X-ray CT

Y S Yang[(a)], A Tulloh[(a)], F Chen[(a)], K Y Liu[(b)], B Clennell[(b)], J Taylor[(c)]

(a): CSIRO Materials Science & Engineering, Locked Bag 33, Clayton, Victoria 3169, Australia
(b): CSIRO Earth Science and Resource Engineering, PO Box 1130, Bentley, WA 6102, Australia
(c): CSIRO eResearch & Computational and Simulation Sciences, GPO Box 664, Canberra ACT 2601, Australia

Sam.Yang@csiro.au

Abstract. A data-constrained non-linear optimization approach has been developed to characterize microscopic distributions of mineral phases and pores in a sandstone sample using X-ray CT data sets acquired at 35keV and 45keV beam energies as constraints. The approach minimizes discrepancy between the expected and measured linear absorption coefficients and maximizes Boltzmann distribution probability. It enables integration of both the 3D X-ray CT data-constraints and global level information, and leads to more accurate predictions of microscopic 3D compositional distributions in material samples. Permeability simulations and comparisons with experimentally measured porosity indicate that DCM characterisation agrees reasonably with experimental observations. However, segmentation of CT images leads to under-estimation of porosity and permeability.

1. Introduction

In oil and gas reservoir rock characterization, and in materials development and processing, it is often the case that one needs to optimize a given set of material performance properties. Material performance properties are closely related to their microstructures. Considerable experimental, theoretical and numerical efforts have been devoted to 3D microscopic characterization [1-7]. X-ray CT has been widely used as a sample non-destructive method for microstructures characterization. However, segmentation of X-ray CT images is not always adequate to discriminate material compositions, as different combinations of material phases may exhibit similar X-ray absorption properties.

A level of success has been achieved in microstructures characterization using a data-constrained modeling (DCM) linear optimization approach with multi-energy (spectrum) X-ray CT [8-17]. However, it is relatively sensitive to experimental noise, and in general not suitable for characterization of materials consisting of more than three compositions. In this article, both multi-energy X-ray CT data and microscopic correlations among material compositions are incorporated into the DCM model. The correlations are formulated as phenomenological self energies for each composition as well as intra- and inter-voxel interaction energies between different compositions. The overall formulation of the problem resembles one of statistical mechanics [18]. It is capable of treating partial occupancy of voxels and can in principle incorporate a wide range of types of data and sources of a priori information about the sample. The methodology is generic and is applicable to a wide range of science and engineering disciplines.

2. DCM nonlinear optimization formulation

The model is defined on a simple-cubic grid with a cubic voxel (volumetric pixel) on each grid point. For a voxel at n ($n = 1,2,\cdots,N$) where N is the total number of voxels in the system, the model attempts to minimize the following objective function:

$$T_n = \sum_{l=1}^{L}\left[\delta\mu_n^{(l)}\right]^{2\zeta} + E_n \tag{1}$$

where L is the number of X-ray CT data sets acquired from the same sample using L different beam energies. This is equivalent to minimizing the discrepancy between the expected and the measured linear absorption coefficients and maximizing Boltzmann distribution probability [18-19]. The parameter ζ (≥ 0) is introduced to approximate relative error weight distribution for the CT reconstructed absorption coefficients. Its default value is $\zeta = 1$.

For each X-ray spectrum (beam energy) l, the relative CT error is expressed as

$$\delta\mu_n^{(l)} = \frac{1}{\mu_{\max}^{(l)}}\left|\sum_{m=0}^{M}\mu^{(m,l)}v_n^{(m)} - \hat{\mu}_n^{(l)}\right|, \tag{2}$$

where $\mu^{(m,l)}$ is the total linear absorption coefficient for material m, $\hat{\mu}_n^{(l)}$ is the CT reconstructed linear absorption coefficient, and $\mu_{\max}^{(l)} = \max\left\{\mu^{(m,l)}; m = 0,1,2,\cdots,M\right\}$. The optimization is achieved by adjusting the volume fraction variables $v_n^{(m)}$ ($m = 0,1,2,\cdots,M$) for each material composition m, where M is the total number of compositions in the system, subject to the constraints:

$$\sum_{m=0}^{M} v_n^{(m)} = 1 \quad \text{where} \quad 0 \leq v_n^{(m)} \leq 1 \qquad \text{for } m = 0,1,2,\ldots,M \tag{3}$$

The dimensionless phenomenological interaction energy is expressed as

$$E_n = \sum_{m=0}^{M} v_n^{(m)} S^{(m)} + \sum_{k=0}^{K}\sum_{j=1}^{N^{(k)}}\sum_{m_1=0}^{M}\sum_{m_2=m_1}^{M}\left(v_n^{(m_1)} v_{n+n_j^{(k)}}^{(m_2)}\right)^{\eta} I_k^{(m_1,m_2)} \tag{4}$$

Where $S^{(m)}$ denotes the self-energy (chemical potential) for composition m, $I_k^{(m_1,m_2)}$ denotes the interfacial energy between neighboring voxel compositions m_1 and m_2 at distance k, where $k = 0$ ($k > 0$) denotes intra- (inter-) voxel. The maximum interaction range is denoted by K. The number of neighboring voxels with neighboring range k is denoted by $N^{(k)}$ and the j th neighboring vector with a neighboring range k is denoted by $n_j^{(k)}$. The parameter η with a default value of $\eta = 0.5$ is used to approximate the volume to interface relation.

Multi-spectrum X-ray CT with a synthetic CIPS (Calcite In-situ Precipitation System) sandstone sample has been carried out at beam energies 35 and 45keV [15]. DCM non-linear optimization has been applied to the data sets which represent a region of $480 \times 480 \times 280$ voxels in the sample. Each voxel represents a sample volume of $5.92 \times 5.92 \times 5.92$ microns. The selected region represents a sample volume of $2.84 \times 2.84 \times 1.66$ mm. A typical CT image, DCM non-linear reconstructed and least-square segmented compositional maps are shown in Figure 1. DCM non-linear reconstruction shows an average porosity of 19.6% which compares favorably with the He gas measured value of 19.5%. The model uses parameters: $I_0^{(\text{pore,calcite})} = 0.015$, $I_0^{(\text{pore,quartz})} = I_0^{(\text{quartz,calcite})} = 1$, $I_1^{(\text{pore,pore})} = I_1^{(\text{quartz,quartz})} = I_1^{(\text{calcite,calcite})} = -0.015$. A multi-energy CT least-square segmentation, which is similar to threshold segmentation for single energy CT, gives a much lower overall porosity of 11.3%. To minimize error near sample boundary and image artifacts slices 1-50 and 215-250 have been excluded in porosity estimations.

Figure 1 shows that calcite exists both as coatings on the quartz grains and as clusters between quartz grains. This is consistent with SEM observations [19]. In DCM generated images, calcite pixels have varying intensities. This indicates the coexistence of calcite and void. In other words, calcite exists in varying degrees of porosity in the sample which is consistent with experimental observations. This partial occupancy of voxels cannot be obtained by image segmentation.

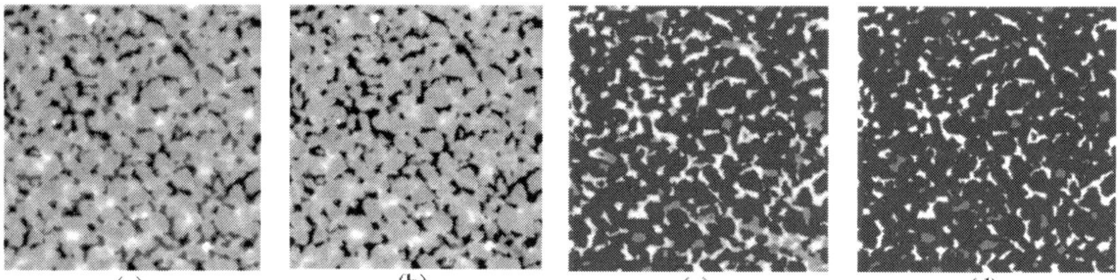

(a) (b) (c) (d)

Figure 1. X-ray CT image and compositional maps at slice 140. (a) X-ray CT image at monochromatic beam energy 35keV. (b) X-ray CT image at monochromatic beam energy 45keV. (c) Microscopic compositional distributions from DCM non-linear optimization. (d) Microscopic compositional distributions from least-square segmentation. In (c) and (d), quartz, calcite and void are represented as blue, red and white respectively. Volume fractions are represented by their colour intensities.

3. Permeability simulations

A random walk method is used to predict permeability defined by Darcy's law [20]. In general, permeability is a function of compositional volume fractions and internal fine structure. For the pre-defined sandstone sample, compositional volume fractions have been derived from the DCM non-linear model or least-square segmentation model. However, the fine structure below the voxel size is unknown. An analytic formula derived in [20] is used to calculate voxel based permeability under an isotropic structure constraint, i.e. compositions mixed randomly. Our numerical study in this section will use an isotropic structure for the effective permeability simulations.

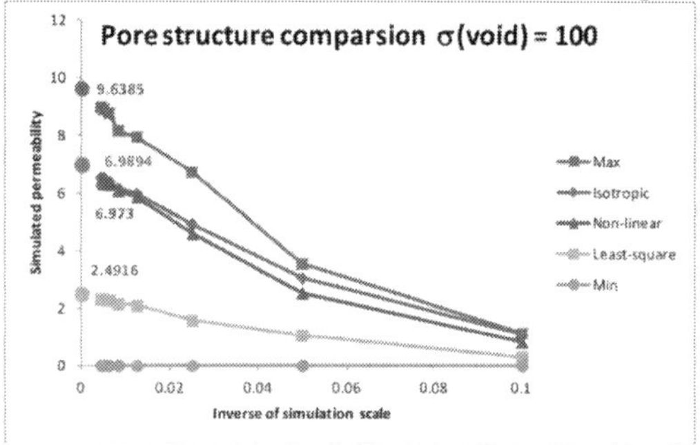

Figure 2. Comparison of the random walk simulation results from different pore structure constraints with given σ (void) = 100. Simulations based on the theoretical maximum and minimum are plotted on the top and bottom respectively. The converging point for each case is represented by a circle on the vertical axis. The simulation result from the DCM non-linear output correlates well with the isotropic microstructure constraint.

Figure 2 shows the permeability simulation results based on different pore connectivity constraints. For each constraint, simulated permeability is plotted against the inverse of simulation physical scale. Circles on the vertical axis show converging values of permeabilities for different pore configurations. Since quartz is not permeable and the permeability of calcite is insignificant compared to void, only void permeability is required to calculate voxel-based permeability. In our numerical simulations, the void phase has been assigned a dimensionless permeability value of 100, and quartz and calcite have been assigned a value of 0. The permeability value of a voxel is calculated based on its compositional volume fractions [20]. The top and bottom plots in the figure represent the simulations based on the theoretical maximum and minimum to define an upper and a lower bound for the effective permeabilities. The middle three plots correspond to three pore structure configurations: isotropic mixture of compositions, from DCM non-linear optimization and from least-square segmentation.

Conclusions to draw from Figure 2 are firstly, the simulated permeabilities based on different pore structure constraints are bounded by the theoretical range, thereby verifying the random walk algorithm. Secondly, the least-square segmentation method leads to a reduction in pore connectivity, which results in a significant underestimation of effective permeability. Thirdly, DCM non-linear optimization improves pore network activity. The simulated permeability from the DCM output shows a quantitative agreement with the permeability simulated by applying an isotropic structure constraint. As can be seen, both limits are very close to each other.

4. Summary

Phenomenological intra- and inter-voxel interactions among material compositions as well as self-energies are incorporated into the data-constrained model (DCM) for characterization of microstructures of various materials. It is possible to incorporate prior knowledge and other measurements. The model is formulated as a constrained non-linear optimization problem. Numerical results indicate that the DCM non-linear optimization approach produces a more robust and accurate prediction of compositional microstructures, and leads to more accurate modeling of permeability properties.

The X-ray CT images and DCM reconstructed compositional maps are available from the CSIRO Data Access Portal [21]. This work is sponsored by CSIRO Computational & Simulation Science Transformational Capability Platform and Shanxi Province One-Hundred Person project.

References

[1] Spanos G. (Eds) 2006 *Scripta Materialia* **55** 1
[2] MacPherson R. D. and Srolovitz D. J. 2007 *Nature* **466** 1053
[3] Mourachov S. 1997 *Computational Materials Science* **7** 384
[4] Raabe D. 2002 *Annu. Rev. Mater. Res.* **32** 53
[5] Rohl A. L. 2003 *Current Opinion in Solid State and Materials Science* **7** 21
[6] Suzudo T. 2004 *Physica A* **343** 185
[7] Yang Y. S., Blake N., Abbott T. B. and McCarthey J. F. 1993 *Scripta Metallurgica et Materialia* **29** 1285
[8] Vinegar H. J. and Wellington S. L. 1987 *Rev. Sci. Instrum.* **58** 96
[9] Yang Y. S., Tulloh A., Cole I., Furman S., Hughes A., 2007 *J. Aust. Ceram. Soc.* **43,** 159
[10] Yang S., Furman S., Tulloh A., 2008 *Advanced Materials Research Vol 32: Frontiers in Materials Science & Technology* (Trans tech Publications) 267
[11] Granton P. V., Pollmann S. I., Ford N. L., Drangova M., Holdsworth D. W. 2008 *Med. Phys.* **35** 5030
[12] Yang Y. S., Gureyev T. E., Tulloh A., Clennell M. B., Pervukhina M. 2010 *Measurement Science & Technology* **21**, 047001 (6pp), doi:10.1088/0957-0233/21/4/047001
[13] Yang S., Gao D., Muster T., Tulloh A., Furman S., Mayo S., Trinchi A. 2010 *Materials Science Forum* **654-656** 1686
[14] Mayo S. C., Tulloh A.M., Trinchi A., Yang Y. S. 2012 *Microscopy and Microanalysis* **18** 524
[15] TrinchiA., Yang Y. S., Huang J. Z., Falcaro P., Buso D. and Cao L. Q. 2012 *Modelling Simu. Mater. Sci. Eng.* **20** 015013
[16] Wang H. P., Yang Y. S, Wang Y. D., Yang J. L., Jia J., Nie Y. H. 2012 *Fuel* http://dx.doi.org/10.1016/j.fuel.2012.11.079
[17] Yang Y. S. 2012 *Lecture Notes in Information Technology* **15** (ISBN: 978-1-61275-015-6, ISSN:2070-1918, Information Engineering Research Institute) pp198-205
[18] Thompson C. J. *Classical Equilibrium Statistical Mechanics*, (Clarendon Press, Oxford, 1988)
[19] Yang Y. S., Wang H. P., Gao J. R. 2012 *Journal of Shanxi University* **35** No. 2 248
[20] Chen F. F., Yang Y. S. 2012 *Modelling Simu. Mater. Sci.Eng.* **20** 045005
[21] Sam Yang, Keyu Liu, Sherry Mayo, Andrew Tulloh, *CSIRO Data Collection* (2012) http://dx.doi.org/10.4225/08/5045B5990B44E

X-ray and EUV micro-imaging systems for laser ICF diagnostics

Yi Shengzhen[1, 2], Mu Baozhong[1, 2,*], Wang Xin[1, 2], Huang Wannian[3], Jiang Li[1, 2], Wang Zhanshan[1, 2]

[1] Tongji University, Shanghai, 200092, P.R. China
[2] Key Laboratory of Advanced Micro-structured Materials (MOE), Tongji University, Shanghai, 200092, P.R. China
[3] School of Political Science and International Relations, Tongji University, Shanghai, 200092, P.R. China

Abstract. Plasma imaging diagnostics plays an important role for laser ICF. Based on the urgent need to carry out high-resolution, high-throughput plasma diagnostics, grazing-incidence X-ray Kirkpatrick-Baez (KB) microscopes and normal-incidence EUV Schwarzschild imaging system were developed. The X-ray multilayer KB microscopes were successfully been applied in the physics experiments of SGII laser facility. Combined with streaked camera, the Mo-backlit implosion flow line of hollow Carbon-Hydrogen (CH) spherical target was obtained in SGII. The 4.75keV single-channel and four-channel KB microscopes were also developed for self-emission and short-pulse backlit imaging diagnostic of CH cylindrical target. In addition, according to the need of ultra-short laser pulse plasma diagnostics, the Schwarzschild imaging system working at 68eV was researched, and the physical experiments of hot electron transport with Schwarzschild imaging system were performed in SILEX-I laser facility.

1. Introduction

Laser inertial confinement fusion (ICF) is of great significance for national economy, national security, and basic research such as plasma physics[1,2]. X-ray imaging diagnostics of implosion area is an important region of laser ICF. Through X-ray imaging of high-temperature and high-density plasma implosion area, and coupled with spectrometry or time-resolved measurement devices, the diagnostics system of time, spatial, and spectrum characteristics can be formed. The basic state parameters such as irradiation uniformity, symmetry of compression, radial convergence, implosion time and velocity, electron temperature and density distribution, and other important information can be obtained to provide experimental references for structural design of the target, the optimization of laser irradiation conditions and simulation programs. The Kirkpatrick-Baez (KB) microscopes provide higher throughout and resolution than pinhole imaging. Several successful implementations such as X-ray backlighting or self-luminescence imaging implosion core area of high-energy X-ray imaging, gated monochromatic X-ray imaging (GMXI) and grating spectral imaging have been made in OMEGA and LMJ etc[3,4]. A series of ICF physics experiments such as hydrodynamic instability growth, ablation evolution, and plasma compression uniformity have been launched to obtain quantitative diagnostic information about plasma temperature $<kT>$ and density $<\rho R>$[5-10]. Based on Chinese present

* Corresponding Author, E-mail:mubz@tongji.edu.cn

situation lacking high-precision X-ray imaging devices and the urgent need to carry out high-quality X-ray space-time diagnostics, the paper carried out the research of X-ray multilayer KB imaging system and EUV Schwarzschild imaging system.

2. Optical design of X-ray multilayer KB microscopes

KB microscope as shown in figure.1 consists of two concave spherical mirrors normal perpendicular to each other[11]. The perpendicular structure plays a role to overcome the strong astigmatism. Rays from object point (A) are successively reflected at grazing angles (θ_1 and θ_2) by two mirrors, and obey the Coddington equations to form a two-dimensional image (A'). The KB microscope has serious off-axis aberration, and the resolution decreases obviously with the increase of object field. We have analysed the influences of geometric aberrations and optical machining tolerances on imaging quality, and the imaging quality criterion of KB microscope was established from spatial resolution, MTF and optical efficiency. The work focused on the influences of grazing incidence angle, radius of curvature, and magnification on MTF in different field of view, and provided theoretical and experimental references for the selection of initial structural parameters.

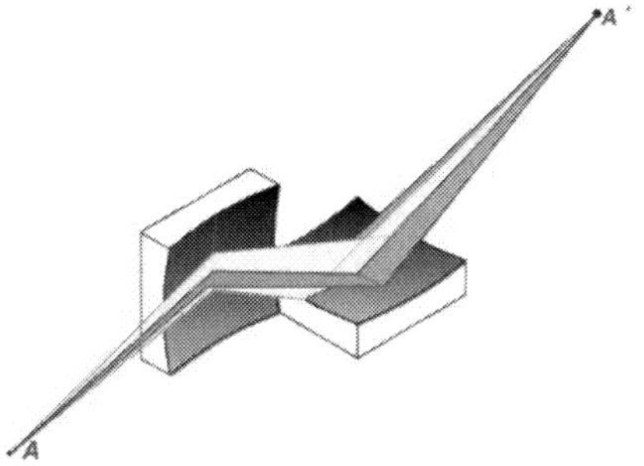

Figure 1. Schematic of single-channel KB microscope

Combing previous researches[12,13], the grazing incidence angle of KB microscopes used in China ICF diagnostics is about 1° to ensure imaging quality, optical efficiency and energy resolution. The radiuses of curvature and length of KB mirrors are about 20m and 10mm, to get better image quality and adequate optical efficiency. The magnification is about 10-15× considering the spatial resolution of image detector, signal response threshold and other special considerations. The spatial resolution decreases near-linearly from 2 microns at central field to near 5 microns at the 200 microns object field. Base on several ICF diagnostic experiments with different types, we designed a series of KB microscopes with detailed optical parameters shown in Table.1. The 4.75keV KB microscope is a four-channel imaging system and will be coupled with X-ray framing camera.

Table 1. Final optical parameters of KB microscopes for different ICF experiments

X-ray energy [a]	Radius	Grazing angle [b]	Magnification	Object distance
2.5keV[c]	18399mm	θ_1=1.0°; θ_2=1.0543°	M_1=15×; M_2=14.12×	171.3mm
4.75keV[d]	20000mm	θ_1=1.0°; θ_2=1.0442°	M_1=8×; M_2=7.56×	196.3mm
8keV[e]	19500mm	θ_1=1.0°; θ_2=1.0479°	M_1=10×; M_2=9.44×	187.2mm

[a] The multilayers were designed and coated for different X-ray energies.
[b] θ_1 stands for the grazing angle of first KB mirror ,and θ_2 for the second mirror.
[c] Used for imaging experiment of implosion flow line of hollow CH spherical target.
[d] Used for self-emission and short-pulse backlit imaging diagnostic of CH cylindrical target.
[e] Used for fast ignition Cu-Kα imaging experiments.

3. Imaging experiments of KB microscope

Imaging experiments of KB microscope is the essential issue for the ICF applications. Along with object field deviates from central field, the spatial resolution of KB microscope decreases rapidly, so the assembly and alignment must reach the positioning accuracy of about 30μm to indicate the best objet field. Firstly, we must find the best object field of KB microscope through X-ray imaging experiments in the laboratory, and then the best object field was replaced by a removable pointer with an Φ200μm reference ball. The ball pointer and KB microscope was connected together with high-precision industrial linear guide which ensured the repeat positioning accuracy up to 20μm.

3.1. Imaging experiments in the laboratory

We have assembled and tested KB microscopes listed in table.1. Figure.2 shows the 8keV KB imaging of 600# Au grid which simulates a resolution pattern. The object was backlit by the copper anode fine-focus of the x-ray tube operated at 45kV and the tube current of 20mA. The image was recorded by a scintillator x-ray CCD (Photonic Science: XDI-50) with 30min exposure time. The CCD pixel size (6.45μm×6.45μm) is small enough to obtain high-resolution imaging results for the magnification M=10×. The image is very clear in central field and gradually blur out with the increase of the field of view. The spatial resolution measured from figure.2, which corresponds to the distance of 20% to 80% of the minimum intensity to the maximum intensity, is approximately 2μm at central field, and better than 5μm within ±200μm object field.The ununiformed brightness is due to the x-ray source, which has been tested by directly projection. The image quality in laboratory experiments was seriously influenced by the background noise of x-ray CCD because of the long time exposure, and it will be improved in the actual ICF experiments in a 2-3ns, short-pulse time.

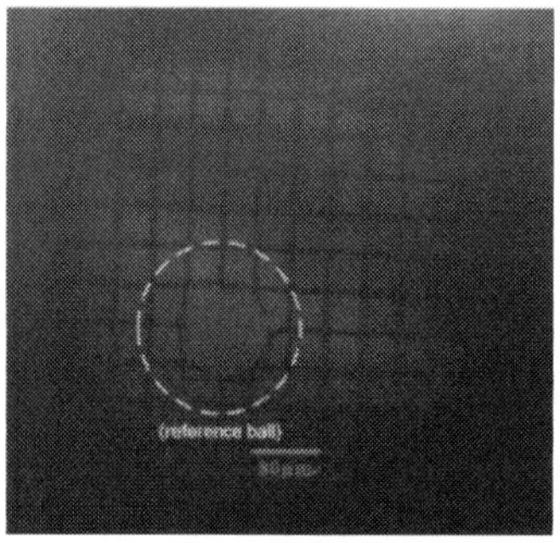

Figure 2. Image of 600# Au grid taken with Copper X-ray tube.

Figure 3. These Image of 2000# Au grid taken with SGII laser.

3.2. Imaging experiments in SGII laser facility

We replaced the Au grid with the ball pointer as shown in figure.2, and tested image characteristics of KB microscope in Chinese SGII laser facility. The reference ball of KB microscope was aligned to the chamber centre, and then replaced by a 2000# Au grid. The backlit X-rays were produced by Cu target driven by the ninth laser beam, and the spectral purity is improved through 4μm thickness Cu filter. The imaging results were shown in figure.3. The separation of the Au grid measured by SEM is 12μm with 5μm width. It is obvious that the clearest grid is located in the brightest region of the total field. For practical applications to ICF fast ignition experiments, the diagnostic target should be placed in this region to obtain the best image resolution and intensity.

Another two KB microscopes list in table.1 were also successfully been applied in the physics experiments of SGII laser facility. Combined with streaked camera, the Mo-backlit implosion flow line of hollow CH spherical target was obtained by 2.5keV KB microscope in SGII. The self-emission and short-pulse backlit imaging of CH cylindrical target was achieved by 4.75keV KB microscope.

4. Schwarzschild microscope for diagnostics of hot electron transport

In addition, according to the need of ultra-short laser pulse plasma diagnostics in fast ignition fusion, we also developed the EUV Schwarzschild imaging system working at 68eV. The spatial distribution of hot electron, caused by ultra-short laser-plasma interaction, can be measured by Schwarzschild imaging system due to it is an achromatic, higher spatial resolution, larger field of view and larger collecting area system[14-17].

The physics experiments of Schwarzschild microscope were performed with 286TW SILEX-I laser facilities. As shown in figure.4, the 30fs, 800nm laser pulses with energy of 4.4J was focused by a *f*/3 off-axis parabolic mirror at 21° from the copper wire axis onto the edge of the target. The EUV intensity emission profile along the copper wire caused by hot electron is found by Schwarzschild microscope as shown in figure.5. As the transport distance increases, the intensity of radiation decreases.

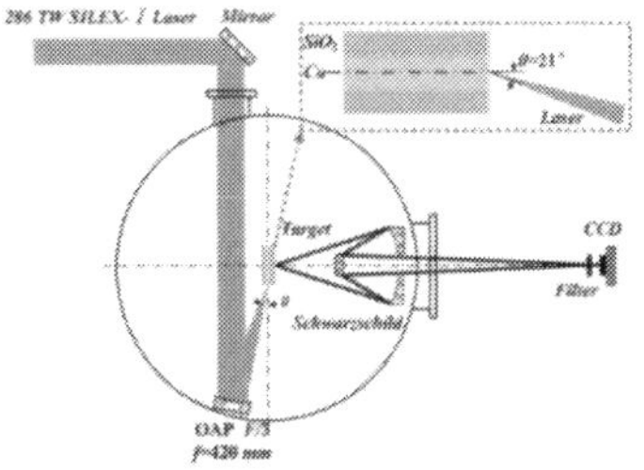

Figure 4. Schematic of hot electron diagnostics on 286 TW SILEX- I laser facility.

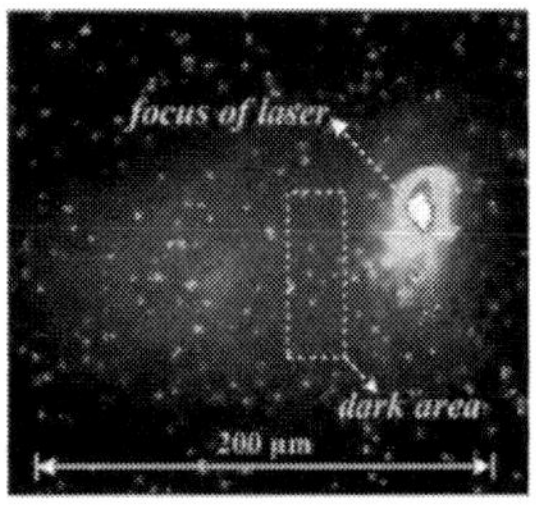

Figure 5. Result of hot electron diagnostics experiment

Acknowledgments

This work was supported by the National Natural Science Foundation of China (Grant No. 11027507,11105098,10825521) and Shanghai Science and Technology Development Funds (No.10QA1406900).

[1] Atzeni S and Meyer-ter-Vehn J 2004 *The physics of inertial fusion: beam plasma interaction, hydrodynamics, hot dense matter* (New York: Oxford University Press)
[2] Lindl J D 1998 *Inertial Confinement Fusion* (New York: AIP Press/Springer-Verlag)
[3] Koch J A, Landen O L and Jr. Barbee T W et al 1998 *Applied Optics.* **37** 1784
[4] Soures J M, McCrory R L and Verdon C P et al 1996 *Physics of Plasmas.* **3** 2108
[5] Gotchev O V, Jaanimagi P A and Knauer J P et al 2004 *Rev Sci Instrum*, **75** 4063
[6] Farshall F J and Qichang S 1995 *Rev Sci Instrum* **66** 725
[7] Marshall F J, Allen M M and Knauer J P et al 1998 *Phys Plasmas*, 1998 **5** 1118
[8] Marshall F J and Bennett G R 1999 *Rev Sci Instrum* **70** 617
[9] Marshall F J, Oertel J A and Walsh P J 2004 *Rev Sci Instrum* **75** 4045
[10] Gotchev O V Hayes L J and Jaanimagi P A et al 2003 *Rev Sci Instrum* **74** 5065
[11] Kirkpatrick P and Baez AV 1948 *Journal of The Optical Society of America* **38** 766
[12] Mu B Z, Yi S Z, Huang S L and Wang Z S 2008 *High Power Laser and Particle Beams* **20** 409
[13] Yi S Z, Mu B Z, Wang X and Huang S L, et al 2009 *High Power Laser and Particle Beams* **21** 1681
[14] Key M H, Cable M D and Cowan 1998 *Physics of Plasmas* **5** 1966
[15] Gremillet L, Amiranoff F and Baton S D 1999 *Physical Review Letters* **83** 5015
[16] Pisani F, Bernardinello A and Batani D 2000 *Physical Review E* 62 R5927
[17] Shealy D L, Hoover R B, Barbee T W and Walker A B C 1990 *Optical Engineering* 29 721

Study of columnar-equiaxed transition and anaxial columnar dendrites growth of hypoeutectic alloy with synchrotron radiation

F G Li [1], J Zhang[1], Q Dong[1], Y B Dai[1], B D Sun[1] and H L Xie[2]

[1]School of Materials Science and Engineering, Shanghai Jiaotong University, Dongchuan Rd. 800#, Minhang District, Shanghai, 200240, P.R. China
[2]Shanghai Synchrotron Radiation Facility, Shanghai Institute of Applied Physics, Shanghai 215600, P.R. China

Email: zj119@sjtu.edu.cn

Abstract. Among solidification processes, the columnar to equiaxed transition (CET) is critical component for formation either of microstructure or macrostructure. Directional solidification of Al-Cu alloy was performed at the BL13W beamline of Shanghai Synchrotron Radiation Facility (SSRF). The imaging results reveal that CET was provoked by external thermal disturbance. The fragment of dendrite tips were come from solute-rich-zone. And they floated to the mushy zone to form the equiaxed dendrites. As the conditions are suitable, a new dendrite morphology sprout and grow up, in which dendrites grow along <110> directions in a binary Al-Cu alloy. These dendrites have no obvious primary arms and were named anaxial columnar dendrites.

1. Introduction

During directional solidification, the thermal flux direction and melt flow are considered that play an important role[1] in the evolution of liquid-solid interface morphology. Among the interface morphology evolutions, the columnar-to-equiaxed transition (CET) is an important part which is still indistinct at some points [2]. Detachment, drift and growth of the fragments of dendrite during solidification, has been widely accepted as the inducement of CET [3-6]. But the internal trig mechanism is still the basic to clarify solidification process.

Based on the imaging observations, the behaviors of fragment in the front of S/L interface were analyzed and discussed. At the same time, the new growth style of dendrite was found. These dendrites have no obvious primary arms and were named anaxial columnar dendrites[7]. They were formed at low thermal gradient (G_T=4.0 K/mm) and solidification speed (v=0.067 mm/s) and are hard to be identified in a traditional metallographic test.

2. Experimental procedure

The solidification was carried out inside a resistance furnace with two opposite windows. 15wt% Al-Cu alloy were prepared by 5N purity Al and Cu in a graphite crucible. Rectangular slices with 60×15×0.7

mm^3 in size were cut and sandwiched between two 120-μm-thick Al_2O_3 plates. The edge of the samples was sealed by high temperature adhesive. Four thermocouples were embedded into the sample with a interval of 8 mm along the edge to measure the temperature gradient during directional solidification.

The synchrotron imaging experiments were performed at beam line BL13W of SSRF. The sample was fixed upright in the holder of the furnace and heated to about 40 K higher than the melting point of the alloy. The cooling rate was kept at R=0.028 K/s and the temperature gradient over the region to be observed was G_T=5.0 K/mm. An incident monochromatic X-ray energy of 20 keV was employed. A detector dead time was 0.2 s, together with an exposure time of 2.5 s. The full image field of view corresponded to 6.8×3.4 mm^2.

3. Results and discuss

3.1. Behavior characteristics of fragment

According to the phase diagram of binary Al-Cu system, the alloy with a nominal composition Al-15 wt%Cu, has a typical hypoeutectic structure. As the eutectic temperature (821.4 K) of Al-Cu alloys is much lower than that (~893 K) of the liquidus, a wider mushy zone would exist during directional solidification. The main dendrite growth and its morphology evolvement happen in the melt in front of the mushy zone (~1-2mm in thickness here), which corresponds to the constitutional supercooling zone.

The whole process of CET during the directional solidification of the alloy was shown in Fig 1. The moment right before CET occurs was set at t=0.0 s (Fig 1a). A slight indication of interface instability was observed at the tips of the dendrites. Meanwhile, a dendrite fragment (named A) detached from the dendritic network was found in a liquid pocket at the right side of the observed area. The interface instability of the columnar dendrites became obvious in Fig 1b. The growth of the columnar dendrites was inhibited and the equiaxed characteristics appeared from the newborn dendrites. The protuberances at the columnar dendrite tips grew up rapidly in different directions (Fig 1b).

It was also noted that the detached dendrite tended to drift upwards away from its original location. Besides, Dendrite A rotated when it floated up in the melt. The rotation angle of the dendrite was measured and shown in Fig 1c. Its axis rotated clockwise 32.2° during 10 seconds floating, indicating that there is a micro melt flow paralleled to the growth interface of the dendrites. Tiny needlelike dendrite fragmentations, named B-D respectively, were observed in Fig 1(c, d, e, g). Their movement and growth processes could be tracked during the observation clearly. According to the initial shape and the size of the fragmentations, it could be inferred that they evolved from the broken dendrite arms, especially the secondary and tertiary arms detached from the dendritic network. From Fig 1f, Fragmentation B had drifted next to the upper edge of the observed area and became a big dendrite. When its movement stopped, the growth space for the dendrites next to it was limited and the solute in the melt between them enriched gradually. A fracture process of a big arm (named G) was recorded by Fig 1g. The detached dendritic arm had two indistinct images to which the arrows pointed at the same figure because of the long exposure time. The result indicated that the fracture of the dendrite was very sudden. The arm moved upwards quickly after fracture and then stopped because of the obstruction of Dendrites B and D, which stood above it, as shown in Fig 1h. From Fig 1h, the growth front of the mushy zone had proceeded out of view. Although the dendrite tips in the view field almost stopped growing, the coarsening of dendrites and the solute enrichment were still found by further observations.

3.2. Anaxial columnar dendrites growth

Anaxial columnar dendrites have special growth characteristics that are completely different from regular dendrites. There is no axis (primary arm) as the trunk of a dendrite. It is composed of a pair of laterally growing dendrites which are split by a linear liquid zone. In this synchrotron imaging work, it is very coincidental to capture the images of the anaxial columnar dendrites at the time of early, as shown in Fig 1(b-d). The two buds of anaxial columnar dendrite, marked 1 and 2, grew up from the stem of the fragment A shown in Fig 2(a,b), in which their growth angles were measured and noted. According to the growth process of fragment A, it can be found that: anaxial columnar dendrite 1 born at the moment while fragment A started to rotate. At that time, the tip of anaxial columnar dendrite 1 just pointed to the opposite direction of heat flow (see Fig 1a). It is not coincidence. With the rotation of fragment A, the anaxial columnar dendrite 2 was born and pointed to the opposite direction of heat flow, too (see Fig 2b). Further, the relative position of anaxial columnar dendrite 2 was similar to dendrite 1 which nucleated at the verge of the solute boundary layer ahead of the growth interface. From these facts, it can be inferred that there is a lot of chances for anaxial columnar dendrite nucleation under suitable thermal gradient and solute concentration.

It is well known that the preferred growth direction of aluminum solid solution which is the primary phase of hypoeutectic alloy is [100][8]. Among the growth direction system, only [110] is angled 45 degrees from [100] direction, as noted in Fig 2. The dendrite fragment A in Fig 2 inherited the favorite growth direction of the original columnar dendrite. Therefore, the anaxial columnar dendrites head to [110].

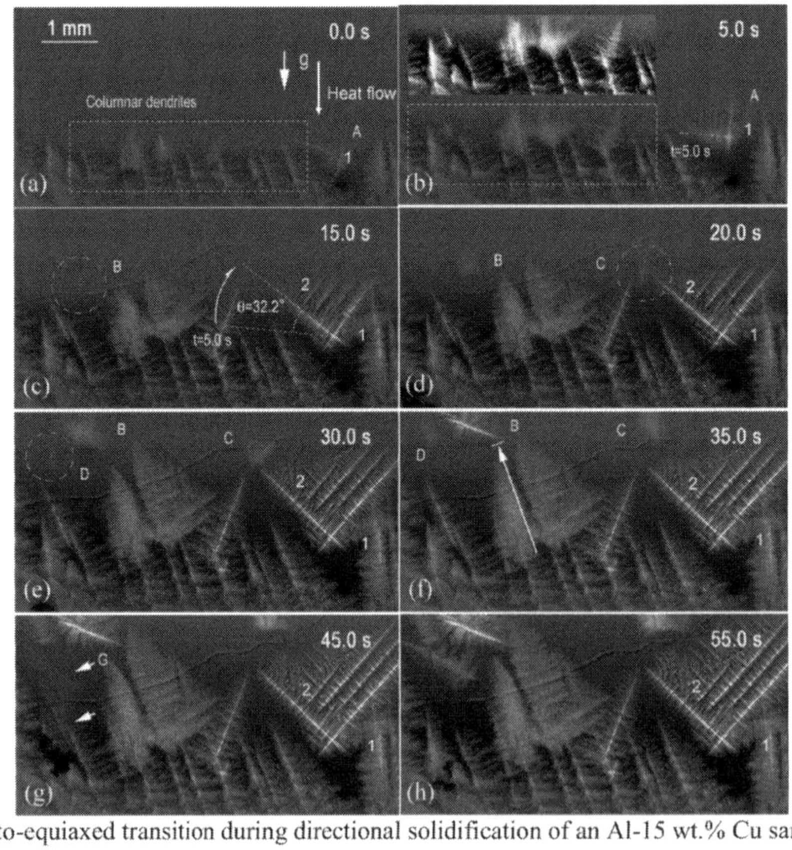

Fig 1. Columnar-to-equiaxed transition during directional solidification of an Al-15 wt.% Cu sample (R=0.028 K/s, G_T =5.0 K/mm). g – direction of gravity, 1 and 2 – anaxial columnar dendrite. (a) columnar dendrites, (b) interface instability, (c) rotation of dendrites, (d) fragmentations, (e) drift and growth of fragmentations, (f) dendrite growth is blocked, (g) fracture of dendritic arm and (h) stagnation of dendrite growth.

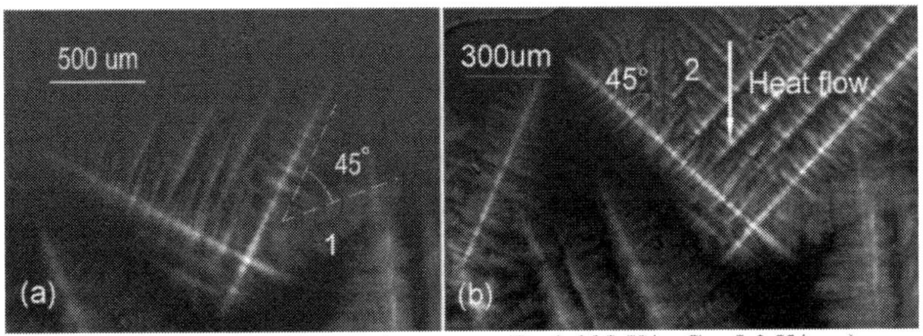

Fig 2. Anaxial columnar dendrite nucleate and grow up (R=0.028 K/s, G_T=5.0 K/mm).

4. Conclusion

The nature convection, caused by external thermal disturbance, leads to growth interface instability and triggers CET. The well-developed columnar dendrites of hypoeutectic alloy can provide the open solute enriched zone. This zone is the only area in which the high order arms of dendrite could detach from matrix.

A new anaxial columnar dendrite morphology was found to appear in directional solidification of the Al-15wt.% Cu alloy. Its growth process was observed using the real-time X-ray imaging technique. The anaxial columnar dendrite is composed of a pair of stems, which are divided by a narrow liquid zone located in its center. This work also provides experimental evidences for the crystalline orientation relationship between the regular and anaxial columnar dendrite. The regular dendrite grows along <100> directions, while the columnar dendrite grows along <110> directions.

Acknowledgments

This work has been performed within the framework of National Science Foundation for Distinguished Young Scholars of China (No.50825401), Hi-Tech Research and Development Program (863) of China (No.2008AA031801) and Project 50821003 supported by NSFC.

Reference

[1] Alain P, Julien D and Marc G 2007 JOM July 71-76
[2] Ruvalcaba D, Mathiesen R H, Eskin D G, Arnberg L and Katgerman L 2007 Acta Mater. **55** 4287-4282
[3] Mathiesen R H, Arnberg L, Bleuet P and Somogyi A 2006 Metall. Mater. Trans. A **37** 2515-2524
[4] Reinhart G, Mangelinck-Noël N, Nguyen-Thi H, Schenk T, Gastaldi J, Billia B, Pino P, Härtwig J and Baruchel J 2005 Mater. Sci. Eng., A **413–414** 384-388
[5] Schenk T, Nguyen Thi H, Gastaldi J, Reinhart G, Cristiglio V, Mangelinck-Noël N, Klein H, Härtwig J, Grushko B, Billia B and Baruchel J 2005 J. Cryst. Growth **275** 201-208
[6] Dong Q, Zhang J, Dong J F, Xie H L, Li Z J, Dai Y B, Liu Y and Sun B D 2011 Mater. Sci. Eng., A **530** 271-276
[7] Dong Q, Zhang J, Dong J F, Dai Y B, Bian F G, Xie H L, Lu Y L and Sun B D 2011 Mater. Lett. **65** 3295-3297
[8] Haxhimali T, KarmaOrientation A, Gonzales F and Rappaz M 2006 nature materials **5** 660-664

Preparation of clay mineral samples for high resolution x-ray imaging

Gennaro Abbati[1], Christian Seim[2], Herbert Legall[2], Holger Stiel[3], Noel Thomas[4] and Thomas Wilhein[1]

[1] Institute for X-Optics, RheinAhrCampus Remagen, Koblenz University of Applied Sciences, Suedallee 2, 53424 Remagen, Germany
[2] Berlin Laboratory for Innovative X-ray Technologies (BLiX), Technical University of Berlin, Hardenbergstrasse 36, 10623 Berlin, Germany
[3] BLiX, Max-Born-Institute, Max-Born-Strasse 2A, 12489 Berlin, Germany
[4] WesterWaldCampus Hoehr-Grenzhausen, Koblenz University of Applied Sciences, Rheinstrasse 56, 56203 Hoehr-Grenzhausen, Germany

E-mail: abbati@rheinahrcampus.de

Abstract. In the development of optimum ceramic materials for plastic forming, it is of fundamental importance to gain insight into the compositions of the clay minerals. Whereas spectroscopic methods are adequate for determining the elemental composition of a given sample, a knowledge of the spatial composition, together with the shape and size of the particles leads to further, valuable insight. This requires an imaging technique such as high resolution X-ray microscopy. In addition, fluorescence spectroscopy provides a viable element mapping technique. Since the fine particle fraction of the materials has a major effect on physical properties like plasticity, the analysis is focused mainly on the smallest particles. To separate these from the bigger agglomerates, the raw material has to pass through several procedures like centrifugation and filtering. After that, one has to deposit a layer of appropriate thickness on to a suitable substrate. These preparative techniques are described here, starting from the clay mineral raw materials and proceeding through to samples that are ready to analyze. First results using high resolution x-ray imaging are shown.

1. Introduction

Insight into the finely particulate nature of plastic clay materials is of fundamental importance for the development of optimum materials for the plastic forming of ceramics. This would provide the basic information required, for example, for the development of special binders to control the degree of agglomeration of the fine particles in an aqueous environment. It would thereby open the door to increasing the sustainable use of natural resources, which would otherwise find no direct economic use. In order to gain this insight, it is necessary to develop a good method for preparing the plastic clay materials.

2. Sample preparation

2.1. Particle size

Depending on which properties of the clay minerals are of interest, it may be necessary to separate the raw materials into several particle size fractions. This can be carried out either with the aid of different filters or by using a centrifuge.

Larger particles and other contaminations of the raw materials are removed by a 32 µm sieve. After liquifying with water, sodium pyrophosphate or ethanol followed by sonication using an ultrasound bath, the sample is centrifuged and then divided into several fractions. For a even better separation, this procedure is repeated a few times with the single fractions, until the desired particle size distribution is reached. These distributions of particle sizes have been measured with a laser particle analyzer, in order to determine influence of the rotational speed and time of the centrifuge.

If, alternatively, the filter-based method of separation is used, and only a few µl of the sample are needed, a syringe filter is a suitable way to achieve this separation. For the comparability of future measurements though, it is important to have more than just a few µl of sample. This points to the superiority of the centrifuge method.

2.2. Sample deposit

After selecting the desired particle size fraction, the sample must be deposited on to a suitable substrate, in this case a silicon nitride membrane with a thickness of 100 nm. This process can be carried out either with the sample in a fluid or in a solid state.

2.2.1. Fluid state For the fluid sample the most obvious way is the application as a droplet with a pipette. After the fluid has evaporated, only the sample particles adhere to the membrane. Owing to the surface tension of the fluid, the particles will not be uniformly arranged but concentrated at the border of the droplet. This leads to a higher layer thickness, which is undesirable because of higher x-ray absorption. By using the technique of spin-coating and hydrophilic membranes, the layer thickness can be decreased, but the non-uniform distribution remains.

2.2.2. Solid state Preparing the sample as a non-fluid has been found to be an appropriate way to deposit it on to the substrate. For this, the sample first has to be dried after the filtering. At this time, the sample is in the state of a fine powder. By poking the box containing this powder, dust is dispersed. Mounting a stage holding the silicon nitride window above this cloud of dust leads to particles adhering to the membrane. Some results of this treatment, which have

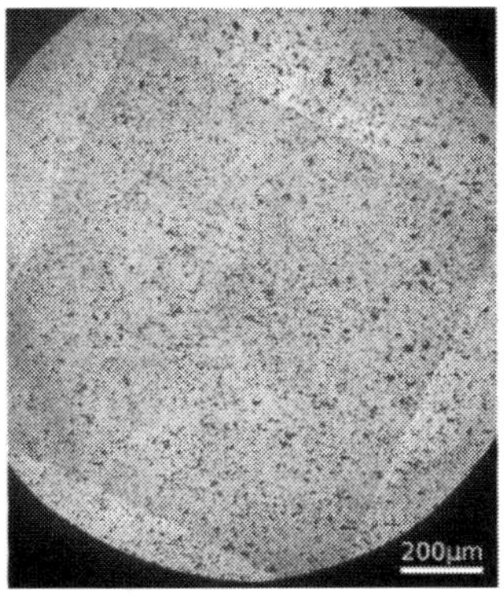

(a) Bright field image

(b) Dark field image

Figure 1: Silicon nitride membranes with sputtered clay minerals imaged by a microscope with visible light been imaged in a light microscope are shown in Fig. 1. In this case, the thickness of the sample depends only on the thickness of the agglomerates. So, in order to obtain a thin sample, the agglomerates have to be split into smaller ones. This can be achieved with dispersants and a sonicator.

3. Results

The sputtered samples have been imaged on the laboratory transmission X-ray microscope (LTXM) at BLiX [2] with an energy of 500 eV (see Fig. 2). The resolution reached down to 30 nm. It can be seen that there are single agglomerates, which can be analyzed independently. However, to visualize the fine clay materials, whose expected shape is that of plates, one has to break up the agglomerates in even smaller particles.

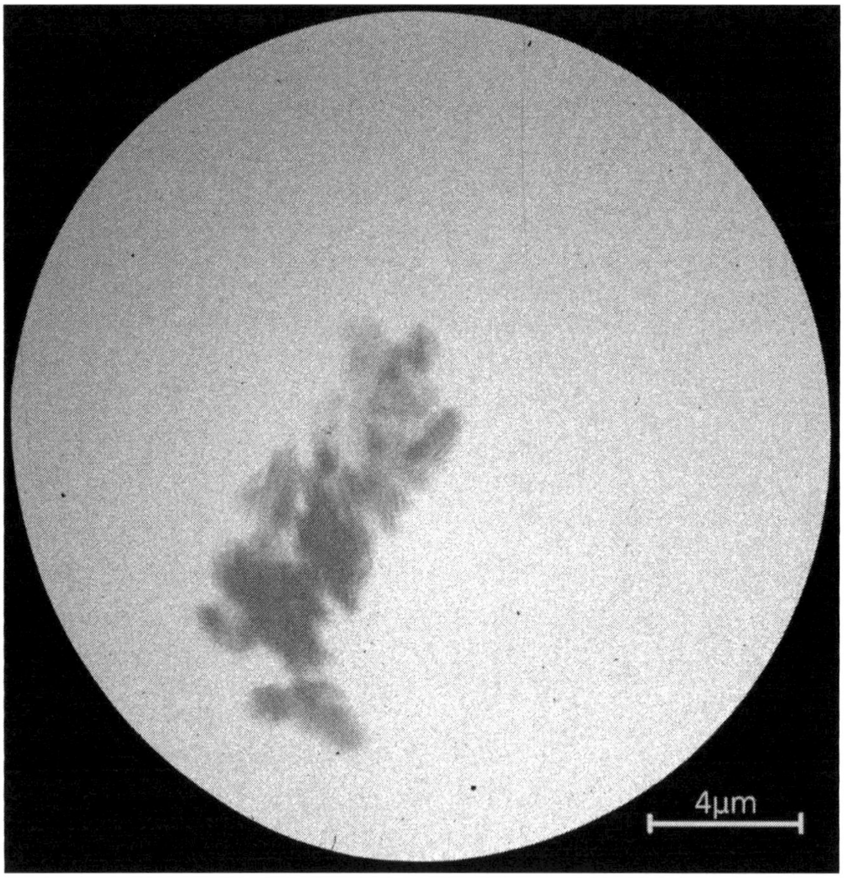

Figure 2: Sputtered clay minerals imaged with the LTXM at BLiX at 500 eV

Further measurements with the purpose of mapping Fe in clay minerals were made at the P04 beamline at PETRA III, DESY. Some results are shown in Fig. 3. For further information about the TXM used for this images, please refer to the proceeding of Philipp Wessels for this conference [3].

These first results show that the mentioned preparation steps can lead to satisfying images. Even more, these separation and filtering steps are essential for visualizing distinct particles.

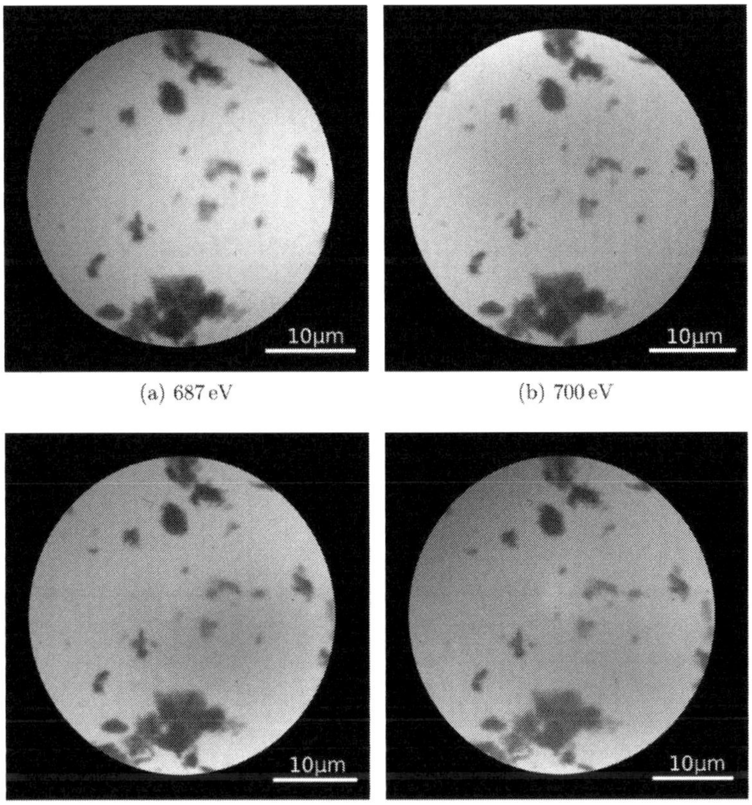

(a) 687 eV (b) 700 eV (c) 712 eV (d) 727 eV

Figure 3: Images of a clay mineral sample near the L-edge of *Fe* (706,8 eV) taken at the soft x-ray beamline P04 at PETRA III, DESY

4. Future Work

The separation of different particle sizes with the centrifuge will be further improved to narrow the size-ranges even more and to increase the accuracy. Furthermore, the use of membrane filters will be investigated as an alternative way of separation.

Further measurements will be done both on the imaging of the fine particles and on element mapping of iron and other elements of which they are composed, e.g. titanium.

Finally, the behaviour of plastic clay materials in an aqueous environment will be analyzed. Therefore a suitable wet cell has to be constructed. A merging of all these techniques using the methods of microtomography will allow the composition of plastic clays to be visualized in three dimensions.

References

[1] Thomas N W 2008 *cfi/Ber. DKG* **85** E29–E40
[2] Legall H et al 2012 *Opt. Express* **20** 18362-18369
[3] Wessels P et al 2012 *J. Phys.: Conf. Series* (see proceedings of this conference)

Journal of Physics: Conference Series **463** (2013) 012052 doi:10.1088/1742-6596/463/1/012052

Soft x-ray imaging of intracellular granules of filamentous cyanobacterium generating musty smell in Lake Biwa

K Takemoto[1], G Mizuta[2], A Yamamoto[3], M Yoshimura[4], S Ichise[5], H Namba[2] and H Kihara[3]

1 Kansai Medical University, 18-89 Uyamahigashi, Hirakata, 573-1136, Japan

2 Ritsumeikan University, 1-1-1 Noji-Higashi, Kusatus, 525-8577, Japan

3 The SR Crenter, Ritsumeikan University, 1-1-1 Noji-Higashi, Kusatus, 525-8577, Japan

4 Nagahama Institute of Bio-Science and Technology, 1266 Tamura, Nagahama, 526-0829, Japan

5 Lake Biwa Environmental Research Institute, 5-34 Yanagasaki, Otsu, 520-0022, Japan

E-mail: takemoto@makino.kmu.ac.jp

Abstract. A planktonic blue-green algae, which are currently identified as *Phormidium tenue*, was observed by a soft x-ray microscopy (XM) for comparing a musty smell generating green strain (PTG) and a non-smell brown strain (PTB). By XM, cells were clearly imaged, and several intracellular granules which could not be observed under a light microscope were visualized. The diameter of granules was about 0.5-1 μm, and one or a few granules were seen in a cell. XM analyses showed that width of cells and sizes of intracellular granules were quite different between PTG and PTB strains. To study the granules observed by XM, transmission in more detail, transmission electron microscopy (TEM) and indirect fluorescent-antibody technique (IFA) were applied. By TEM, carboxysomes, thylakoids and polyphosphate granules were observed. IFA showed the presence of carboxysomes. Results lead to the conclusion that intracellular granules observed under XM are carboxysomes or polyphosphate granules. These results demonstrate that soft XM is effective for analyzing fine structures of small organisms such as cyanobacterium, and for discriminating the strains which generates musty smells from others.

1. Introduction

Since 1969, musty odor problems have frequently occurred in the water supply at Lake Biwa. 2-methylisoborneol (MIB) and geosmin were identified as the causative substances. At present, three species of planktonic blue-green algae from Lake Biwa have been identified as organisms playing important roles in such problems. From the unialgal culture of each alga, *Anabaena macrospora Klebahn* was shown to produce geosmln, and both *Phormidium tenue* (Meneghini) Gomont and *Oscillatoria tenuis* Agardh to produce MIB [1].

A planktonic blue-green algae, which is currently identified as *P. tenue*, is a small filamentous cyanobacterium with a firm sheath, and the trichome is consisted of cells 1.0–1.6 μm in width, and 2.5 –5 μm in length. The sheath is too thin to be visualized by a light microscope (LM). *P. tenis* is a well known MIB producer and is closely associated with the occurrence of musty odor in Lake Biwa. The presence of two-color plankton, green and brown species, have been reported [2]. Some green species produces MIB, and some green species do not produce MIB [2]. On the other hand, all brown species do not produce MIB [2]. Because of a small cell size, it is difficult to know whether there are morphological differences between green strain (PTG) and brown strain (BTB) or between musty smell generating and non-generating strains under the light microscopic examination. Self fluorescence microscopy is a unique method to discriminate such strains from others [2]. However, it is reported that application of this method is restricted to some specific strains.

Soft x-ray microscope (XM) is a powerful tool to investigate whole cells up to 10 μm thickness with much higher resolution than LM. We observed the cyanobacterium by a soft XM to image fine structure in the cells, and compared the morphological difference between the strains. To identify an intracellular structure which was observed by XM, transmission electron microscopy (TEM) and indirect fluorescent-antibody technique (IFA) were also applied.

2. Experimental

2.1. X-ray microscopy (XM)

Suspension of the laboratory-cultured musty smell generating green *P. tenue* strain (PTG) and musty smell non-generating brown strain (PTB) (see Figure 1) was dropped onto a polyimide thin film and air-dried [3]. XM images were acquired by using a soft X-ray microscopy beamline (BL-12) at the SR Center of Ritsumeikan University. Two wavelengths, 2.0 nm and 2.4 nm (below and above the wavelength of the oxygen K-edge threshold), were used for imaging. A resolution at 2.0 nm and 2.4 nm was ca. 80 nm and 88 nm, respectively, judging from the knife-edge estimation (20-80 %).

Figure 1. Photograph of filamentous cyanobacteria, which are currently identified as *P. tenue*. Left: PTG; it generates musty smell. Right: PTB; it does not generate musty smell.

2.2. Transmission electron microscopy (TEM)

Laboratory-cultured PTG suspension was dropped onto a polyimide thin film and air-dried. After fixation with glutaraldehyde and subsequently with osmium tetroxide, cells were embedded in epon. Thin sections are doubly stained with uranium and lead. TEM images were acquired by using Hitachi H-7600 TEM operating at 80 kV accelerating voltage.

2.3. Indirect fluorescent-antibody technique (IFA)

Laboratory-cultured PTB suspension was dropped onto a polyimide thin film and air-dried. Cells were fixed in pre-chilled 100 % methanol were rinsed in PBS and were incubated for 60 min at 4°C in blocking solution containing PBS and 5 % bovine serum albumin (BSA). Cells were then incubated with a polyclonal anti-RuBisCO antibody (Agrisera, Prod. ID AS03 037) diluted 1:500 in blocking solution overnight at 4°C. Cells were washed in PBS and then incubated with an anti-rabbit Oregon green conjugated antibody (Invitrogen) diluted 1:300 in blocking solution for 1 hr at 30°C. Cells were washed in PBS and imaged [4]. IFA image was acquired by using Zeiss Axiovert 200M epifluorescence microscope.

3. Results and discussion

Figure 2 shows x-ray microscopic images of PTG and PTB. Each cell was clearly observed. Granules were also clearly recognized. All granules were not observed by 2.4 nm observations [5]. These granules were also observed in wet condition in our previous XM observations. Thus, we reconfirm that granules were observed both in dry and wet condition [5, 6]. Results also show that oxygen is important constituent element of the granule. Cellular size of PTG was larger than that of PTB. Using x-ray micrographs, average size of cells were estimated. The average width of PTG and PTB was 1.35 μm and 1.08 μm, respectively. Although the number of intracellular granules is about the same (~2 / cell), the size distribution of the granules was quite different between PTG and PTB. The average diameter of PTG and PTB was 0.77 μm and 0.42 μm, respectively.

To investigate the granules in more depth, TEM and IFA were applied to the PTG cells. Figure 3 shows a TEM image of a representative PTG cell on the ultra-thin section. Thylakoid membranes appeared as thin layers around the cell in cross section. The cell included carboxysomes and polyphosphate body. Carboxysomes are spatially ordered in a linear fashion in the cell. Carboxysomes are bacterial microcompartments that contain enzymes, RuBisCO, involved in carbon fixation. Next, IFA was applied to visualize the carboxysomes using anti- RuBisCO antibody. By the IFA, the carboxysomes were also spatially ordered in a linear fashion (Figure 4).

These results show that soft XM is effective for analyzing fine structure of small organisms such as cyanobacterium, and discriminating the strains which generates musty smells from others.

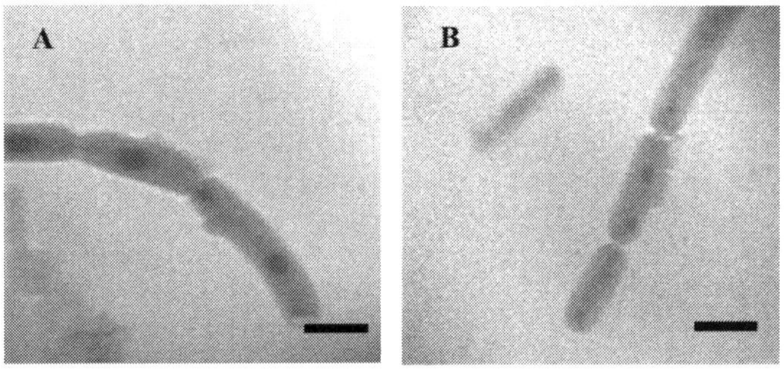

Figure 2. XM images of PTG (A) and PTB (B). Scale bar is 2 μm. Observation wavelength was 2.0 nm. Each exposure time was 120 s.

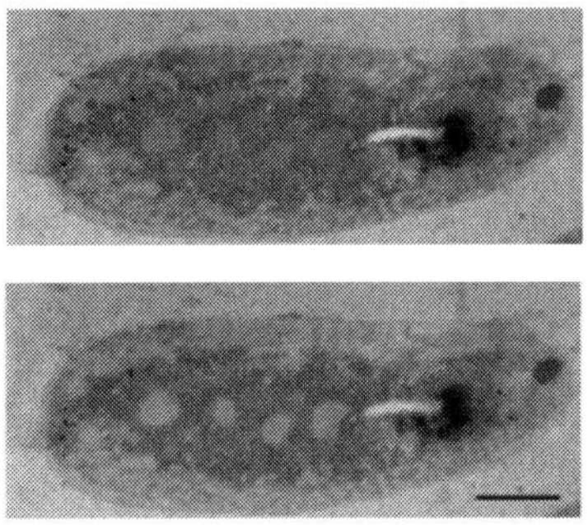

Figure 3. TEM images of PTG. Scale bar is 500 nm.
Green area: thylakoid membranes, membrane system where the fully functional electron transfer chains of photosynthesis and respiration reside. Yellow area, carboxysomes, bacterial microcompartments that contain enzymes involved in carbon fixation. Blue area: polyphosphate body, intracellular nutrient storage granules.

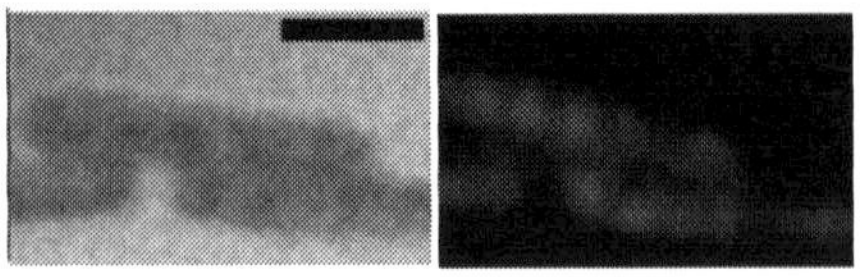

Figure 4. Micrographs of PTB. Left image is XM. Right image is IFA image of the corresponding cells. Scale bar is 5 μm.

4. Conclusions

We observed a small filamentous cyanobacterium, *P. tenue*, by the soft XM for comparing a musty smell generating strain and a musty smell non-producing strain.

Experimental results led to the following conclusions: (1) soft XM showed morphological difference between a musty smell generating strain (PGB) and a musty smell non-producing strain

(PGB). (2) Soft XM is one of optimal tools to distinguish different cyanobacteria strains such as PTG and PTB. (3) TEM and IFA showed that intracellular granules observed under XM may be carboxysomes or polyphosphate granules.

Acknowledgement
This work was supported in part by Open Research Center Project for Private Universities matching fund subsidy (2007-2011), Grant-in-Aid for Scientific Research (C) 24510039 from the Ministry of Education, Culture, Sports, Science and Technology, Japan, and Aid for plant research (2012) from The New Technology Development Foundation.

References

[1] Negoro T, Ando M and Ichikawa N 1988 *Wat. Sci. and Tech.* **20** 117
[2] Nakamura T 1988 *Journal of Water and Waste.* **30** 24 (in Japanease)
[3] Takemoto K, Watanabe N, Hirai A, Nakayama Y and Kihara H 1998 *X-Ray Microscopy and Spectromicroscopy* ed J. Thieme, G. Schmahl, D. Rudolph and E. Umbach (Springer-Verlag, Heidelberg) pp I129-133
[4] Savage D F, Afonso B, Chen A H and Silver P A 2010 *Science* **327** 1258
[5] Takemoto K, Ichise S, Ohigashi T, Namba H and Kihara H 2010 *Memoirs of SR Center Ritsumeikan Univ.* **12** 169
[6] Takemoto K, Ichise S, Ohigashi T, Ikegaya H, Namba H and Kihara H 2011 *Jpn. J. Wat. Treat. Bio.* **47** 131 (in Japanease)

Correlative Imaging of Structural and Elemental Composition of Bacterial Biofilms

Y Yang[1,5], R Heine[1], F Xu[1], H Suhonen[2], L Helfen[1,2], A Rosenhahn[3,4], T Gorniak[3,4], S Kirchen[4], T Schwartz[4], T Baumbach[1]

[1] Institute for Photon Science and Synchrotron Radiation /ANKA, Karlsruhe Institute of Technology (KIT), 76344 Eggenstein-Leopoldshafen, Germany

[2] European Synchrotron Radiation Facility (ESRF), 38043 Grenoble, France

[3] Analytical Chemistry-Biointerfaces, University of Bochum, 44801Bochum, Germany.

[4] Institute of Functional Interfaces (IFG), Karlsruhe Institute of Technology (KIT), 76344 Eggenstein-Leopoldshafen, Germany

[5] Institute of Geoscience, University of Freiburg, 79104 Freiburg, Germany.

thomas.schwartz@kit.edu

Abstract. Synchrotron-based phase contrast tomography (holotomography) and scanning hard X-ray fluorescence microscopy (SXFM) are combined to characterize the three-dimensional (3D) structural and corresponding elemental distribution of bacterial biofilms of *Pseudomonas aeruginosa*. Samples were fixed without contrast agents or microtomal sectioning. Within an intact microbial community single bacteria are clearly resolved, and their morphology can be directly visualized together with the elemental content. Such 3D set of complementary information at cellular level is essential for gaining a deeper understanding of biofilm evolution aiming to develop potential strategies on biofilm growth control and prevention.

1. Introduction

Bacterial biofilms are broadly defined as structured, multicellular communities of bacteria that are enclosed in an exopolymeric matrix and attached to a surface [1]. Many bacteria are well-known for their robustness against antimicrobial agents. This manifests itself particularly in health care where contamination of medical implants and wounds poses a severe threat to human health. But also industrial equipments like ventilation and oil wells suffer from biofilm formation [2]. Microbial organization has been investigated on the macro-scale and micro-scale [3-5] with special emphasis on their architecture and elemental composition. Biofilms show remarkable resistance to wetting [6], gas penetration [6], and uptake of heavy metals like copper, lead and zinc [7]. In this context it is important to understand which mechanisms allow this complex matrix to be robust to such an extent.

More insight into this matter is expected from the investigation of the structural organization of the bacteria in the biofilm and the measurement of the content of relevant trace elements and their distribution within the biofilm. Here we propose the application of a correlative imaging approach which enables both structural and elemental visualization of intact fixed biofilms. Due to the low transmission of natural multilayer biofilms in the visible spectral range, conventional optical microscopy is hampered. Electron microscopy requires a large number of very thin sections due to their low penetration depth. For maintaining the integrity of the microbial communities, noninvasive multiple contrast hard X-ray imaging might be the method of choice.

Hard X-ray phase contrast tomography (holotomography) has outstanding performance in visualizing 3D cellular architectures of thick and non-transparent biological samples. The real part of the refractive index, i.e., the local variation in electron density, can be retrieved from the recorded Fresnel diffraction patterns based on a contrast transfer function approach [13]. For hard X-rays, the sensitivity of reconstructed phase maps can be two orders of magnitude better than conventional absorption contrast images [14]. Often, high resolution with good contrast is accessible without the use of sectioning or contrast agents, increasing the scientific relevance of the results.

For an elemental analysis of biological samples, synchrotron-based scanning hard X-ray fluorescence microscopy (SXFM) is a suitable label-free probe. SXFM with nano-scale resolution has emerged a few years ago [8] as a powerful chemical analysis tool in life sciences [9], material sciences [10] and environmental sciences [11]. By detecting characteristic fluorescence signals, this method is capable of mapping multiple elements simultaneously down to nano-scale resolution without labeling by dyes or heavy metals. Furthermore, low requirements for sample size and preparation, the high penetration power of hard X-rays and trace element sensitivity (sub-femtogram absolute detection limits) [12] renders this technique to be an excellent non-invasive way to investigate the elemental distribution within both sensitive and valuable small bulk samples.

Here we report about the results of a feasibility study aiming to image bacterial biofilms by a correlative approach [e16] combining 2D fluorescence mapping and 3D tomographic reconstruction of the electron density.

2. Materials and Methods

Gram-negative, aerobic rod bacteria *Pseudomonas aeruginosa* (environmental isolate) were selected as the representatives of the most-studied single species to form biofilms on substrates (Fig.1). A standard electron microscopy sample support, namely silicon nitride membranes (Si_3N_4) was chosen as substrate on which the bacteria were cultured in BM2 mineral medium (62 mM potassium phosphate buffer (pH 7), 7 mM (NH4)2SO4, 2 mM MgSO4, 10 μM FeSO4, 0.4% (wt/vol) glucose) for 48 h. Planktonic, non adherent bacteria were removed by washing with phosphate buffered saline (pH 7.5) and subsequently the biofilms were fixed with 2.5 % glutaraldehyde solution for 1 h.

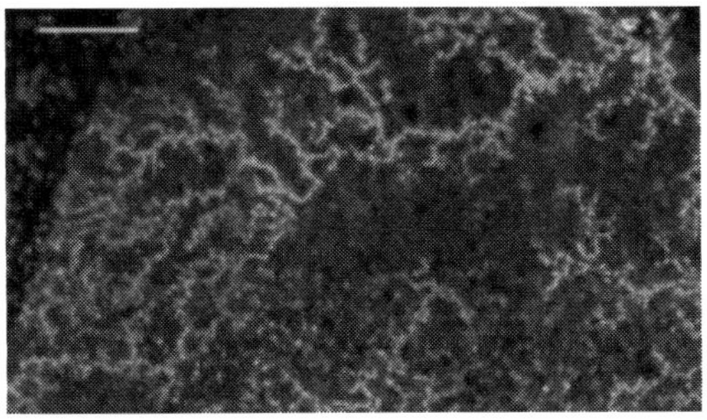

Figure 1. Optical micrograph (Nikon, Plan Fluor) of a fixed bacterial biofilm on a Si_3N_4 window. Magnification is 400 ×. Scale bar is 50 μm.

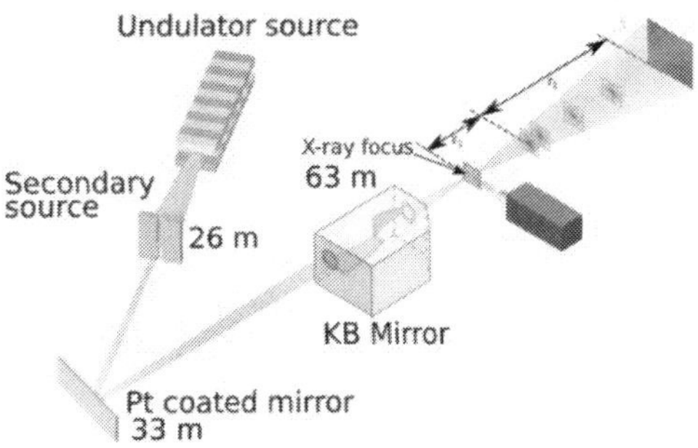

Figure 2. Sketch of the experimental setup at ID 22.

Both, the SXFM and holotomography measurements have been performed at the nano-imaging end station of beamline ID22 at the European Synchrotron Radiation Facility (ESRF). The end station operates at energies between 17-29 keV and is dedicated to 2D and 3D nano-imaging with multiple contrast mechanisms (see Fig. 2): propagation-based full-field phase contrast imaging (x-ray holography), scanning-probe fluorescence (covering atomic numbers Z>13), and scanning-probe diffraction [17]. In our experiments, an undulator source produced X-rays of 17 keV incident energy. The rays were subsequently reflected from a Pt coated mirror to remove higher harmonics and, then focused by a pair of elliptically bent, graded multilayer mirrors (Kirkpatrik-Baez configuration). The measured focal spot size was 60 × 65 (vertical × horizontal) nm^2 FWHM with a photon flux of 10^{12} photons/s at 17 keV. The high photon flux rendered a satisfying signal to noise ratio in a relatively short time.

For holotomography, 1999 phase contrast projection micrographs per tomographic rotation were recorded at four different propagation distances respectively. Each projection was acquired with 0.2 s exposure time. The pixel size of 25 nm in the sample is directly related to the physical detector pixel size of 0.756μm and the geometrical magnification (30.2 ×) given by the sample position between the focal spot and the detector. Subsequently performing SXFM at the same sample position, the specimen was moved into the focus and raster scanned with a scanning pitch of 100 nm for 2D elemental mapping. Fluorescence spectra were collected with a dwell time of 0.1 s by a silicon drift detector (Vortex-EX, SII Nanotechnology). The spectra were normalized and fitted against the measured standards using PyMca for quantification of elemental contents [18].

3. Results and discussion

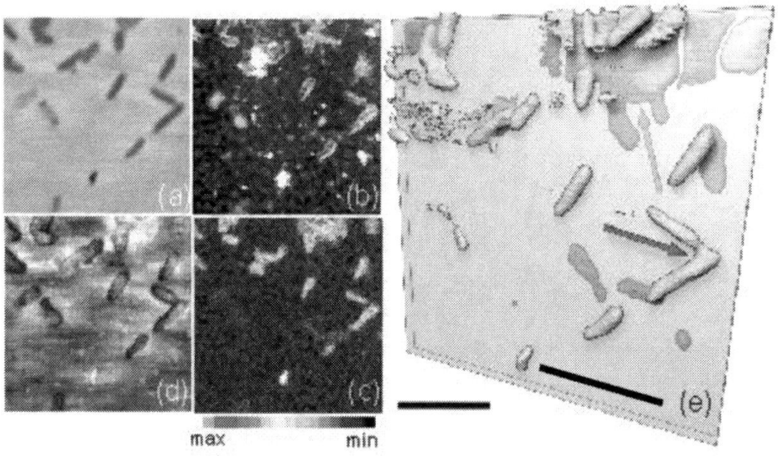

Figure 3. (a) Cross section of the reconstructed holotomogram showing single bacteria. (b)-(c) X-ray fluorescence images of the bacterial biofilm, with contrast adjusted independently (b: Fe, c: P). Scanned area 10 × 10 μm^2. (Scanning pitch: 100 nm/pixel, dwell time: 0.2 s/pixel, FWHM of beam: 100 nm). (d) Superposed image of 2D elemental compositions and the reconstructed cross section (red: Fe, green: P, gray: cross section (a)). (f) Corresponding volume rendering of the same area. Scale bar is 5 μm

Fig. 3 shows exemplary results. The bacteria are well resolved in the fluorescence micrographs (see Figs. 3 (b) and (c) for the distribution of two selected elements) and their morphology is reproduced by the holotomogram (Figs. 3 (a) and (e)). In the volume rendering (Fig. 3 (e)) we observe two layers of bacteria that formed on both sides of the membrane, where different development states can be distinguished. On the foreside we find bacteria which are still in initial attachment (red arrow), on the back side of the membrane there are also bacteria in the state of irreversible attachment (blue arrow) with bacterial cells cemented to the substrate and forming nascent cell clusters [17]. Furthermore, when combined with the fluorescence images, the elemental composition of bacterial cells and the clusters can be determined without ambiguity (compare Fig. 3 (d)). Phosphate is an indispensable component of many macromolecules, like proteins, nucleic acids, and lipids [2], so P appears spread over the whole cell as can be seen in Figs. 3 (c) and (d). The map of the Fe content (Figs. 3 (b) and (d)) shows higher concentrations along the contours of the cells indicating that Fe is richer along the cell membranes. One explanation for this distribution comes from the strong Fe binding agents called siderophores. They are suspected to be secreted by the bacteria to tightly bind Fe from the exterior of the cells and transport it into the cells as Fe is an essential trace nutrient for microbial life processes [18].

Conclusion

X-ray holotomography in combination with fluorescence microscopy provide a comprehensive protocol for biofilm research to simultaneously and correlatively investigate morphology and elemental composition. It is demonstrated as a versatile probe which benefits from both high resolution down to presently sub 100 nm pixel size and trace element sensitivity. Overall, this protocol enables visualization of spatial structures and elemental distribution of individual bacteria without inducing changes to the original microbial organization by slicing or chemical staining. This is especially useful for maintaining the natural state of a biofilm within its original habitat, opening the door for possible *in-situ* microbial research in the future.

Acknowledgements

We want to thank European Synchrotron Radiation Facility for providing the beamtime, and also thank Dr. Thomas van de Kamp for the help on volume rendering.

References

[1] O'Toole G, Kaplan H B and Kolter R. 2000 *Annu. Rev. Microbiol.* **54** 49-79
[2] Madigan M T, Martinko J M, Dunlap P V and Clark D P. 2009 *Pearson Press.* 677-679
[3] Iltis G C, Armstrong R T, Jansik D P, Wood B D, and Wildenschild D. *Water Resour. Res.* **47** W02601
[4] Lawrence J R, Swerhone G D W, Leppard G G, Araki T, Zhang X, West M M and Hitchcock A P. 2003 *Appl. Environ. Microbiol.* 69 5543-5554
[5] Neu T R, Manz B, Volke F, Dynes J J, Hitchcock A P, and Lawrence J R. 2010 *FEMS Microbiol Ecol.* **72** 1-21
[6] Epsteina A K, Pokroya B, Seminarab A, and Aizenberga J. 2011 *PNAS.* **108** 995-1000
[7] Teitzeil G M and Parsek M R. 2003 *Appl. Environ. Microbiol.* **69** 2313-2320.
[8] Vogt S, Maser J, Paunesku T, Stojićević N, Lai B and Woloschak G E. 2004 *Microsc Microanal.* **10** 1028-1029
[9] Ortega R, Cloetens P, Devès G, Carmona A, Bohic S. 2007 *PLoS ONE.* **2** e925
[10] Buonassisi T, Istratov A A, Heuer M, *et al.* 2005 *J. Appl. Phys.* **97** 074901
[11] Endo S, Terada Y, Kato Y and Nakai I. 2008 *Environ Sci Technol.* **42** 7152-8
[12] Bacquart T, Devès G, Carmona A, *et al.* 2007 *Anal Chem.* **79** 7353-9
[13] Cloetens P, Ludwig W, Baruchel J, *et al.* 1999 *J. Phys. D.* **32** A145-A151
[14] Cloetens P, Mache R, Schlenker M, and Lerbs-Mache S. 2006 *PNAS.* **103** *14626-14630*
[15] Bleuet P, Cloetens P, Gergaud P, *et al.* 2009 *Rev. Sci. Instrum.* **80** 056101
[16] Xu F, Helfen L, Suhonen H, *et al.* 2012 *PLoS ONE.* **7(11)**:e50124
[17] Martinez-Criado G, Tucoulou R, Cloetens P, *et al.* 2011 *J Synchrotron Radiat.* **19** *10-18*
[18] Solé V A, Papillon E, Cotte M, Walter Ph, Susini J. 2007 *Acta Part B* **62** 63-68
[19] Sauer K, Camper A K, Ehrlich G D, *et al.* 2002 *J. Bacteriol.* **184**:1140-1154.
[20] Cox, C D and Adams P. 1985 *Infect. Immun.* **48**:130-138.

The Stockholm laboratory cryo x-ray microscope: towards cell-cell interaction studies

E. Fogelqvist, M. Selin, D.H. Martz, A. E. Christakou, and H. M. Hertz

Biomedical and X-Ray Physics, KTH Royal Inst. of Technology/Albanova, 10691 Stockholm, Sweden

E-mail: emelie.fogelqvist@biox.kth.se, marten.selin@biox.kth.se

Abstract. We describe recent improvements in the Stockholm laboratory x-ray microscope and the first experiments aiming towards studies of cell-cell interaction. The shorter exposure time due to a higher brightness laser-plasma source will become of large importance for tomography while the reproducible cryo preparation of few-cell samples is essential for the interaction studies.

1. Introduction

X-ray microscopy (XRM) in the water window (λ=2.3-4.4 nm; E=284-540 eV) has the capability to image whole unstained cells with unprecedented resolution and contrast [1,2]. Results with high biological relevance are presently emerging [3,4] from microscopes located at high-average spectral brightness synchrotron sources. Typically exposure times are in the seconds range. Also at the laboratory microscopes, synchrotron-quality images are now produced but with longer exposure times due to the lower source brightness [5,6]. With the latest upgrade of our compact soft x-ray source we have demonstrated stable operation with a brightness of $>1.5\times10^{12}$ ph/(s$\times$mm$^2\times$mrad$^2\times$line) and 10-second exposure-time imaging [7], i.e. approaching the performance of microscopes at early synchrotrons.

The primary goal of our laboratory XRM is to study cryo-fixated cells in their near-native environment. In such studies the high-brightness source will be of particular importance for tomography, significantly reducing the exposure time compared to in the early proof-of-principle experiments [8], and, thus, making studies of biological relevance possible. In the present paper we describe the current state of the microscope and some early attempts towards studies of cell-cell interactions.

2. Microscope arrangement

Figure 1 shows the experimental microscope arrangement, with the source, the condenser and imaging optics, the cryo sample holder and the detector. Below we briefly describe these sub-systems, with emphasis on the source due to its recent improvement.

The laser-plasma source is based on a regenerative liquid-nitrogen jet target allowing high-repetition-rate operation. The 20 μm diameter jet is formed by driving liquid nitrogen through a fused-silica capillary nozzle with a backing pressure of 10-20 atm, resulting in a jet velocity of approximately 60 m/s. The beam from a 2 kHz, 100 mJ, 600 ps, Nd:YAG master-oscillator power-amplifier (MOPA) laser [9] is focused onto the jet to form the plasma. The nitrogen plasma emits strong peaks in the water-window corresponding to electron transitions in the highly ionized nitrogen.

The emission line utilized in our microscope is the $\lambda/\Delta\lambda>500$ hydrogen-like (NVII, 1s-2p) line at 2.478 nm [10]. The source size is typically 14×20 μm^2 FWHM, and it radiates the flux 1.1×10^{15} ph/(s×sr×line) of the 2.48 nm line uniformly into 4π sr, resulting in an average spectral brightness of $>1.5\times10^{12}$ ph/(s×mm^2×mrad^2×line) [7].

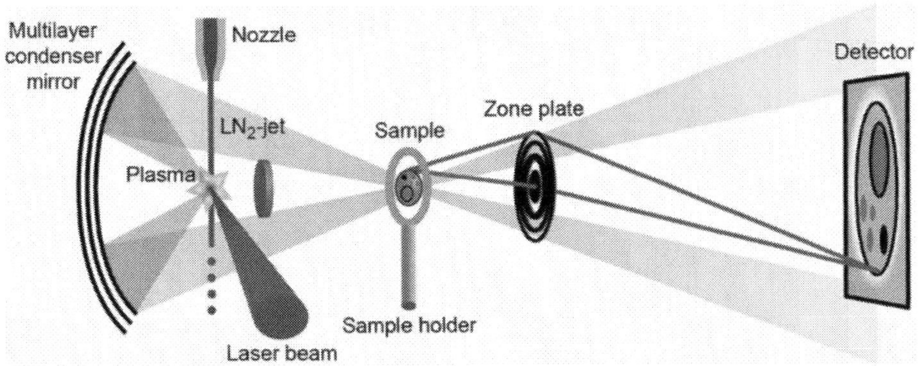

Figure 1. The Stockholm soft x-ray microscope with the high-power liquid-jet source, a Cr/V multilayer condenser mirror, cell sample in sample holder, nickel zone plate and CCD detector.

The source is imaged by a Cr/V multilayer condenser mirror onto the sample plane with a magnification of 1.6×. The narrow mirror bandpass ($\lambda/\Delta\lambda\approx300$) matches the 2.48 nm emission line so that it also serves as monochromator. At the sample plane cryo-prepared samples on TEM-grids are inserted using a modified TEM-sample-holder stage with both cryo and tilt capabilities. Thus, it facilitates cryo tomography operation. The illuminated object is imaged onto the detector by a 25-50 nm outermost zone-width, 50-100 μm diameter nickel zone-plate. The image is recorded by a cooled back-illuminated CCD detector.

3. Towards cell-biological studies

The goal of our XRM activity is to allow high-resolution studies of cells and cell-cell interactions. For such studies cryo sample preparation is necessary to avoid radiation damage on the samples, see, e.g. [5]. Thus, a robust protocol of cryo preparation for the laboratory microscope has been developed. Sample quality and reproducibility have been the key goals. The method is demonstrated on parasites and for immune system natural killer (NK) cell-target cell interaction studies.

3.1. Cryo preparation for laboratory XRM

The method used for cryo fixation is plunge-freezing, i.e., fast immersion of the sample into liquid ethane in order to obtain a cooling rate higher than 10^4 K/s. At this cooling rate vitreous ice is formed, thereby avoiding damage to the sample substructures. We use 3 mm gold TEM grids as sample carriers in the laboratory x-ray microscope. To avoid ice crystal formation it is essential that the water layer on the grid is thin. The removal of excess liquid is done by blotting the sample carefully from the sides. With non-adherent cells or cells that are not given proper time to adhere to the carriers this method often results in removal of a majority of the cells. However, with the use of adherent cells the number of cells remaining on the grid after blotting and cryo fixation is reasonably reproducible and reasonably high.

In the next two sections we show two examples of imaging based on the protocol. First single-cell images of two types of parasites are demonstrated. They were grown according to standard procedure and seeded onto the TEM grid before plunge freezing [5] but not given the proper time to adhere to the carrier. Still, they remained in sufficient numbers to allow imaging.

The second example is aimed at cell-cell interaction studies. Here the cancer target cell was the adherent cell line HEK 293T. The target cells were cultivated on 3 mm gold TEM grids with a thin carbon support layer. After 24 hours incubation at 37 °C in an atmosphere of 5% CO_2 the NK cell solution was added to the grid. Subsequently, after another hour of incubation, the NK-target cell sample was imaged in the x-ray microscope or in a confocal laser scanning microscope. Figure 2 shows two light microscope images taken just before the NK cells were added in order to investigate the plunge freezing. Here the HEK cells can be identified on the grid before and after cryo fixation.

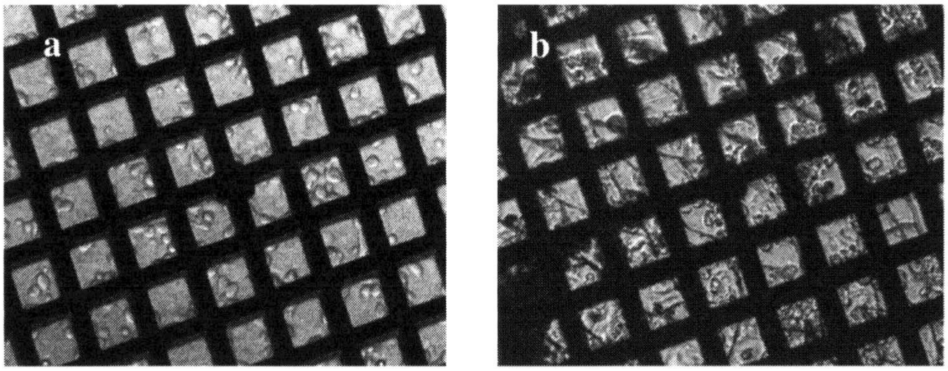

Figure 2. Light microscopy image before (a) and after (b) cryofixation.

3.2. Cryo imaging of parasites, 2D and 3D

Figure 3 shows recent images of parasites [5]. Figure 3a shows the protozoan parasite *Spironucleus Salmonicida*. The flagella as well as subcellular structures in the carbon-rich cell body are clearly visible. Vesicular structures, not seen in light microscopy, are clearly visible, as are the flagellar canals. Figure 3b shows another protozoan parasite, *Giardia intestinalis*. Several flagella and an intact adhesive disc and nuclei are observed.

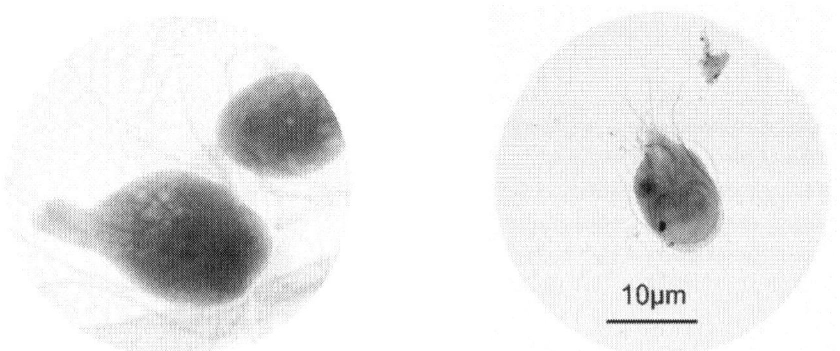

Figure 3. Parasites *Spironucleus Salmonicida* (a) and *Giardia intestinalis* (b) (from [5]).

3.3. Towards cell-cell interaction studies

Natural killer cells are cells of the immune system that have the ability to release toxins and kill virus-infected cells and cancer cells [11]. NK cells play an important role in host defence and immune regulation. When a NK cell encounters another cell, it forms an immune synapse in order to investigate whether the cell is a threat. If there is a downregulation of major histocompatibility complex (MHC) class I molecules on the target cell surface this indicates that the cell is malignant and the NK cell will kill the target cell. It delivers granules carrying cytolytic effector molecules to the target cell, inducing apoptosis. This process is of great interest in order to improve our understanding of the immune system. However, the resolution of light microscopy is not high enough to resolve small structures such as the lytic granules since they typically have a diameter smaller than 100 nm. Therefore x-ray microscopy potentially could shed light on this transport process and perhaps even show the reorganization of the microtubule organization centre (MTOC) that is responsible for the intracellular transport.

Figure 4 shows the first XRM image of a NK and a HEK 293T target cell in close proximity to each other. The image shows promise for studies of the interaction area, especially in tomographic imaging.

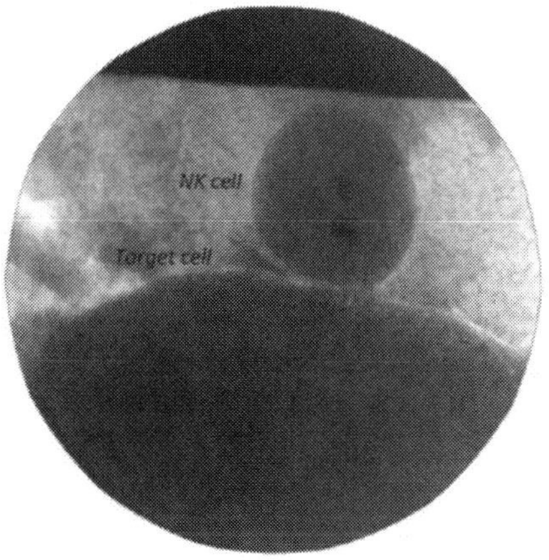

Figure 4. XRM image of NK cell with adjacent target cell. FOV = 25 μm.

References

[1] A. Sakdinawat and D. Attwood, "Nanoscale x-ray imaging", Nature Phot. **4**, 840-848 (2010).
[2] J. Kirz, C. Jacobsen, and M. Howells, "Soft-x-ray microscopes and their biological applications", Quart. Rev. Biophys. **28**, 33 (1995).
[3] M. Uchida et al. "Soft X-ray tomography of phenotypic switching and the cellular response to antifungal peptoids in Candida albicans", Proc. Natl Acad. Sci. USA **106**, 19375 (2009).
[4] G. Schneider et al., "Three-dimensional cellular ultrastructure resolved by X-ray microscopy", Nature Meth. **7**, 985 (2010).
[5] H.M. Hertz et al., "Laboratory cryo soft X-ray microscopy", J. Struct. Biol. **177**, 267-272 (2012).
[6] H. M. Hertz et al, "Laboratory soft x-ray microscopy for high resolution imaging of environmental colloidal structures", Chem. Geol., in press (2012).
[7] D. H. Martz et al, "High-brightness liquid-jet laser-plasma source for water-window microscopy", Opt. Lett., in press (2012).
[8] M. Bertilson et al, "Laboratory soft-x-ray microscope for cryotomography of biological specimens", Opt. Lett. **36**, 2728-2730 (2011).
[9] Institute for Laser Technology, www.ilt.fhg.de
[10] P.A.C. Jansson, U. Vogt, and H.M. Hertz, "Liquid-nitrogen-jet laser-plasma source for compact soft x-ray microscopy", Rev. Sci. Instrum. **76**, 43503 (2005).
[11] N. J. Topham and E. W. Hewitt, "Natural killer cell cytotoxicity: how do they pull the trigger?", Immunology **128**, 7-15 (2009).

Development of Soft X-ray Contact-type Microscope and Application to Micro-spectroscopy in Water-window

T. Ejima, Y. Neichi, F. Ishida, and M. Yanagihara

IMRAM, Tohoku University: 2-1-1, Katahira, Aoba-ku, Sendai, 980-8577 JAPAN

ejima@tagen.tohoku.ac.jp

Abstract. A contact-type microscope for the observation of organelles in a culture solution is fabricated with the use of a scintillator plate which shows high quantum efficiency and linearity in Water-window wavelength region. The fabricated microscope was applied to micro –spectroscopy, and an absorption spectrum of polystyrene beads was obtained from SX images taken by the microscope.

1. Introduction

Light in soft X-ray (SX) wavelength region can excite core-level electrons of a material, then chemical bonding states and/or electronic structures of materials can be clarified by core-level spectroscopies using SX light [1]. Especially, the wavelength region between C and O absorption edges (λ=2.3 ~ 4.4 nm) is called Water-window, because the light in this wavelength region is transparent to water comparing with the other wavelength region and organelles will be observed in cultures. If spectra of organelles in bio-cells can be measured at around C, N, and O absorption edges, chemical bonding states of the organelles will be clarified.

When organelles in a bio-cell are observed by use of a SX microscope, the organelles look different from the same organelles observed by a visible (VI) microscope. This is because the contrasts of the organelles in a SX image are different from the ones in a VI image. Then the identification of each organelle in SX images is difficult by itself. In VI wavelength region, several microscopic techniques were developed to observe the organelles [2]. If these techniques are simultaneously used with the SX observation, the organelles will be exactly identified. Therefore, we think that comparison of a SX image with VI images at the same position is necessary to identify organelles in a bio-cell.

As an easy microscopic method to compare a SX image with a VI one, a contact-type microscope was fabricated on a basis of the two-dimensional detector, SXV-IC, which we have developed [3]. In the developed microscope, a transparent SX image is converted to a VI one by a scintillator plate. Moreover the scintillator plate is transparent in VI wavelength region, and then transparent VI images can be observed by the change of the incident wavelength from SX light to VI light. Flexibility of incident wavelength enables us to identify organelles. Application of the microscope to spectroscopic measurements will be presented in this paper.

2. Evaluation of SCOM in Water-window

Optical layout of the fabricated microscope, Soft x-ray COntact Microscope (SCOM), is represented in Figure 1. SX light irradiates a measurement sample through a SiN membrane. The shadow of the incident SX light is converted to a VI image by a scintillator plate placed at the back of the sample. The converted VI image is magnified by a VI optics through the transparent scintillator plate.

2.1. Sample-cell for bio-cells in culture solution

In the fabricated microscope, a SX shadow of a sample is converted to a VI image by a scintillator plate. Therefore it is desirable that the sample is firmly attached on the scintillator plate, even if the sample is bio-cells in a culture solution. For this purpose, we have fabricated a sample cell for observation of bio-cells in a culture solution.

As represented in Figure 1, a fabricated sample-cell was designed as follows. A sample being measured is pressed on a scintillator plate by a SiN membrane. The SiN membrane adhered on a stainless-steel plate was 100 nm thick and 0.25×0.25 mm^2 in window size. The scintillator plate was adhered on another stainless-steel plate and was 0.2 mm thick and 4.5 mm in diameter. Both the stainless-steel plates were set to a base with O-rings not to leak the solution from the cell.

To check a leak-rate from the sample-cell, the sample-cell with and without water was placed into a vacuum chamber. Then partial pressure of H_2O was measured. The measurement was made by quadrupole mass filter (CANNON-ANELVA, M-101QA-TDM) before and after the placement of the sample-cell into the vacuum chamber. Base pressure was 2.3×10^{-6} Pa before the placement, and that was 4.4×10^{-6} Pa after the placement. Difference between the values was caused by the pumping time before and after the placement. After the measurements, the water remained in the sample-cell.

The change of the partial pressure rate is plotted in Figure 2. Zero point of the horizontal axis in Figure 2 is 30 minutes after the placement into the vacuum chamber to avoid an initial fluctuation of the pressure caused by the introduction of the sample-cell. As it can be seen from the figure, the change of the pressure rate was nearly equal to each other.

2.2. Quantum efficiency of the scintillator

Scintillator material is a key component in the system because of the quality of SX images depending on specifications of the scintillator plate. The specification that relates to SCOM is quantum efficiency $q_{scin}(\lambda_{SX})$ for a wavelength λ_{SX} in SX region. Using quantum efficiency $q_{scin}(\lambda_{SX})$ of the scintillator plate, numerical aperture of VI objective lens NA, transmittance of VI microscope $T_{VIM}(\lambda_{scin})$ at the luminescence wavelength λ_{scin} of the scintillator material, and quantum efficiency $q_{CCD}(\lambda_{scin})$ of the CCD image sensor, total quantum efficiency $Q(\lambda_{SX})$ of SCOM is represented as

$$Q(\lambda_{SX}) \propto q_{scin}(\lambda_{SX})\, T_{VIM}(\lambda_{scin})\, q_{CCD}(\lambda_{scin})\left(1-\sqrt{1-NA^2}\right). \qquad (1)$$

Except for the quantum efficiency $q_{scin}(\lambda_{SX})$ of the scintillator, all parameters are known: transmittance of the VI microscope is more than 0.8 and dispersion of the transmittance is flat [4], quantum efficiency of the CCD image sensor is 0.63 at the luminescence wavelength λ_{scin} [5]. Therefore, total quantum efficiency $Q(\lambda_{SX})$ can be obtained from measurement results.

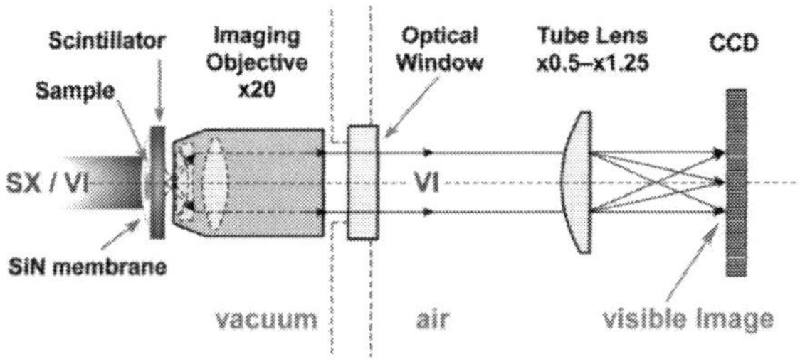

Figure 1. Schematic representation of SCOM.

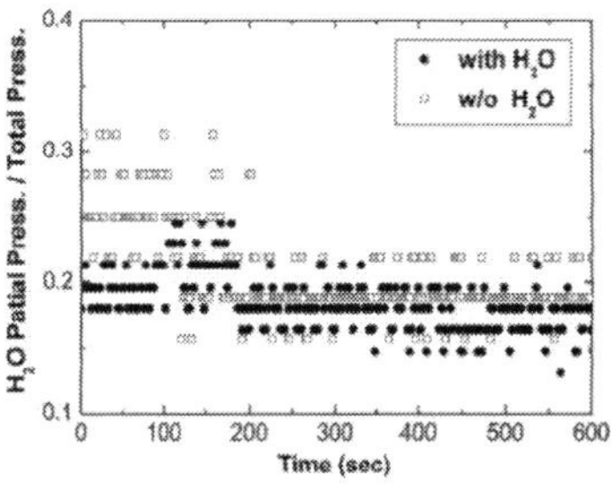

Figure 2. Change of partial pressure H_2O.

We chose Ce:LYSO (Oxide Co., Japan) as a material of the scintillator plate, because its quantum efficiency is high in SX region and it has no deliquescence [3]. Both sides of the scintillator plate were optical-polished and its quantum efficiency was measured with the use of SCOM. The values of relative quantum efficiency were determined by Equation (1). All measurements were carried out at beamline BL11D in Photon Factory, KEK, Japan under the following conditions: wavelength resolution was 500, wavelength of incident light, 2 ~ 5 nm, and exposure time depending on the photon intensity, 5 ~ 30 sec [6].

Measured spectrum of relative quantum efficiency is plotted in Figure 3. The efficiency decreases slowly as the increase of the wavelength, and show a local minimum value at 2.3 nm. The efficiency decreases again from 2.4 nm and shows the minimum value at around 4.4 nm. The efficiency increases from 4.4 to 5 nm. The small dip structure at 2.3 nm is O-K absorption edge, which is originated from the component atoms in the mother material of the scintillator. The minimum value at 4.4 nm will be caused by C-K absorption edge, and its origin will be C atoms remained on the surface of the scintillator plate, because the plate was rinsed out by ethanol and acetone solutions.

2.3. Spatial resolution

Spatial resolution was measured by an edge response of a polystyrene bead in water. The measurement was made by the same setup of the quantum efficiency measurement under the following conditions: measurement wavelength region was 2.1 ~ 4.4 nm, wavelength resolution, 500, and exposure time depending on the photon intensity, 240 ~ 300 sec. An image of polystyrene beads in water taken at wavelength 2.30 nm is presented in Figure. 4 (a). Black points were the polystyrene beads in the figure, and the other area was water. The black points showed no movement during the exposure, therefore the beads were considered to be fixed on the scintillator plate.

Intensity profile of the beads was made as follows: the several points were selected and intensity profiles were averaged. The intensity profile was fitted by Gaussian curve to estimate the HWHM value as the spatial resolution. Obtained profile is designated as black circles in Figure 4 (b), the fitting result is as a solid curve. The HWHM value was 0.72 μm, it is comparable with the ideal value, 0.69 μm, under the conditions that numerical aperture of objective lens was 0.4 and observation wavelength, 450 nm [7].

3. Application to micro-spectroscopy

When the fabricated microscope was applied to micro-spectroscopy measurement, polystyrene beads in water were chosen as a standard sample. Observation conditions were the same as the evaluation of spatial resolution, and one of the SX images has been presented in Figure. 4(a). Spectrum of a region of interest (ROI) was obtained as follows: **1.** SX images of the irradiation light were observed by SCOM changing the incident wavelength. **2.** SX images of the sample were observed as well as the irradiation light. **3.** Sample image was normalized by the exposure time and then normalized by the irradiation light image at the same incident wavelength. Obtained images are transmittance ones. **4.** A ROI to obtain a spectrum is determined referring to the same ROI of VI

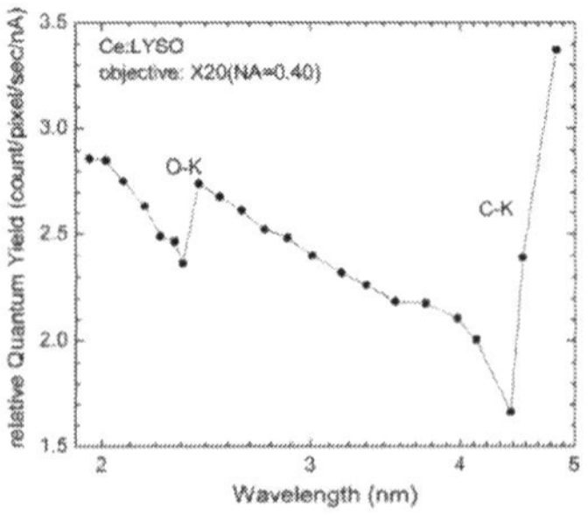

Figure 3. Relative quantum efficiency of Ce:LYSO.

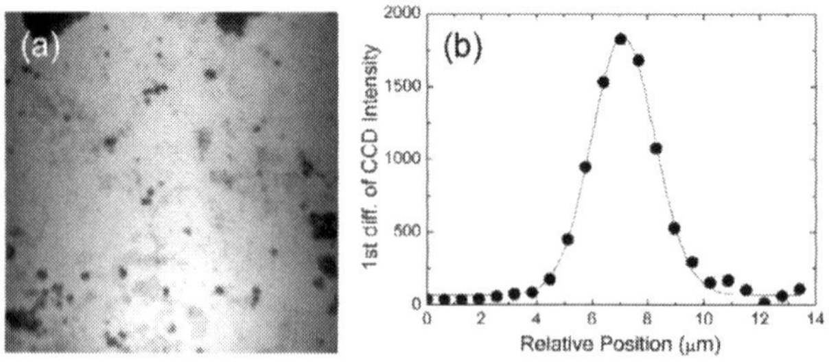

Figure 4. a) Polystyrene beads image at a wavelength 2.30 nm. b) Edge response of the beads in water. HWHM was 0.72 μm.

images, and intensities of pixels in the ROI are averaged. The intensity-change according to the wavelength was the transmittance of the ROI. **5.** Finally, absorptance of the ROI is obtained on the assumption that there is no interference in the sample.

Beads and water areas were selected as ROIs, and the spectra of both ROI were obtained. Thickness of the sample was less than 1μm, which was estimated from the absorptance of the water. Therefore optical density spectrum of the beads can be calculated. The obtained spectra are plotted in Figure. 5. Peak structures observed at 2.3 nm, 3.1 nm, and 4.3 nm are O-K, N-K, and C-K absorption edges, respectively.

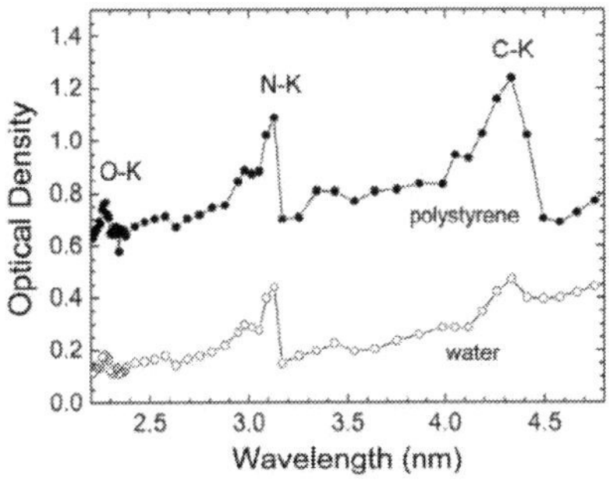

Figure 5. Optical density spectra of both polystyrene beads and water.

The N-K absorption edge peaks were observed in both the beads area and the water one, therefore the peaks are originated from the SiN membrane. As same as the N-K edge peaks, the O-K absorption peaks of both ROIs showed the same shape and almost the same intensity. Therefore, the peaks will be originated from the water in the cell. The C-K absorption edge peaks show the different shape between the beads and the water. The origin of the peak in the polystyrene beads is understood by two major peaks of π and σ bands in polystyrene [8]. The polystyrene-beads peak is broadened by the wavelength resolution and rough wavelength interval, therefore the shape of the peak looks a long tail toward the short wavelength side. The peak intensity in water is small and its width is narrow comparing with the peak of the polystyrene beads. The different shape of the peak suggests that the origin of the peak is the contamination of the SiN window.

4. Summary

A soft X-ray contact-type microscope, SCOM, for taking SX images and obtaining spectra from the images was fabricated and evaluated. The evaluation results showed that transmittance spectrum of polystyrene beads can be obtained after taking SX images. Then spatial resolution of SCOM was measured by edge response method, and its value coincides with the ideal diffraction-limit one. These results suggest that a small organelles identified by visible-light observation can be measured using SX wavelength region by SCOM. It will be a powerful analytical tool to determine not only chemical changes in a biological sample but also electronic structures of nanostructures in a solution.

References

[1] F de Groot & A Kotani, Core level spectroscopy of solids, (CRC Press (Taylor & Francis G.), 2008, Boca Raton) p. 23
[2] for example, B Alberts, *et al.*, *Molecular biology of the cell*, (Garland Science, Taylor & Francis Group, New York, 2010) Chap. 9.
[3] T Ejima, *et al.*, AIP CP **1234**, (2010) 811.
[4] Datasheet of Objective lens, UV Series for ultraviolet rays, Union Optical Co. Ltd.
[5] Datasheet of CCD sensor, CCD47-10, e2v technologies Ltd.
[6] Photon Factory activity Report 1997 #15 (1997) A101., http://pfwww.kek.jp/users_info/station_spec/bl11/bl11d.html
[7] for example, D Attwood, *Soft X-rays and extreme ultraviolet radiation*, (Cambridge University Press, Cambridge, 2000) Chap. 9.
[8] D A Fischer, G E Mitchell, A T Yeh, J L Gland, Appl. Surf. Sci., **133** (1998) 58.

Journal of Physics: Conference Series **463** (2013) 012056 doi:10.1088/1742-6596/463/1/012056

In situ observation of cellular organelles with a contact x-ray microscope

M Kado[1], M Kishimoto[1], S Tamotsu[2], K Yasuda[2] and K Shinohara[1, 3]

[1]Japan Atomic Energy Agency, Umemidai, Kizugawa-shi, Kyoto, 619-0215, Japan
[2]Nara Women's University, Kitauoyahigashi-machi, Nara-shi, Nara, 630-8506, Japan
[3]Waseda University, Okubo, Shinjuku-ku, Tokyo, 169-8555, Japan

E-mail: kado.masataka@jaea.go.jp

Abstract. A contact x-ray microscope coupled with a highly intense laser plasma soft x-ray source has been developed and in situ observations of cellular organelles have been conducted. The soft x-rays were generated by irradiating a high power laser pulse onto a thin-foiled gold target and about 1.3×10^{15} photons/sr were obtained, which allowed the inner structures of live wet biological cells to be imaged. Single shot flash imaging is a key technique to image cellular organelles of live biological cells avoiding degradation of the spatial resolution of the images resulting from motion blur and radiation damage. The use of a fluorescence microscope to identify cellular organelles in conjunction with the soft x-ray microscope has allowed several cellular organelles to be identified precisely in the soft x-ray images. Combining the fluorescence microscope with the soft x-ray microscope will be very useful and will establish the technique as a powerful tool to analyze living function of biological cells.

1. Introduction

Laser plasma soft x-ray sources have advantages such as high intensity and short pulse duration and a contact x-ray microscope coupled with a laser plasma soft x-ray source has the potential to capture an x-ray image of biological specimens in a single shot flash imaging without any artificial treatment such as staining, fixation, and slicing. However in order to realize the single shot flash imaging of hydrated living biological cells extremely high intense laser plasma x-ray sources are required and success has been limited with the images only showing the surface structure of cells [1].

In order to realize the single shot flash imaging of wet live cells it is also very important to avoid reduction of x-ray photon flux caused by low efficiency of x-ray optics. In this respect, the contact type x-ray microscopy that requires no x-ray optics has an advantage over x-ray microscopes with x-ray optics. Hence, combining the highly intense x-ray source and contact-type x-ray microscopy allows in situ observation of cellular organelles of live biological cells to be realized. According to the estimation by Sayre et al. [2], a photon flux of more than 1×10^5 photons/μm^2 is required on specimens to obtain 100 nm spatial resolution or more than 1×10^{15} photons/sr at the source.

It is also important to establish how to identify cellular organelles in x-ray images for biological applications since the inner structures of cells are complicated and it is difficult to distinguish each organelle in the x-ray images. We propose an in situ method of a soft x-ray microscope in conjunction with an optical fluorescence microscope to identify the cellular organelles of live hydrated cells in the soft x-ray images obtained of biological cells. Direct comparison of both the soft x-ray images and the fluorescence images allows cellular organelles to be identified in the soft x-ray images.

Journal of Physics: Conference Series **463** (2013) 012056 doi:10.1088/1742-6596/463/1/012056

2. Flash imaging with highly intense laser plasma soft x-rays

The laser plasma x-ray sources are suitable for the single-shot flash imaging to observe live biological cells since they have advantages such as bright and short pulse duration. There have been several studies using a laser plasma soft x-ray source with an x-ray microscope [3]. However most of them did not achieve a single shot flash imaging due to the difficulty of obtaining enough photon flux [2].

The high power Nd:glass laser system, which can generate over 20 J output energy in a 600 ps pulse duration, was employed to generate enough photon flux to capture an x-ray image with a single shot flash exposure. The pulse duration of the soft x-rays emitted from the laser plasma source is also expected to be about 600 ps [4]. The 600 ps duration is shorter than the time scale of any artificial effect on image blurring including radiation damage and Brownian motion [5].

The laser pulse was focused onto a thin-foil gold target at a laser intensity of about 1×10^{14} W/cm^2. The x-ray flux was measured by a soft x-ray plasma camera with Fresnel zone plate [6] and as ~1.3×10^{15} photons/sr over the wavelength range 2.3 nm to 4.4 nm in the so-called "water window" [7]. Figures 1 (a) and (b) show the soft x-ray images obtained using the soft x-ray plasma camera of the laser plasma source with gold target thickness of 50 μm and 20 nm, respectively. It is clear that the x-ray intensity obtained using the 20 nm thick target in (b) is much brighter than that with the 50 μm thick target in (a).

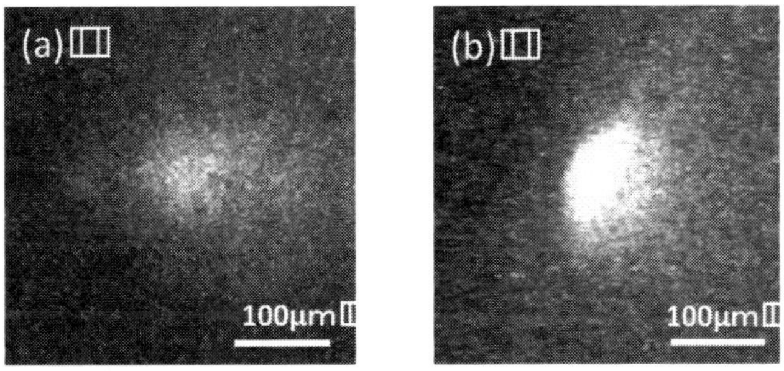

Figure 1. Soft x-ray images of the laser plasma source with target thickness of (a) 50 μm and (b) 20 nm. The image of the laser plasma with 20 nm target thickness is brighter than that with 50 μm target thickness.

3. In-situ observation with contact-type x-ray microscope

Leydig cells from mouse testicle were placed directly onto an x-ray photoresist and were cultured for two to three days to establish good contact onto the surface of the photoresist and the thickness of the cells were measured to be about 5 μm. Poly methyl methacrylate (PMMA) with thickness of 500 nm was used as the x-ray photoresist and it was coated onto a thin glass plate. The biological cells on the PMMA were covered with a silicon nitride membrane with thickness of 200 nm and were enclosed in a stainless steel holder specifically designed to protect the biological cells from vacuum. The cells were stained with fluorescent dyes of Phalloidin, Mito-tracker, and/or DAPI, placed in the specimen holder and observed by a fluorescence microscope to obtain images for the cellular organelles of cytoskeleton, mitochondria, and/or chromatin, respectively. After observation by the fluorescence microscope the specimen holder containing the cells was placed in the vacuum chamber for soft x-ray imaging. The time interval between the observation by the fluorescence microscope and the irradiation with soft x-rays was typically 10 minutes. After exposure to soft x-rays the PMMA was rinsed with sodium hypochlorite (NaClO) to remove the cells and developed with a mixed solution of methyl isobuthyl ketone (MIBK) and isopropyl alcohol (IPA). Before the development of the PMMA the surface of it was carefully observed by differential interference microscopy to make sure that no cells were left on the surface. Soft x-ray images of the cells were obtained by scanning the surface of the developed PMMA using an atomic force microscope. Shown in Figure 2 is the typical soft x-ray image of the wet Leydig cells. Fine structures surrounding nucleus were clearly visible and structures in the nucleus were also visible. The structures surrounding the nucleus were identified to be mitochondria and the structures in the nucleus were identified to be chromatin. The identification of the structures was done using a fluorescence microscope, which is described in the next section. The spatial resolution was measured to be about 90 nm [8], which equates well to that estimated as achievable with an x-ray flux on the specimens of 4.4×10^5 photons/μm^2 [2].

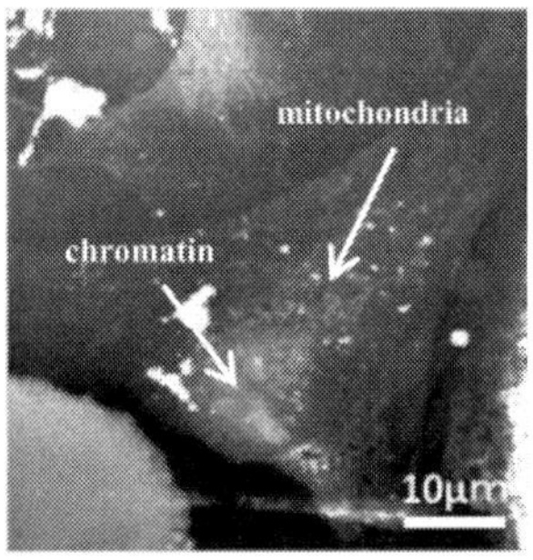

Figure 2. A typical soft x-ray image of wet Leydig cells. Fine structures surrounding nucleus are clearly visible and structures in the nucleus are also visible.

4. Identification of cellular organelles using a fluorescence microscope

The soft x-ray images of Leydig cells were directly compared with corresponding fluorescence images and each cellular organelle was identified [9, 10]. Shown in Figures 3 (a) and (b) are the fluorescence image and soft x-ray image of the cells, respectively. As the cells had been stained with Mito-tracker to label the mitochondria, the structures that appeared red in (a) were identified as mitochondria. The same structures as in (a) were also found in (b). Lattice-like structures were also seen in the soft x-ray image, which were not seen in the fluorescence image. These structures were identified as cytoskeleton according to the previous results observing the cells stained with Phalloidin [8].

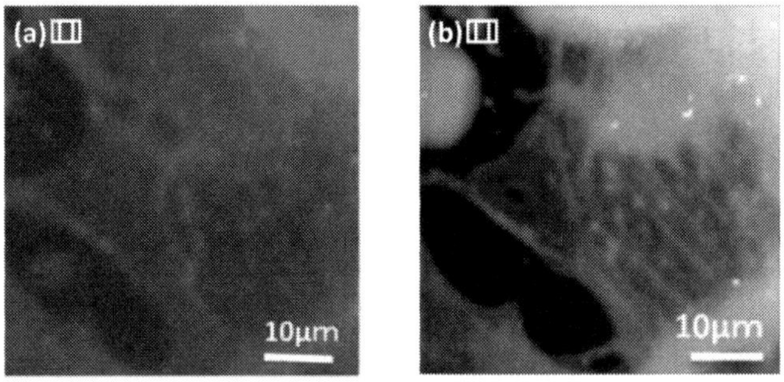

Figure 3. Fluorescence image (a) and a soft x-ray image (b) of Leydig cells. The structures appearing bright red dots in (a) are mitochondria and the same structures are also seen in (b).

Shown in Figure 4 is an enlarged area of the soft x-ray image shown in Figure 3 (b). Mitochondria and cytoskeleton were clearly recognized and those structures were found to be in close proximity. This is the first result where it was possible to identify the cellular organelles in the soft x-ray images obtained by in-situ observation of wet biological cells.

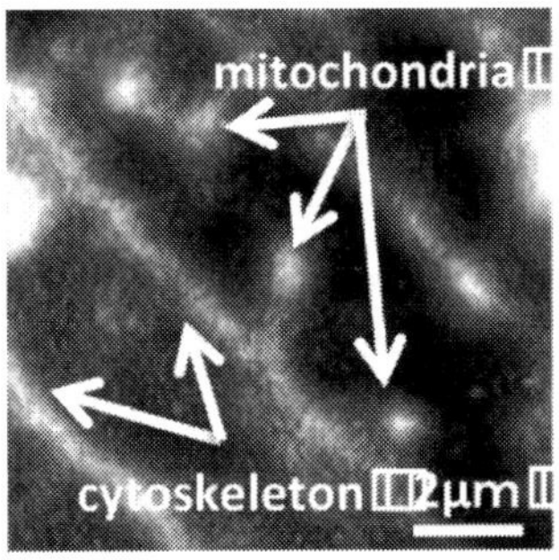

Figure 4. An enlarged image of the soft x-ray image shown in Figure 3 (b). Mitochondria and cytoskeleton were clearly recognized and those structures were found to be in close proximity.

5. Summary

In summary a contact soft x-ray microscope with an intense laser plasma soft x-ray source has been developed and live Leydig cells were observed. Using a high intensity laser plasma source with the soft x-ray microscope made it possible to image wet biological cells with a single shot flash exposure to avoid motion blur and radiation damage. The specially designed specimen holder made it possible to obtain fluorescence images and soft x-ray images of the same cells in the same environment. Comparing the fluorescence images with the soft x-ray images for the same cells allowed cellular organelles in the soft x-ray images to be clearly identified for the first time and fine structures of the cellular organelles to be analyzed.

The soft x-ray microscope has been expected to be a powerful tool for life science research for a long time as it allows the observation of live biological specimens with high spatial resolution [8]. However the lack of the brightness of the x-ray source prevented the ability to image live biological specimens. The difficulty in identifying the inner structures of biological cells was also a barrier for use of the soft x-ray microscope for life science. In the present work, solving these problems using the high intensity laser plasma soft x-ray source and the hybrid microscope system combined with fluorescence microscope, in-situ observation of cellular organelles in a live cell has been realized.

Acknowledgements

This research was partially supported by the Ministry of Education, Science, Sports and Culture, Grant-in-Aid for Scientific Research (A), 23241038, 2012.

References

[1] Kado M, Richardson M C, Rajyaguru J M, Muszynski M J and Yamamoto Y 1999 Proc. of Experimental Biology and Medicine **220** 27
[2] Sayre S, Kirz J, Feder R, Kim D M and Spiller E 1977 Ultramicroscopy **2** 337
[3] Johansson G A, Holmberg A, Hertz H M and Berglund M 2002 Rev. Sci. Instrum. **73** 1193
[4] Mochizuki T, Mima K, Ikeda N, Kodama R, Shiraga H, Tanaka K A and C. Yamanaka 1987 Phys. Rev. A **36** 3279
[5] Shinohara K and Ito A 1991 J. Microscopy **161** 463
[6] Kado M, Mori M, Nishiuchi M, Ishino M and T. Kawachi 2009 Journal of Physics: Conference Series **186** 012031
[7] Kado M, Ishino M, Kishimoto M, Tamotsu S, Yasuda K, Kinjo Y and Shinohara K 2011 Proc. of SPIE **8139** 81390O1
[8] Kado M, Ishino M, Tamotsu S, Yasuda K, Kishimoto M, Nishikino M, Kinjo Y and Shinohara K 2010 JEEE **130** 1774
[9] Kado M, Ishino M, Tamotsu S, Yasuda K, Kishimoto M, Nishikino M, Kinjo Y and Shinohara K 2011 AIP Conf. Proc. **1365** 391
[10] Kado M, Kishimoto M, Ishino M, Tamotsu S, Yasuda K and Shinohara K 2012 AIP Conf. Proc. **1465** 246

AUTHOR INDEX

AUTHOR INDEX

AUTHOR INDEX